Solar Technologies for Buildings

Solar Technologies for Buildings

Ursula Eicker
University of Applied Sciences, Stuttgart, Germany

Originally published in the German language by B.G. Teubner GmbH as *"Ursula Eicker: Solare Technologien für Gebäude. 1. Auflage (1st edition)".*

Telephone (+44) 1243 779777

Email (for orders and customer service enquiries): cs-books@wiley.co.uk
Visit our Home Page on www.wileyeurope.com or www.wiley.com

Other Wiley Editorial Offices

John Wiley & Sons Inc., 111 River Street, Hoboken, NJ 07030, USA

Jossey-Bass, 989 Market Street, San Francisco, CA 94103-1741, USA

Wiley-VCH Verlag GmbH, Boschstr. 12, D-69469 Weinheim, Germany

John Wiley & Sons Australia Ltd, 33 Park Road, Milton, Queensland 4064, Australia

John Wiley & Sons (Asia) Pte Ltd, 2 Clementi Loop #02-01, Jin Xing Distripark, Singapore 129809

John Wiley & Sons Canada Ltd, 22 Worcester Road, Etobicoke, Ontario, Canada M9W 1L1

Wiley also publishes its books in a variety of electronic formats. Some content that appears in print may not be available in electronic books.

British Library Cataloguing in Publication Data

A catalogue record for this book is available from the British Library

ISBN 0-471-48637-X

Produced from files supplied by the Author.
Printed and bound in Great Britain by Antony Rowe Ltd, Chippenham, Wiltshire
This book is printed on acid-free paper responsibly manufactured from sustainable forestry in which at least two trees are planted for each one used for paper production.

Contents

Preface

The heating energy requirement of buildings can be reduced from today's high levels to almost zero if buildings are thoroughly insulated, passive solar gains through windows are used efficiently, and the supply of fresh air takes place via a heat-recovery system. However, all buildings still have an energy requirement for electricity and warm water provision which cannot be met by passive measures. Active solar technologies are especially appropriate for meeting this energy requirement, as the elements can be integrated into the shell of the building, thus substituting classical building materials and requiring no additional area. Solar modules for photovoltaic electricity production can be built like glazing into all common construction systems, and are characterised by a simple, modular system technology. Thermal collectors with water and air as heat conveyors are installed for warm water provision and heating support, and can replace complete roof covers if the collector surface is large. For today's increasing air-conditioning and cooling demand, especially in office buildings, thermally driven low-temperature techniques are interesting; these can use not only solar energy but also waste heat. Apart from electricity production, solar heating and cooling, solar energy is used in the form of daylight and thus contributes to a reduction in the growing electricity consumption. The intention of this book is to deal with all solar technologies relevant to meeting the energy requirements of buildings, so that both the physical background is understood and also concrete approaches to planning are discussed. The basic precondition for the sizing of active solar plants is a reliable database for hourly recorded irradiance values. New statistical procedures enable the synthesis of hourly radiation data from monthly average values, which are available world-wide from satellite data, and also partly from ground measurements. For the use of solar technologies in urban areas, an analysis of the mutual shading of buildings is particularly relevant.

Solar thermal systems with air- and water-based collectors are a widely used technology. For the engineer-planner, the system-oriented aspects such as interconnecting, hydraulics and safety are important, but for the scientific simulation of a thermal system the heat transfer processes must also be examined in detail. The fact that with thermal collectors not only heat can be produced, is pointed out in the extensive section on solar cooling. The current technologies of adsorption and absorption cooling as well as open sorption-supported air conditioning can all be coupled with thermal collectors and offer a large energy-saving potential, particularly in office buildings.

Photovoltaic generation of electricity is then discussed, with the necessary basics for current-voltage characteristics as well as the system-oriented aspects. Since photovoltaics offers particularly interesting building integration solutions, new procedures for the calculation of thermal behaviour must be developed. For these new elements, component characteristic values are derived, which are needed for the building's heating-requirement calculations.

The book concludes with a discussion of passive solar energy use, which plays an important role in covering heat requirements and in the use of daylight. What is crucial for the efficiency of solar energy is the effective storage capability of the components, which must also be known in cases where transparent thermal insulation is used. Linking an outline of basic physical principles with their applications is designed to facilitate a sound knowledge of innovative solar-building technologies, and to contribute to their being accepted in planning practice.

This book is due to the initiative of my now retired colleague Professor Jenisch, who was always interested in solar technology, while working on classical thermal building physics. His contacts with the German publishing house Teubner led first to a German version of the current book, which appeared in 2001.

Within the Department of Building Physics at the Stuttgart University of Applied Sciences, solar technology is now very important. Without the support of the research group on solar energy, the broad subject range of the book would not have been possible. I would like to thank my co-workers Christa Arnold, Uwe Bauer, Volker Fux, Martin Huber, Guenther Maendle, Uli Jakob, Dieter Schneider, Uwe Schuerger and Peter Seeberger for their input and support with many of the book's graphics. I also wish to thank my theory colleague, Professor Kupke, for his various suggestions regarding the section on passive solar energy use.

I am, however, most indebted to Dr Juergen Schumacher for his continuous support, which extends beyond the book's content. It was with his simulation environment INSEL, at an intensive conference in Barcelona, that most of the calculation results were obtained.

Ursula Eicker, Stuttgart
January 2003

Abbreviations in the text

A	area [m^2]
A_r	receiver surface [m^2]
A_s	sender surface [m^2]
A_c	cross-section surface [m^2]
C	charge concentration [kg_{H2O}/kg_{sor}]
E_{gap}	band gap [eV]
G	irradiance [W/m^2]
G_{eh}	extraterrestrial irradiance on horizontal plane [W/m^2]
G_{en}	extraterrestrial irradiance on plane oriented normally to beam [W/m^2]
G_{sc}	solar constant [W/m^2]
G_h	irradiance on horizontal plane (global irradiance) [W/m^2]
G_b	direct beam irradiance [W/m^2]
G_{dh}	diffuse irradiance on horizontal plane [W/m^2]
G_t	irradiance on tilted plane [W/m^2]
I	current [A]
I_0	saturation current [A]
I_{SC}	short circuit current [A]
K	extinction coefficient [m^{-1}]
L	characteristic length [m]
L_e	irradiance density [W/m^{2}sr]
L_v	luminance [lm/m^{2}sr]
Nu	Nußelt number [–]
Pr	Prandtl number [–]
P_{AC}	AC power [W]
P_{DC}	DC power [W]
Q_u	useful energy [kWh]
R	resistance [Ω]
R_{a-g}	heat resistance between absorber and glas cover [m^{2}K/W]
R_{g-o}	heat resistance between glas cover and environment [m^{2}K/W]
Re	Reynolds number [–]
R_p	parallel resistance [Ω]
R_s	series resistance [Ω]
S_d	shading factor for diffuse irradiance [–]
T	temperature [K]
T_a	absorber temperature [K]
T_{ads}	adsorber temperature [K]
T_b	back side temperature [K]
T_{des}	desorber temperature [K]
T_e	evaporator temperature [K]
T_f	fluid temperature [K]
$T_{f,in}$	fluid inlet temperature [K]
$T_{f,out}$	fluid exit temperature [K]
T_g	glas temperature [K]
T_G	generator temperature [K]
T_{sky}	sky temperature [K]
T_o	outside/exterior temperature [K]

T_i	room temperature [K]
T_c	condensor temperature [K]
T_{dp}	dew point temperature [K]
U	heat transfer coefficient (U-value) [W/m²K]
U_{eff}	effective U-value [W/m²K]
U_f	heat transfer coefficient front [W/m²K]
U_b	heat transfer coefficient back [W/m²K]
U_l	heat transfer coefficient per metre length [W/mK]
U_s	heat transfer coefficient side wall [W/m²K]
U_t	total heat transfer coefficient [W/m²K]
V	voltage [V]
$V(\lambda)$	spectral sensitivity of the eye [–]
V_c	volumetric content collector [litre]
V_{cc}	volume collector circuit [litre]
V_{oc}	open-circuit voltage [V]
W	width [m]
c_a	heat capacity of air [J/kgK]
c_v	heat capacity of water vapour [J/kgK]
d_h	hydraulic diameter [m]
g	gravity constant [m/s²]
q	elementary charge [C]
h	heat transfer coefficient [W/m²K]
h_c	convective heat transfer coefficient [W/m²K]
$h_{c,w}$	convective heat transfer coefficient due to wind forces [W/m²K]
h_r	radiative heat transfer coefficient [W/m²K]
h_e	evaporation enthalpy [kJ/kg]
h_o	enthalpy outside air [kJ/kg]
h_l	enthalpy liquid [kJ/kg]
h_v	enthalpy vapour [kJ/kg]
k_{max}	photometric equivalent [lm/W]
m	avalanche coefficient [–]
p_a	pressure dry air [Pa]
p_w	pressure water vapour [Pa]
v	velocity[m]
Φ	luminous flux [lm], heat recovery efficiency [–]
α	optical absorptions coefficient [–]
α_I	temperature coefficient of current [K^{-1}]
α_V	temperature coefficient of voltage [K^{-1}]
β	orientation angle from horizontal [°]
β'	heat expansion coefficient [K^{-1}]
δ	sheet thickness [m]
ε	emission coefficient [–]
ν	kinematic viscosity [m²/s]
ξ_r	concentration of rich solution [–]
ξ_p	concentration of poor solution [–]
η	efficiency [–]
ρ	reflection coefficient
τ	transmission coefficient [–]

1 Solar energy use in buildings

1.1 Energy consumption of buildings

Buildings account today for about 40% of the final energy consumption of the European Union, with a large energy saving potential of 22% in the short term (up to 2010). Under the Kyoto protocol, the European Union has committed itself to reducing the emission of greenhouse gases by 8% in 2012 compared to the level in 1990, and buildings have to play a major role in achieving this goal. The European Directive for Energy Performance of buildings adopted in 2002 (to be implemented by 2005) is an attempt to unify the diverse national regulations, to define minimum common standards on buildings' energy performance and to provide certification and inspection rules for heating and cooling plants. While there are already extensive standards on limiting heating energy consumption (EN832 and prEN 13790), cooling requirements and daylighting of buildings are not yet set by any European standard. The reduction of energy consumption in buildings is of high socio-economic relevance, with the construction sector as Europe's largest industrial employer representing an annual investment of 868×10^9 € (2001) corresponding to 10% of gross domestic product. Almost two million companies, 97% of them small and medium enterprises, employ more than 8 million people (European Commission, 1997).

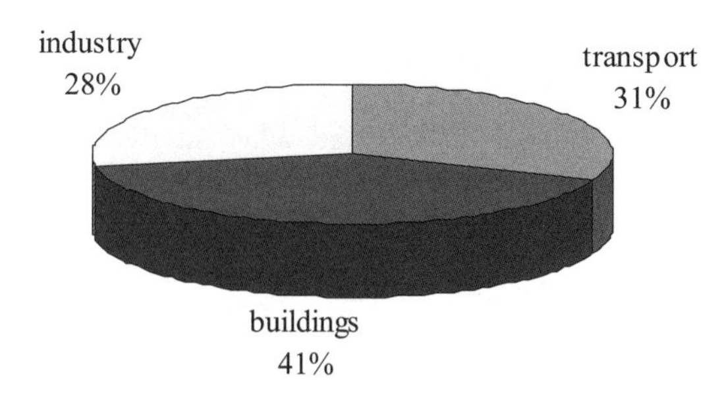

Figure 1.1: Distribution of end energy consumption within the European Union with a total value of 10^{12} MWh per year (Deschamps, 2001).

The distribution of energy use varies with climatic conditions. In Germany, where 44% of primary energy is consumed in buildings, 32% is needed for space heating, 5% for water heating, 2% for lighting and about 5% for other electricity consumption in residential buildings (Diekmann, 1997). The dominance of heat-consumption, almost 80% of the primary energy consumption of households, is caused by low thermal insulation standards in existing buildings, in which today 90% and even in 2050 60% of residential space will be located (Ministry for Transport and Buildings, Germany, 2000).

Since the 1970s oil crisis the heating energy requirement, particularly of new buildings, has been continuously reduced by gradually intensified energy legislation. With high heat insulation standards and the ventilation concept of passive houses, a low limit of heat consumption has meanwhile been achieved, which is around 20 times lower than today's values. A crucial factor for low consumption of passive buildings is the development of new glazing and window technologies, which enable the window to be a passive solar element and at the same time cause only low transmission heat losses. In new buildings with low heating requirements other energy consumption in the form of electricity for lighting, power and air conditioning, as well as in the form of warm water in residential buildings, is becoming more and more dominant. Electricity consumption within the European Union is estimated to rise by 50% by 2020. In this area renewable sources of energy can make an important contribution to the supply of electricity and heat.

1.1.1 Residential buildings

Due to the wide geographical extent of the European Union of nearly 35° geographical latitude difference (36° in Greece, 70° in northern Scandinavia), a wide range of climatic boundary conditions are covered. In Helsinki (60.3° northern latitude), average exterior air temperatures reach –6°C in January, when southern cities such as Athens at 40° latitude still have averages of +10°C. Consequently the building standards vary widely: whereas average heat transfer coefficients (U-values) for detached houses are 1 W/m²K in Italy, they are only 0.4 W/m²K in Finland. The heating energy demand determined using the European standard EN 832 is comparable in both cases at about 50 kWh/m²a.

If existing building standards are improved to the so-called passive building standard, heating energy consumption can be lowered to less than 20 kWh/m²a. The required U-values for the building shell are listed below for both current practice buildings and passive buildings.

Table 1.1: U-values in residential buildings according to national building standards and the requirements of passive buildings construction (Truschel, 2002).

	Rome		*Helsinki*		*Stockholm*	
U-values	*Current standard* [W/m²K]	*Passive building* [W/m²K]	*Current standard* [W/m²K]	*Passive building* [W/m²K]	*Current standard* [W/m²K]	*Passive building* [W/m²K]
Wall	0.7	0.13	0.28	0.08	0.3	0.08
Window	5	1.4	2.0	0.7	1.7	0.7
Roof	0.6	0.13	0.22	0.08	0.28	0.08
Ground	0.7	0.23	0.36	0.08	0.21	0.1
Mean U-value	1.0	0.33	0.43	0.16	0.36	0.17

The resulting heating energy requirement for current building practice varies between 55 kWh/m²a in Stockholm/Sweden and 93 kWh/m²a in Helsinki/Finland. These values can be lowered by nearly 80% when applying better insulation to the external surfaces and reducing ventilation losses.

Independent of the standard of insulation, water heating is always necessary in residential buildings, and this lies between about 220 (low requirement) and 1750 kWh per person per year (high requirement), depending on the pattern of consumption. For the middle requirement range of 30–60 litres per person and day, with a warm-water temperature of 45°C, the result is an annual consumption of 440–880 kWh per person, i.e. 1760–3520 kWh for an average four-person household. Related to a square metre of heated residential space, an average value of 25 kWh/m²a is often taken as a base.

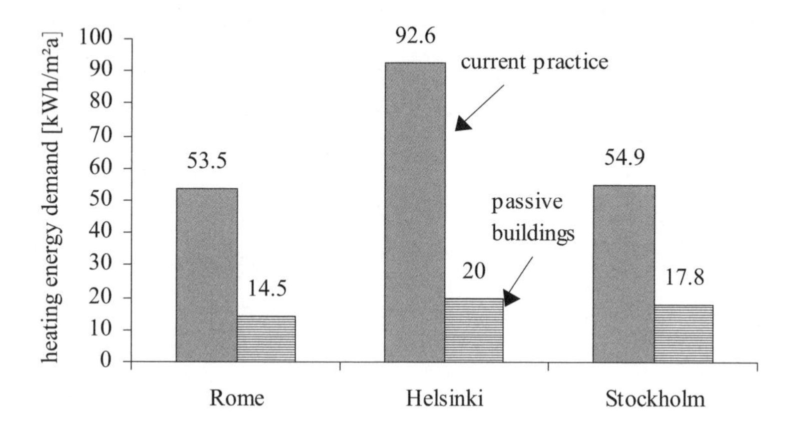

Figure 1.2: Heating energy demand for residential buildings in three European climates with current practice constructions (high values) and passive building standards (low values).

The average electricity consumption of private households, around 3600 kWh per household per year, is of a similar order of magnitude. Related to a square metre of heated residential space, an average value of 31 kWh/m²a is the result. An electricity-saving household needs only around 2000 kWh/a. In a passive building project in Darmstadt (Germany), consumptions of between 1400 and 2200 kWh per household per year were measured, which corresponds to an average value of 11.6 kWh/m²a. Low energy buildings today have heat requirements of between 30 and 70 kWh/m²a.

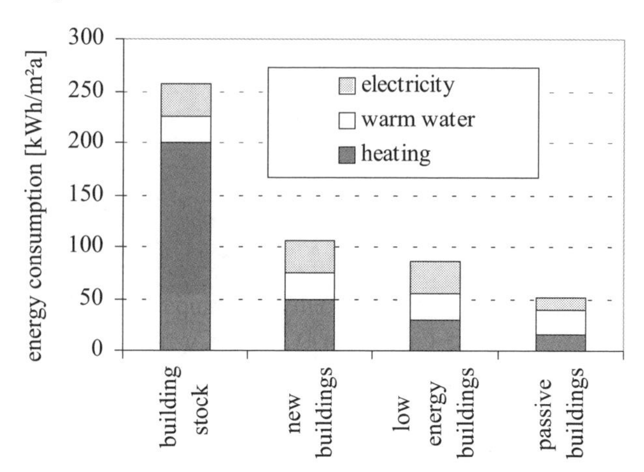

Figure 1.3: End energy consumption in residential buildings per square metre of heated floor space in Germany.

1.1.2 *Office and administrative buildings*

Existing office and administrative buildings have approximately the same consumption of heat as residential buildings and most have a higher electricity consumption. According to a survey of the energy consumption of public buildings in the state of Baden-Wuerttemberg in Germany the average consumption of heat is 217 kWh/m²a, with an average electricity consumption of 54 kWh/m²a. The specific energy consumption of naturally ventilated office buildings in Great Britain is in a similar range of 200–220 kWh/m²a for heating and 48–85 kWh/m²a for electricity consumption (Zimmermann, Andersson, 1998). If the final energy consumption for heat and electricity is converted to primary energy consumption, comparable orders of magnitude of both energy proportions result. Still more important are the slightly higher costs of electricity.

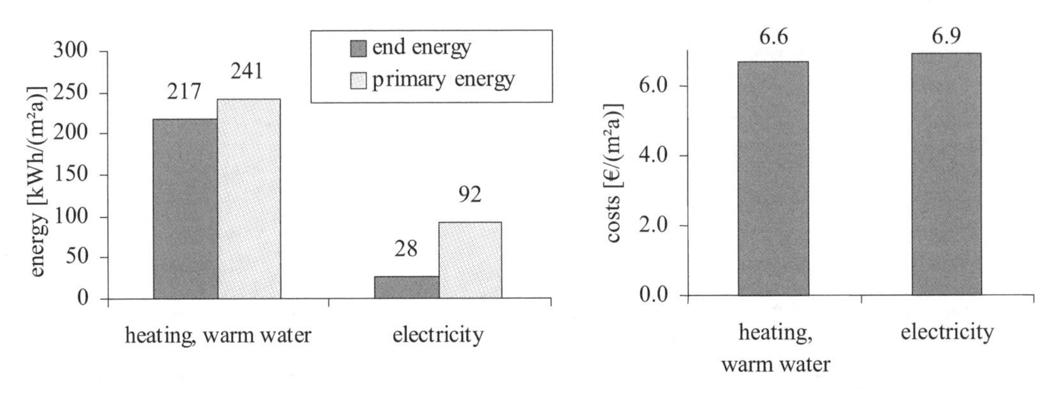

Figure 1.4: Annual energy consumption and operating costs of public buildings in Baden-Wuerttemberg (an area of 4.4 million square metres).

Both heat and electricity consumption depend strongly on the building's use. In terms of the specific costs, electricity almost always dominates.

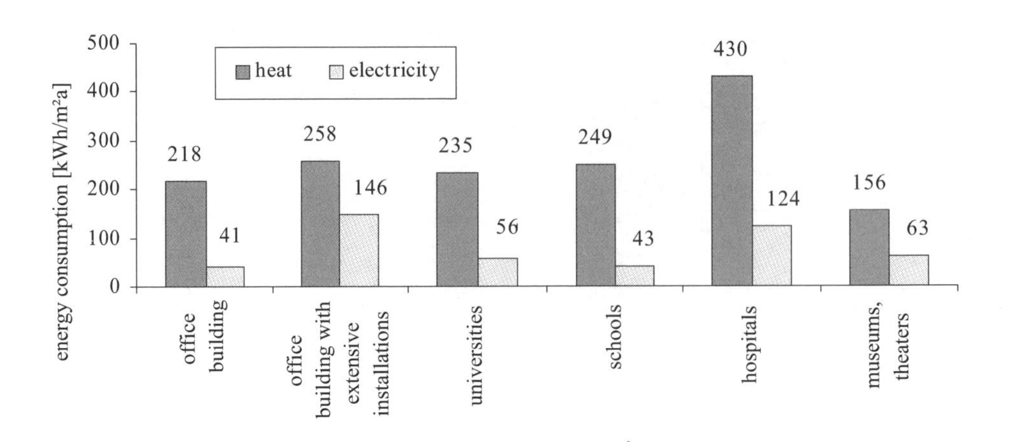

Figure 1.5: Final energy consumption by building type in Baden-Wuerttemberg.

If one compares the energy costs of commercial buildings with the remaining current monthly operating costs, the relevance of a cost-saving energy concept is also apparent here: more than half of the running costs are accounted for by energy and maintenance. A large part of the energy costs is due to ventilation and air conditioning.

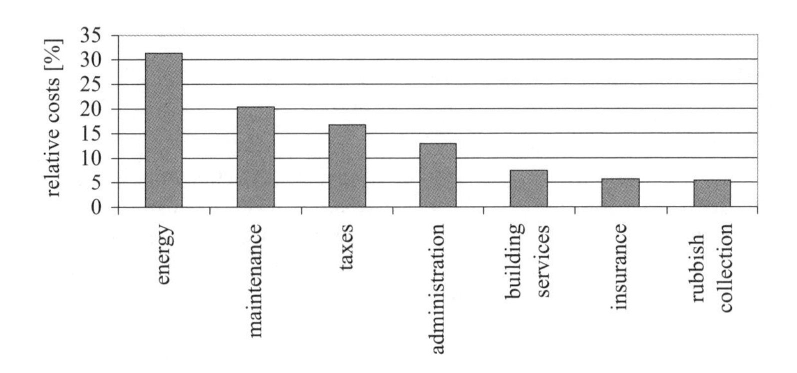

Figure 1.6: Percentage distribution of operating costs of office buildings per square metre of net surface area.

Heat consumption in administrative buildings can be reduced without difficulty, by improved thermal insulation, to under 100 kWh/m²a, and even to a few kWh per square metres and year in a passive building. Related to average consumption in the stock, a reduction to 5–10% is possible. Electricity consumption dominates total energy consumption where the building shell is energy-optimised and can be reduced by 50% at most. Even in an optimised passive energy office building in southern Germany, the electricity consumption remained at about 33 kWh/m²a, mainly due to the consumption of energy by office equipment such as computers.

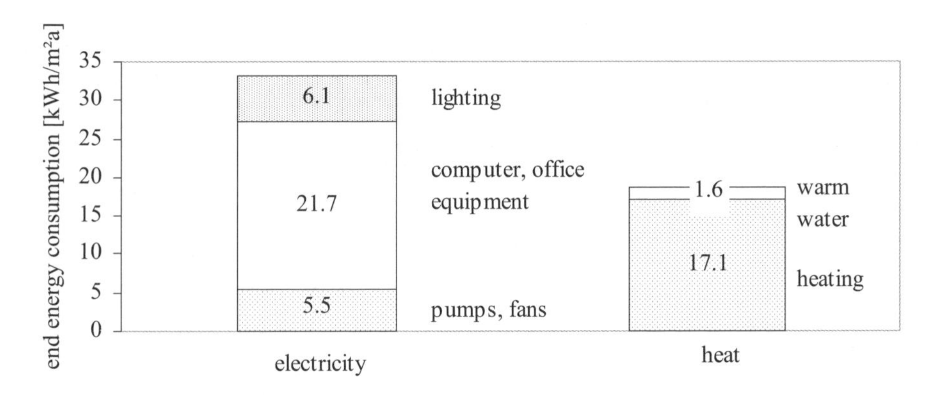

Figure 1.7: Measured consumption of electricity, heat and water heating in the first operational year of an office building with a passive house standard in Weilheim/Teck, Germany (Seeberger, 2002).

While the measured values for heat consumption correspond well with the planned values, the measured total electricity consumption exceeds the planned value of 23.5 kWh/m²a by 42%.

A survey of good practice in office buildings in Britain showed that the electricity consumption in naturally ventilated offices is 36 kWh/m²a for a cellular office type, rising to 61 kWh/m²a for an open plan office and up to 132 kWh/m²a for an air-conditioned office (Zimmermann, 1999).

1.1.3 Air conditioning

In Europe energy consumption for air conditioning is rising rapidly. This is due to increased internal loads through electrical office appliances, but also to increased demand for comfort in summer. Summer overheating in highly glazed buildings is often an issue in modern office buildings, even in northern European climates. This unwanted and often unforeseen summer overheating leads to the curious fact that air conditioned buildings in northern Europe sometimes consume more cooling energy than those in Southern Europe that have a more obvious architectural emphasis on summer comfort. According to an analysis of a range of office buildings, an average of 40 kWh/m²a was obtained for southern climates, whereas 65 kWh/m²a were measured in northern European building projects (Mat Santamouris, University of Athens, private communication, 2002).

The largest European air conditioning manufacturer and consumer is Italy, accounting for nearly half of all European production (Adnot, 1999). Sixty-nine per cent of all room air conditioner sales are split units, with total annual sales of about 2 million units. In 1996 the total number of air conditioning units installed in Europe was about 7 500 000 units. Between 1990 and 1996 the electricity consumption for air conditioning in the European Union has risen from about 1400 GWh/year to 11 000 GWh/year and further increases up to 28 000 GWh/year are predicted by Adnot for the year 2010. Without any policy intervention or technological change for solar or waste heat-driven cooling machines, the associated CO_2 emissions will rise from 0.6 million tons in 1990 to 12 million tons in 2010. The average coefficients of performance for all cooling technologies is currently about 2.7 (cooling power to electricity input), with a target of about 3.0 for 2015.

Cooling energy is often required in commercial buildings, with the highest consumption world-wide in the USA. In Europe the cooling energy demand for such buildings varies between 3 and 30 MWh/year. Very little data is available for area-related cooling energy demand. Breembroek and Lazáro (1999) quote values between 20 kWh/m²a for Sweden, 40–50 kWh/m²a for China and 61 kWh/m²a for Canada.

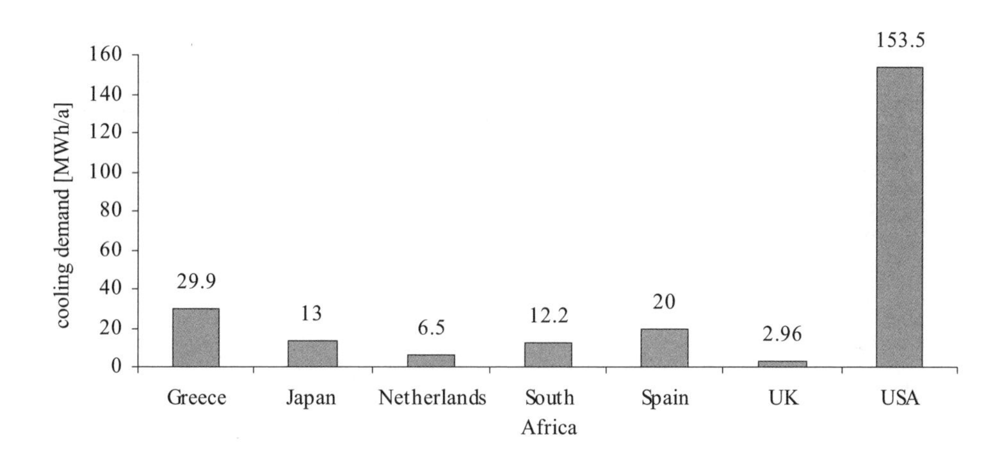

Figure 1.8: Cooling energy demand for new commercial buildings (Breembroek and Lazáro, 1999).

Under German climatic conditions, demand for air conditioning exists only in administrative buildings with high internal loads, provided of course that external loads transmitted via windows are reduced effectively by sun-protection devices. In such buildings, the average summer electricity consumption for the operation of compression refrigerant plants is about 50 kWh/m²a, i.e. the primary energy requirement for air-conditioning is 150 kWh/m²a, higher than the heating energy consumption of new buildings (Franzke, 1995).

In Southern Europe, the installed cooling capacity is often dominated by the residential market. Although in Spain less than 10% of homes have air conditioning systems, 71% of the installed cooling capacity is in the residential sector (Granados, 1997).

About 50% of internal loads are caused by office equipment such as PC's (typically 150 W including the monitor), printers (190 W for laser printers, 20 W for inkjets), photocopiers (1100 W) etc., which leads to an area-related load of about 10–15 W/m². Modern office lighting has a typical connected load of 10–20 W/m² at an illuminance of 300–500 lx. The heat given off by people, around 5 W/m² in an enclosed office or 7 W/m² in an open-plan one, is also not negligible. Typical mid-range internal loads are around 30 W/m² or a daily cooling energy of 200 Wh/m²d, in the high range between 40–50 W/m² and 300 Wh/m²d (Zimmermann, 1999).

Table 1.2: Approximate values for nominal flux of light, and specific connected loads of energy-saving lighting concepts (Steinemann *et al.*, 1992).

Room type	*Required illuminance levels* [lx]	*Specific electrical power requirement* [W/m²]
Side rooms	100	3–5
Restaurants	200	5–8
Offices	300	6–8
Large offices	500	10–15

External loads depend greatly on the surface proportion of the glazing as well as the sun-protection concept. On a south-facing facade, a maximum irradiation of 600 W/m² occurs on a sunny summer day. The best external sun-protection reduces this irradiation by 80%. Together with the total energy transmission factor (g-value) of sun-protection glazing of typically 0.65, the transmitted external loads are about 78 W per square metre of glazing surface. In the case of a 3 m² glazing surface of an enclosed office, the result is a load of 234 W, which creates an external load of just about 20 W/m² based on an average surface of 12 m². This situation is illustrated in the Figure 1.9 for south, east and west-facing facades in the summer:

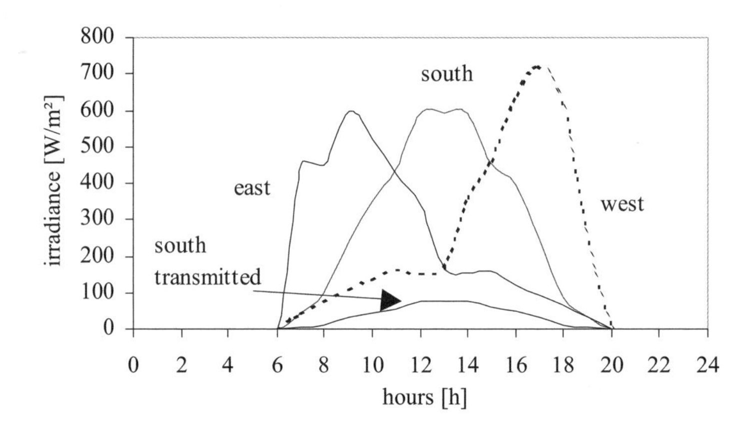

Figure 1.9: Diurnal variation of irradiance on different facade orientations and transmitted irradiance by a sun-protected south facade on a day in August (Stuttgart).

The reducing coefficients of sun-protection devices depend particularly on the arrangement of the sun protection: external sun protection can reduce the energy transmission of solar radiation by 80%, whereas with sun protection on the inside a reduction of at most 60% is possible.

Table 1.3: Energy reduction coefficients of internal and external sun protection (Zimmermann, 1999).

Sun shading system	*Colour*	*Energy reduction coefficient* [–]
External sun shades	Bright	0.13 – 0.2
External sun shades	Dark	0.2 – 0.3
Internal sun shades	Bright	0.45 – 0.55
Reflection glazings	–	0.2 – 0.55

The total external and internal loads leads to an average cooling load in administrative buildings of around 50 W/m².

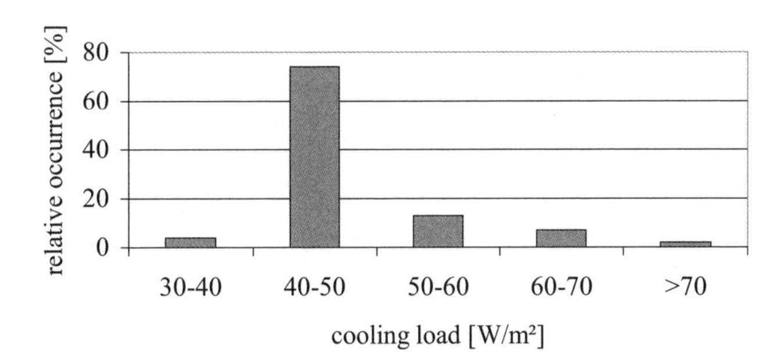

Figure 1.10: Occurrence of typical loads of administrative buildings in Germany.

With a cooling load of 50 W/m² the loads are typically distributed as shown in Figure 1.11.

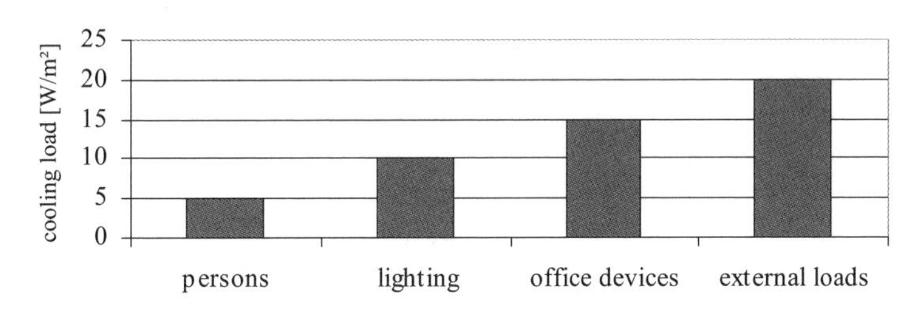

Figure 1.11: A typical breakdown of the cooling load at a total load of 50 W/m².

1.2 Meeting requirements by active and passive solar energy use

1.2.1 Active solar energy use for electricity, heating and cooling

Active solar-energy use in buildings today contributes primarily to meeting electricity requirements by photovoltaics, and to warm water heating by solar thermal collectors. Meeting the space heating requirement by solar thermal systems is recommended if conventional heat insulation potential is fully exhausted or if special demands such as monument protection or facade retention do not permit external insulation. Support heating with thermal collectors, with small contributions of approximately 10–30%, is always possible without significant surface-specific losses. Outside-air pre-heating with thermal air collectors can also make a significant contribution to reducing ventilation heat losses.

In air-conditioned buildings, thermal cooling processes such as open and closed sorption processes can be powered by active solar components.

When considering the potential solar contribution to the different energy requirements in buildings (heating, cooling, electricity), it is necessary to analyse the solar irradiance, the transformation efficiency of the solar technology in question and the available surface potential in buildings as well as the economically usable potential.

For a first design of a solar energy system, it is usually sufficient to consider the annual solar energy supply on the receiver surface. The maximum annual irradiance is achieved in the northern hemisphere on south-facing surfaces inclined at an angle of the geographical latitude minus about 10°. In Stuttgart the maximum irradiation on a 38° inclined south-facing surface is 1200 kWh/m²a. A deviation from south orientation of + or – 50° leads to an annual irradiation reduction of 10%. A south-facing facade receives about 72% of the maximum possible irradiation G_{max} (defined in Figure 1.12 as 100%).

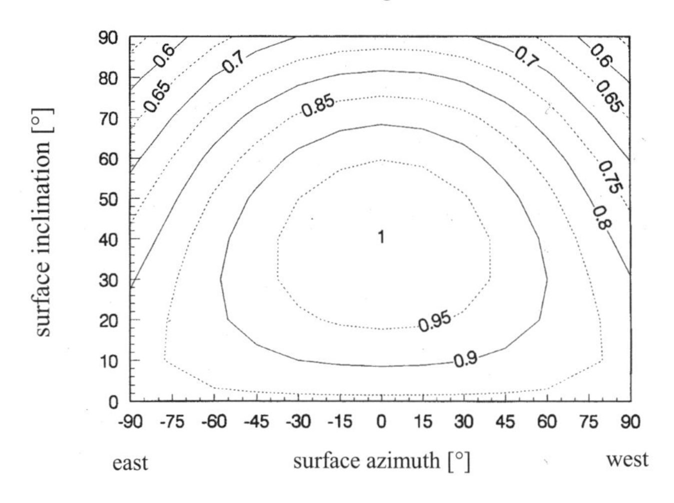

Figure 1.12: Annual irradiation depending on surface azimuth and angle of inclination in Stuttgart (Staiß, 1996).

An azimuth of 0° corresponds here to south-orientation. From surface orientation and system efficiency of the selected solar technique, the annual system yield can be estimated. Thus for example a photovoltaic solar system with an efficiency η_{PV} of 10% on a south-facing surface inclined at 40° from the horizontal, at an annual irradiation G of 1200 kWh/m²a, produces an annual system yield of

$$Q_{PV} = \eta_{PV}\, G = 0.1 \times 1200 \frac{kWh}{m^2 a} = 120 \frac{kWh}{m^2 a}$$

and accordingly a thermal solar plant for water heating with 35% solar thermal efficiency η_{st} produces about

$$Q_{st} = \eta_{st}\, G = 0.35 \times 1200 \frac{kWh}{m^2 a} = 420 \frac{kWh}{m^2 a}$$

For an economical electricity consumer with a yearly consumption of 2000 kWh, a

17 m² PV system would be sufficient to meet annual requirements (this corresponds to an installed performance of about 2 kW). Accordingly, in administrative buildings with an electricity requirement of between 25 and 50 kWh/m²a, a PV system with 20–40% of the effective area would have to be used to fully cover requirements.

On this calculation basis, for a medium-range warm water requirement of 2500 kWh per year, a surface of 6 m² would be sufficient for 100% cover.

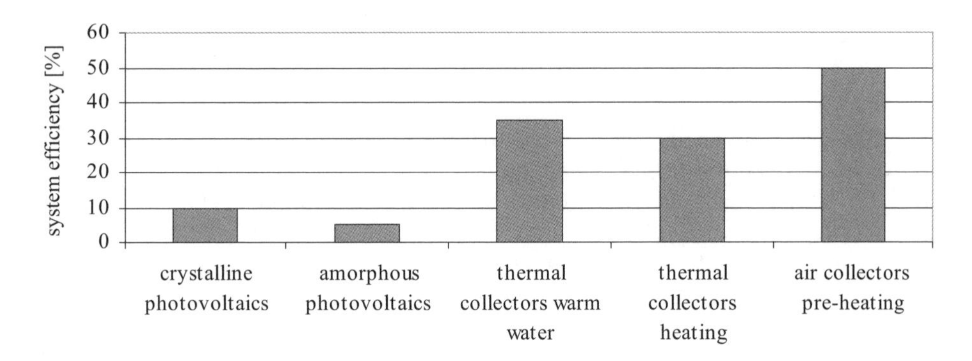

Figure 1.13: Average annual system efficiencies of active solar technologies.

However, because of low irradiance levels in winter, the annual requirements with this surface are covered to 60–70% at most. With heating-supporting systems it is assumed that all-year use of the thermal collectors is possible through warm water heating in the summer. Due to the oversizing of the collector surface in the summer, however, the specific yield drops.

For more specific uses of solar technology for heating only (for example, air collectors for fresh air pre-heating) or cooling, the irradiance must be divided into at least the two periods of summer and winter, in order to make possible a rough estimation of yield.

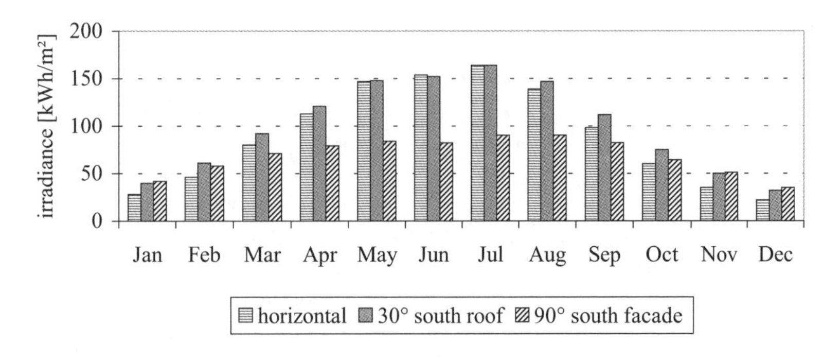

Figure 1.14: Monthly irradiation of differently inclined surfaces in Stuttgart.

If, for example, an air collector system, displaying a high efficiency of 50% with small rises in temperature and no heat exchange losses, is used on a south-facing facade for fresh air pre-heating, then an energy yield of 200 kWh/m² can be obtained during a heating season irradiation (October–April) of 400 kWh/m².

With solar thermal applications for air-conditioning, the system efficiency is calculated as the product of the solar yield η_{st} and the performance figure of the cooling machine. With open or closed sorption systems with low-temperature heat drive, the performance figures are, for instance, 0.7–0.9. If, for example, the summer irradiation (June–September) on a south-facing roof area is 575 kWh/m², and the thermal efficiency η_{st} of the solar plant is on average 40%, a surface-related energy quantity for air-conditioning of

$$Q_{\text{cooling}} = \eta_{st}\, G \times COP = 0.40 \times 575 \frac{kWh}{m^2} \times 0.8 = 184 \frac{kWh}{m^2}$$

Thus in the case of an average cooling power requirement of 125 kWh per square metre of effective area, the result is an area requirement of 0.7 m² collector surface per square metre of effective area. Although irradiation and cooling requirements clearly correlate better in summer than in winter, a possible phase shift between supply and requirement cannot be considered in the rough estimation. For this, dynamic system simulations based on the physical models described in the following chapters are necessary.

1.2.2 Meeting heating energy requirements by passive solar energy use

The most important component of passive solar energy use is the window with which short-wave irradiation can be very efficiently converted into space heating, and daylight made available. The total energy transmission factor of the glazing corresponds to efficiencies of active solar components, about 65% with today's double glazed coated low-emissivity windows. Thus an energy quantity of 260 kWh/m² per square metre of window area can be obtained on a south-facing facade in the heating season, as long as no space overheating occurs in the transition period due to over-large window areas.

$$Q_{heating\ period} = gG = 0.65 \times 400 \frac{kWh}{m^2} = 260 \frac{kWh}{m^2}$$

For a net energy balance, transmission heat losses must be deducted from solar gains, which for a thermally insulated glazing with a heat transfer coefficient (U-value) of 1.1 W/m²K are about 90 kWh/m². The result is a net maximum energy gain of some 170 kWh/m².

A further element in passive solar energy use is transparent thermal insulation of solid external walls. With similar values as good thermally insulated glazing (U-values around 1 W/m²K and g-values between 0.6–0.8, depending upon thickness and structure), similar energy savings to windows can be made with transparent thermal insulation. Here, too, overheating problems are crucial in the spring and autumn transition period for the total yield, which in practice lies between 50 and 150 kWh/m².

2 Solar irradiance

A square metre of horizontal Earth surface receives, under German climatic conditions, between 925 kWh/m² in the north to 1170 kWh/m² in the south solar irradiance annually, i.e. a daily average of around 3 kWh/m^2. Direct solar irradiance accounts for just about 50%; the remainder is diffuse irradiance from the atmosphere. In southern Europe the annual irradiance on a horizontal surface can reach up to 1770 kWh/m² (Almeria/Spain at 37° northern latitude). In northern Europe with geographical latitudes between 60–70° the irradiance drops from 990 kWh/m² in Helsinki to a low of 700 kWh/m² in northern Norway, where there is no sun for four months in winter.

For the dimensioning and yield prognoses of active and passive solar technology in buildings, it is often not sufficiently exact to determine only the monthly or annual solar irradiance on a roof or a facade surface and to then multiply this by the system efficiency. In particular with the thermal use of solar energy (active and passive), the dynamic storage behaviour of building components and heat stores is decisive for the solar contribution to energy requirements.

For system simulation a temporal resolution of the solar irradiance of one hour has proved a good compromise between computational accuracy and computational time, so in what follows, time series of hourly average irradiance will be presented. On the basis of the purely geometry-dependent hourly irradiance on a surface outside the atmosphere, extraterrestrial irradiance, weakening and dispersion in the atmosphere are taken into account using statistical methods. Allocation of the irradiance into a direct proportion of irradiance diffused by atmospheric dispersion enables the conversion of horizontal irradiance on surfaces oriented at any angle. The effect of shading, which plays a significant role particularly in urban space, can be determined from the geometrical relationship between recipient surfaces and points in the sky.

2.1 Extraterrestrial solar irradiance

2.1.1 Power and spectral distribution of solar irradiance

The radiating power of the sun results from a process in which four hydrogen nuclei fuse into a helium nucleus. The resulting loss of mass, totalling 4.3 million tons per second, is transferred into a freed power of 3.845×10^{26} W. This energy, released at extremely high temperatures (> 10^7 K), is transferred by irradiance and convection to the outer photosphere.

Extraterrestrial irradiance develops predominantly in the photosphere, which is composed of inhomogeneous gases of low density. The photosphere consists of strongly ionised gases which constantly recombine with free electrons and whose kinetic energy is transferred into a continuous irradiance spectrum. Over it is a reversal layer of some 100 km thickness, which contains almost all the elements in the earth's crust. The chromosphere, consisting of hydrogen and helium and about 2500 km thick, forms the sun's atmosphere together with the reversal layer. The corona, which expands far into the solar system, is a gas layer much hotter than the chromosphere (Iqbal, 1983). If the sun is regarded as a black body, an equivalent irradiance temperature, T_s, can be calculated from the specific radiant emittance M, on the basis of the Stefan–Boltzmann Law (with the Boltzmann constant $\sigma = 5.67051 \times 10^{-8}$ W/(m^{2}K^4)). The specific radiant emittance is defined as the relationship of total radiating power ϕ and sun surface A_s (6.0874×10^{12} km^2).

$$M(T) = \frac{\Phi}{A_S} = \sigma T_S^4 = \frac{3.845 \times 10^{26} W}{6.0874 \times 10^{18} m^2} = 63.11 \frac{MW}{m^2}$$
$$\Leftrightarrow T_S = \sqrt[4]{\frac{M}{\sigma}} = 5777 K \qquad (2.1)$$

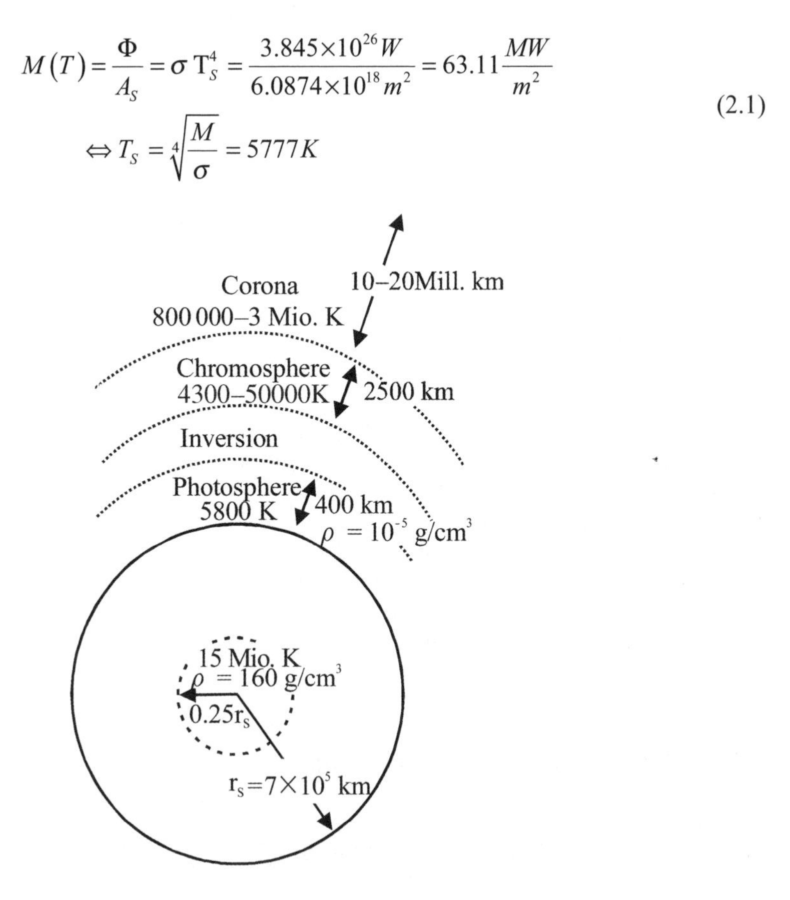

Figure 2.1: Structure of the sun (not to scale).

The irradiance G_{sc} outside the terrestrial atmosphere, known as the solar constant, can be calculated from the entire radiant emittance of the sun (the product of specific radiant emittance and sun surface MA_s), by relating this radiant emittance to a square metre of the spherical surface A_{se}, which is formed by the radius of the distance earth–sun. The average distance between the earth and the sun of $r_0 = 1.496 \times 10^{11}$ m is called an astronomical unit (AU).

$$G_{sc} = M\frac{A_s}{A_{se}} = M\left(\frac{r_s}{r_0}\right)^2 = 63.11\left(\frac{6.9598\times10^8}{1.4959789\times10^{11}}\right)^2\frac{MW}{m^2} = 1367\frac{W}{m^2} \quad (2.2)$$

The sun's radiating power of 3.845×10^{26} W is diluted by the squared relationship of its radius r_s to the sun-earth distance r_0 (factor 2.16×10^{-5}).

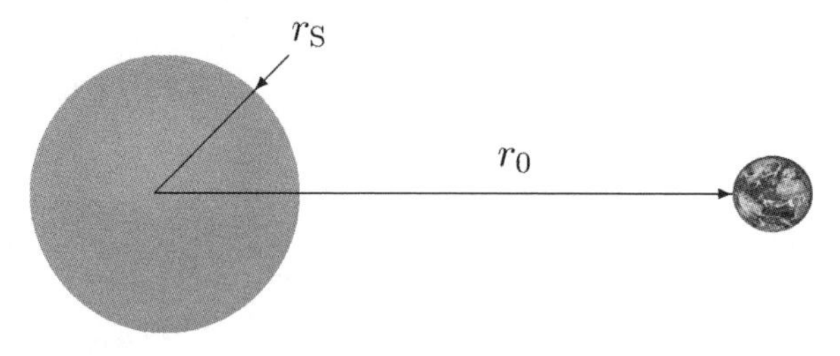

Figure 2.2: Sun radius and sun–earth distance.

The deviations between the spectral distribution of a black emitter and measured extraterrestrial irradiance are caused by absorption and dispersion in the outer, cooler layers of the photosphere, which apart from hydrogen and helium contains about 2 % by mass of heavy elements. Altogether 20 000 accurately measured absorption lines in the solar spectrum can be observed.

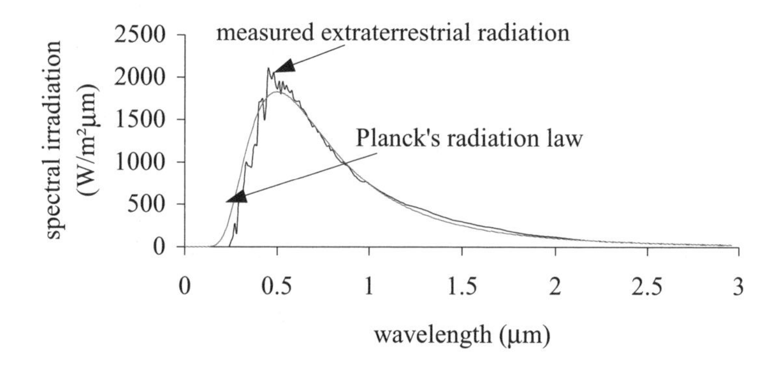

Figure 2.3: Measured and calculated spectral distribution of the sun.

The spectral radiant emittance of a thermal emitter G_λ [in $Wm^{-2}\mu^{-1}$] is calculated, using Planck's irradiance law, as a function of the temperature T [K] and wavelength λ [µm]:

$$G_\lambda = \frac{C_1}{\lambda^5\left(\exp\left(C_2/(\lambda T)\right)-1\right)} \quad (2.3)$$

with constants $C_1 = 3.7427 \times 10^8$ W μm^4 m^{-2} and $C_2 = 1.4388 \times 10^4$ µm K. At a sun surface temperature of 5777 K the calculated spectrum is illustrated in Figure 2.3. If the extraterrestrial irradiance is integrated over the wavelength, the cumulative irradiance power can be calculated. Only 48% of the extraterrestrial irradiance intensity is in the visible range of 380–780 nm (1 nm = 10^{-9} m). Apart from ultraviolet irradiance (< 380 nm), which accounts for 6.4% of the total intensity, 45.6% is given off in the upper infrared. Above 3000 nm the irradiance is negligible.

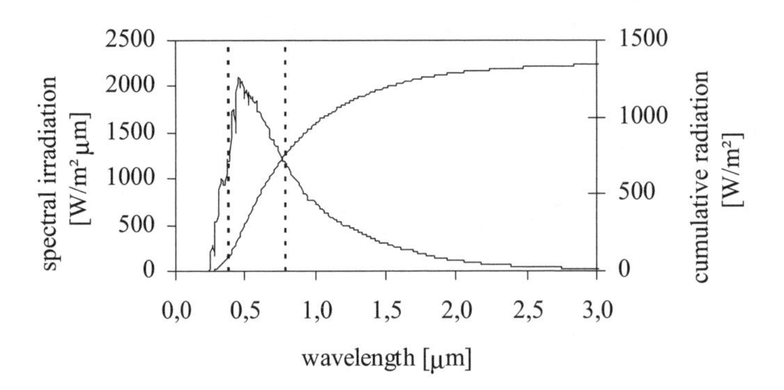

Figure 2.4: Spectral intensity and cumulative power of extraterrestrial irradiance.

Total ultraviolet irradiance below 0.38 µm is about 92.6 W/m^2. The visible area within the broken lines has a total power of 660 W/m^2, and the remainder of the total irradiance of 1367 W/m^2 is infrared.

2.1.2 Sun–Earth geometry

The orbit of the earth around the sun in the so-called ecliptic plane is slightly elliptical with a minimum distance of 0.983 AU on 3 January and a maximum distance of 1.017 AU on 4 July.

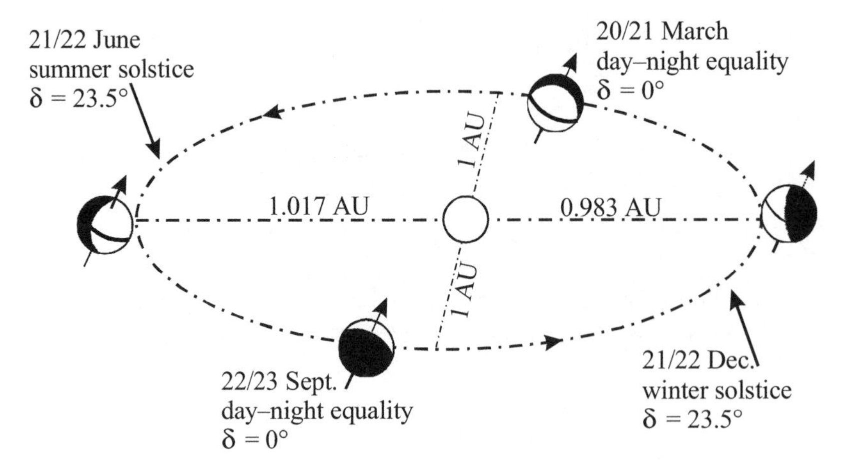

Figure 2.5: Ecliptic plane and position of the earth at the winter and summer solstices, and also at the spring and autumn equinoxes.

The changing distance leads to a fluctuation of the extraterrestrial irradiance on a normal surface G_{en} of approximately ±3%. For a given day-number n the irradiance can be calculated by a simple approximation formula (Duffie and Beckmann, 1980) (error < 0.3%) or more exactly by a Fourier series expansion (Spencer, 1971).

$$G_{en} = G_{sc}\left(\frac{r_0}{r}\right)^2 = G_{sc}\left(1+0.033\cos\left(\frac{360\,n}{365}\right)\right) \tag{2.4}$$

$$G_{en} = G_{sc}\begin{pmatrix}1.000110+0.034221\cos B+0.001280\sin B\\ +0.000719\cos 2B+0.000077\sin 2B\end{pmatrix} \tag{2.5}$$

$$\text{with} \qquad B = 360^\circ\,\frac{n-1}{365}$$

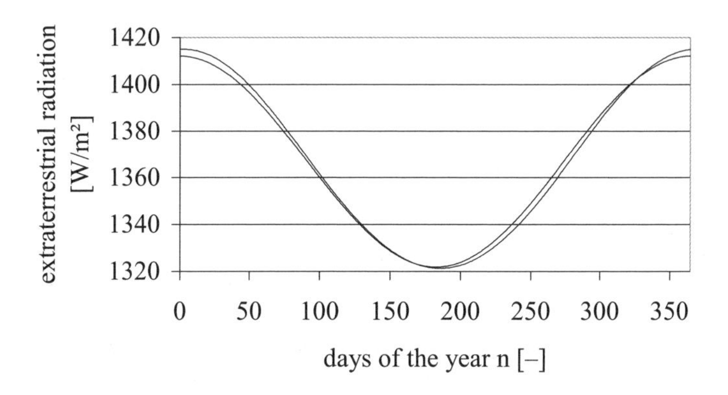

Figure 2.6: Variation of the extraterrestrial irradiance on a normal surface calculated by Fourier expansion (thick curve) and by the approximation equation (thin curve).

2.1.2.1 Equator coordinates

The earth rotates around its polar axis, which is inclined at 23.45° to the normal of the ecliptic plane. The daily irradiance fluctuations are caused by the rotation on the polar axis, the seasonal ones by the inclination of the polar axis relative to the sun. Let us first analyse the seasonal modifications of the irradiance, which result from the movement of the Earth around the sun with the polar axis in a constant position in space.

Declination

If the sun is theoretically 'observed' from the centre of the earth during the annual circulation of the earth around the sun, the angle between sun direction and equatorial plane constantly changes. This angle is called the declination and is positively defined in the northern hemisphere. Positive declination angles (sun above the equator) to a maximum of

23.45° characterise the summer in the northern hemisphere, negative declination angles to –23.45° the winter. At the autumn and spring equinoxes, the declination is 0°.

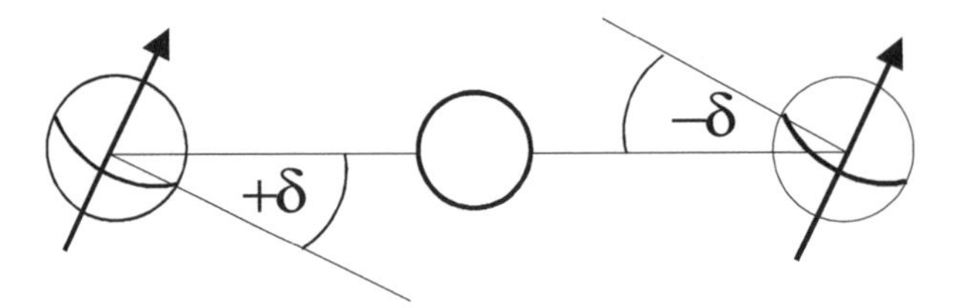

Figure 2.7: Positive and negative declination angles in summer and winter in the northern hemisphere.

The change in declination in one day is at most 0.5° and is often ignored. The declination can be calculated by a simple approximation equation or by a Fourier series:

$$\delta = 23.45 \sin\left(\frac{360}{365}\,(284 + \mathrm{n})\right) \tag{2.6}$$

$$\delta = \begin{pmatrix} 0.006918 - 0.399912\cos B + 0.070257\sin B - 0.006758\cos 2B \\ + \ 0.000907\sin 2B - 0.002697\cos 3B + 0.00148\sin 3B \end{pmatrix}\frac{180°}{\pi} \tag{2.7}$$

Hour angle and equation of time

While the declination clearly determines the position of the earth in the ecliptic plane relative to the sun, the hour angle characterises the daily irradiance fluctuations due to the earth's rotation. The hour angle is defined as the angle between local longitude and the longitude at which the sun is at its zenith. Hour angle $\omega = 0$ means the highest position of the sun during the day (the so-called upper culmination), i.e. in the northern hemisphere the time when the sun is exactly in the south. The true solar time or the true local time *TLT* is 12.00 h, at the highest position of the sun.

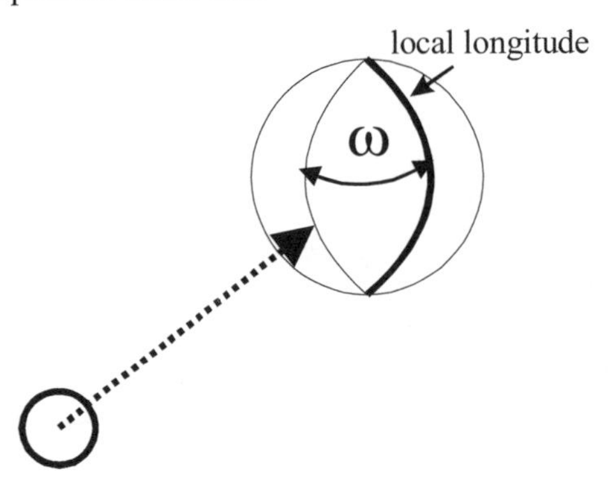

Figure 2.8: Hour angle ω of the sun as an angle deviation between local longitude and longitude with the sun at its zenith.

A sun's day of ω = 0 to ω = 360° is the period between two consecutive days in which the sun crosses the local meridian (degree of longitude of the observer) in each case. Since the earth does not move at a constant rate in its elliptical orbit, one sun-day is not exactly 24 h long. At the nearest point to the sun (perihelion on 3 January), the velocity is highest, according to Kepler's second law, i.e. the sun-day is at its shortest; at the furthest point (aphelion on 4 July) it is smallest. This velocity-determined whole-year period is overlain by a second deviation with biannual periodicity, which results from the projection of sections of orbit of equal length on the ecliptic plane onto sections of orbit of different lengths on the equatorial plane.

In order to calculate the hour angle and thus the true solar time *TLT* from the local time-of-day with a day length fixed to 24 h, the local time must be corrected by this temporal deviation (equation of time E_t in minutes).

$$E_t = 229.2 \begin{pmatrix} 0.000075 + 0.001868 \cos B - 0.032077 \sin B \\ -0.014615 \cos 2B - 0.040849 \sin 2B \end{pmatrix} \tag{2.8}$$

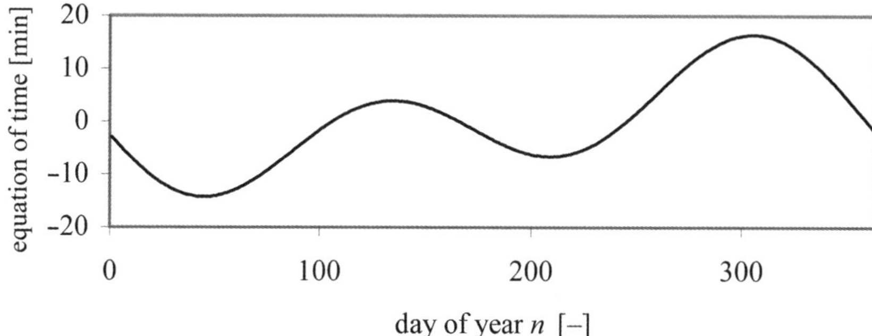

Figure 2.9: Equation of time as a deviation of the true solar time from universal time.

A further correction of local time is necessary, as only zones of around 15° longitude have their own standard time; there is not one for each degree of longitude. In most Western European countries, Central European Time applies, which corresponds to 15° longitude east. For each degree of deviation between local (L_{local}) and Standard Time Meridian (L_{Zone}), a longitudinal correction L_c of 4 minutes must be taken into account (360°/(24 h × 60 min/h)).

$$L_c = 4\left(L_{Zone} - L_{local}\right) \quad [\text{min}] \tag{2.9}$$

The true local time *TLT* (or solar time) therefore results from the standard time corrected by the constant longitude correction as well as by the equation of time. During Central European Summer Time a further hour must be deducted.

$$\begin{aligned} TLT &= CET - L_c + E_t \\ &= CET_s - 1h - L_c + E_t \end{aligned} \tag{2.10}$$

The hour angle ω results directly from the solar time. An hour angle of 15° corresponds to a one-hour deviation from solar noon (in the morning negatively, in the afternoon positively).

$$\omega = (TLT - 12.00h) \times \frac{15°}{h} \tag{2.11}$$

Example 2.1

Calculation of the hour angles of the sun at 12.00 h local time in Stuttgart (longitude 9.2° east) on 1 February, 1 July and 1 October (taking summer time into consideration).

For Stuttgart the result is a constant longitude correction of L_c = 4 min/° × (15° – 9.2°) = 23.2 min = 0.387 h.

Date	*day number n*	*B* [°]	E_t [min]	E_t [h]	*TLT*	ω [°]
1.2.	32	30.57	–13.1	–0.218	11.39h	–9.08
1.7. (CET_S)	182	178.52	–3.5	–0.058	10.55h	–21.68
1.10. (CET_S)	274	269.26	+10.5	0.175	10.79h	–18.18

For the calculation of the *TLT* both longitude correction and equation of time were converted decimally into hours.

2.1.2.2 Horizon coordinates

From the declination and hour angle the sun's height α_s and azimuth γ_s can be determined for each location with latitude ϕ. For these angles, described as horizon coordinates, the azimuth γ_s for the north direction is defined as 0°, east +90°, south +180° and west 270°, and the elevation angle α_s is determined from the horizontal plane. The complementary angle to the elevation angle is the zenith angle θ_z, which at the same time represents the angle of incidence of direct solar irradiance on a horizontal surface.

$$\cos\theta_z = \sin\delta\sin\Phi + \cos\delta\cos\Phi\cos\omega = \sin\alpha_s \tag{2.12}$$

$$\gamma_s = \begin{cases} 180° - \arccos\left(\dfrac{\sin\alpha_s \sin\Phi - \sin\delta}{\cos\alpha_s \cos\Phi}\right) & \text{for} \quad TLT \leq 12.00\,\text{h} \\ 180° + \arccos\left(\dfrac{\sin\alpha_s \sin\Phi - \sin\delta}{\cos\alpha_s \cos\Phi}\right) & \text{for} \quad TLT > 12.00\,\text{h} \end{cases} \tag{2.13}$$

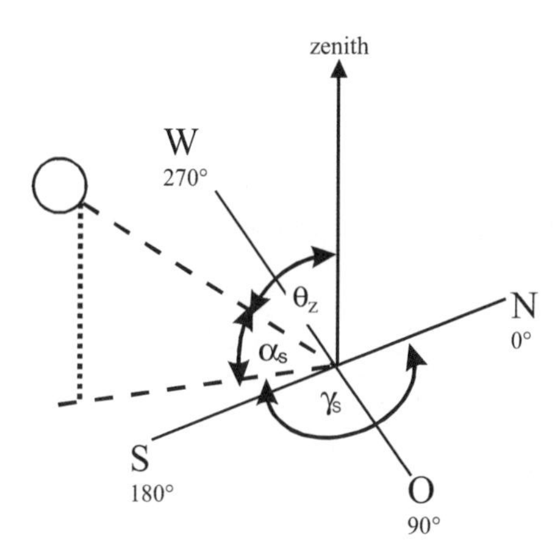

Figure 2.10: Zenith angle θ_z, elevation angle α_s and azimuth angle γ_s of the sun.

Example 2.2

a) Determination of the angles of the solar azimuth between the autumn and spring equinoxes in the Northern Hemisphere:

Between 23.9 and 21.3, the solar azimuth always lies between east and west, i.e. 90° and 270°. This results directly from Equation (2.13), as the declination is $\delta = 0°$ and the elevation angle $\alpha_s = 0°$ at sunrise and sunset, and thus arccos (0) = 90°.

b) Determination of the solar azimuth at the equator at 12.00 h TLT at the winter and summer solstice in the northern hemisphere:

21 December:

Declination angle –23.45°, i.e. the sun is below the equator and thus exactly in the south: $\gamma_s = 180°$.

21 June:

Declination angle +23.45°, i.e. the sun is above the equator and thus exactly in the north: $\gamma_s = 0°$.

Sunrise time and length of day

At sunrise and sunset the zenith angle θ_z is exactly 90°. From it the hour angle ω_s results, based on Equation (2.12).

$$\cos\omega_s = -\frac{\sin\phi\sin\delta}{\cos\phi\cos\delta} = -\tan\phi\tan\delta \qquad (2.14)$$

The number of daylight hours N can be calculated from the hour angle, since the hour angle changes by 15° every hour. The factor 2 results from taking into consideration morning and afternoon hours.

$$N = \frac{2}{15^\circ} \arccos\left(-\tan\phi \ \tan\delta\right) \tag{2.15}$$

Example 2.3

Calculation of the hour angles of sunrise, and of the number of daylight hours for the 1 February, 1 July and 1 October in Stuttgart (48.8° northern latitude).

	1.2	1.7	1.10
Declination δ [°]	–17.5	23.1	–4.2
ω_S [°]	–68.4	–119.1	–85.2
N [h]	9 h 11 min	15 h 57 min	11 h 22 min

2.1.2.3 Sun-position diagrams

For the illustration of the sun's height and azimuth angles over the course of the year, at a given location, sun-position diagrams with either cartesian or polar coordinates can be used. Cartesian coordinates, with the elevation angle as a function of the azimuth, are particularly suitable for the representation of shading horizons. Shaded objects can easily be entered into the sun-position diagram with the respective azimuth and elevation angles, and the times of shading can be directly seen.

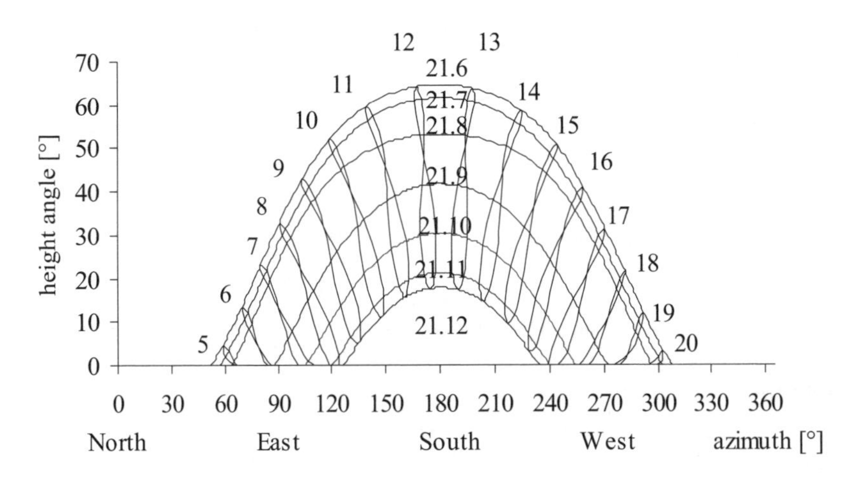

Figure 2.11: Sun-position diagrams in cartesian coordinates for Stuttgart.

The elevation angles calculated for the second half of the year can be used symmetrically for the first half (21 July corresponds to 21 May etc.). From the equation of time arise the lines of same local time, here *CET* (the so-called analemma). While cartesian coordinates illustrate the elevation angles of the sun and shading objects, polar diagrams make clear the azimuth angles of the sun and the position of further buildings.

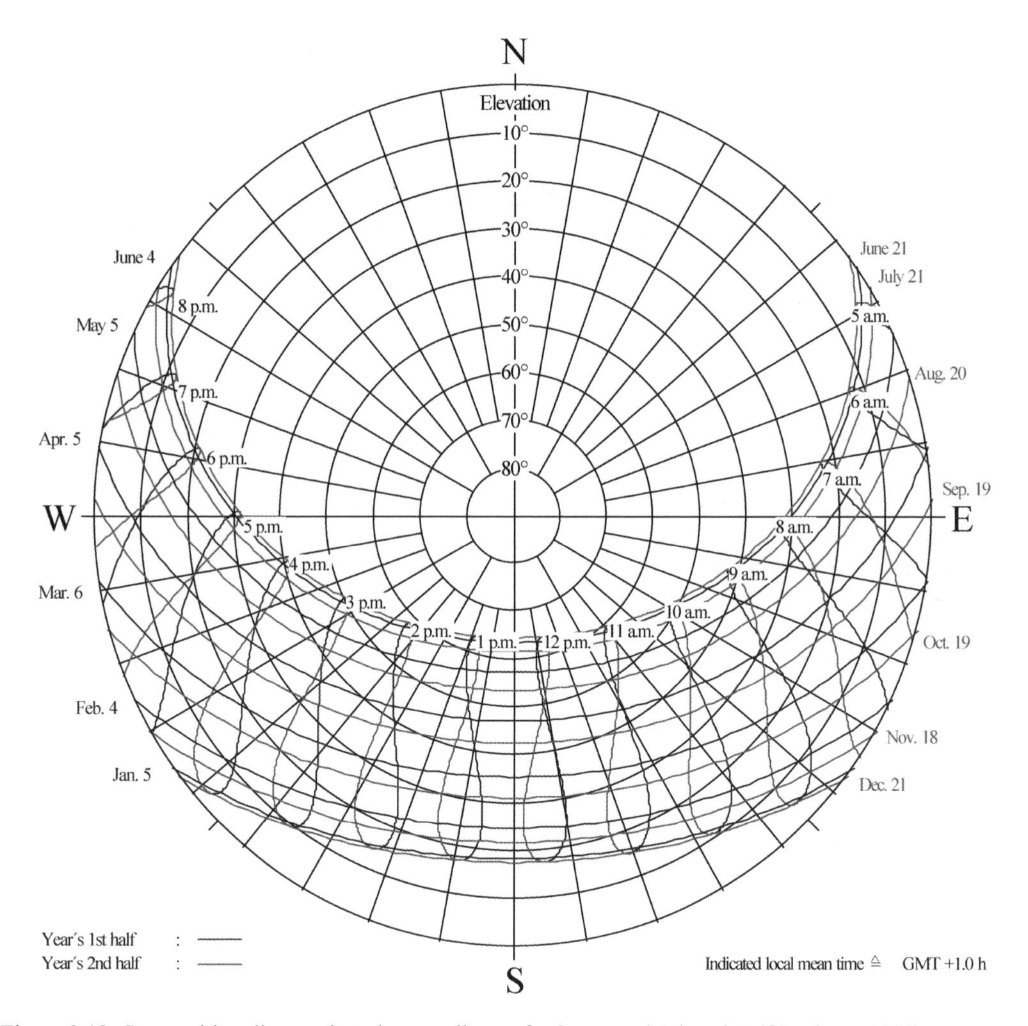

Figure 2.12: Sun-position diagram in polar coordinates for Stuttgart (University of Bochum, 1999)

The outer circle corresponds to an elevation angle of zero, i.e. the horizon, the centre of the polar diagram to the zenith.

Angle of incidence on randomly inclined surfaces

The angle of incidence of direct rays on an inclined receptor surface depends on the angle of inclination β of the surface to the horizontal, and also on the surface azimuth γ. From a given position of the sun (zenith angle θ_z and solar azimuth γ_s), the angle of incidence can be calculated from the horizon coordinates θ_z and γ_s:

$$\cos\theta = \cos\theta_z \cos\beta + \sin\theta_z \sin\beta \cos(\gamma_s - \gamma) \qquad (2.16)$$

or, when using equator coordinates δ and ω,

$$\begin{aligned}\cos\theta &= (\cos\beta\sin\Phi + \cos\Phi\cos\gamma\sin\beta)\sin\delta \\ &+(\cos\beta\cos\Phi - \sin\Phi\cos\gamma\sin\beta)\cos\delta\cos\omega \\ &- \sin\gamma\sin\beta\cos\delta\sin\omega\end{aligned} \tag{2.17}$$

Simplifications of the formula result e.g. from horizontal surfaces with $\beta = 0°$ (Equation (2.12) of the zenith angle) or vertical and south-oriented facade surfaces (β = 90°, γ = 180°):

$$\cos\ \theta = -\cos\Phi\sin\delta + \sin\Phi\cos\delta\cos\omega \tag{2.18}$$

At solar noon with $\omega = 0°$ the following formula applies for a south-oriented surface:

$$\theta_{\text{noon}} = |\Phi - \delta - \beta| \tag{2.19}$$

For a horizontal surface with $\beta = 0°$ the maximum and minimum positions of the sun in a year can thus be quickly determined. The highest position in Stuttgart on 21 July with a declination of 23.45° results for a zenith angle θ_z of 48.8° – 23.45° = 25.35°, which corresponds to a sun height of 64.65°. The lowest position of the sun at solar noon is with a declination of 23.45° on 21 December and a zenith angle of 48.8 – (–23.45) = 72.25°, i.e. a sun height of 17.75°.

Example 2.4

Calculation of the angle of incidence on a 10° inclined surface with a surface azimuth of 160° on 1st October at 11.00 h *TLT* in Stuttgart.

The zenith angle is θ_z = 54.6°, the sun azimuth γ_s = 161.5°. From this an angle of incidence of 44.6° results. The same result is obtained by use of the equator coordinates δ = –4.2° and ω = –15°.

2.2 The passage of rays through the atmosphere

While extraterrestrial irradiance values on randomly oriented surfaces are solely geometry-dependent and thus simple to calculate, the passage of rays through the atmosphere is so complex that simple procedures like the use of monthly turbidity factors (still used in the German daylighting standard DIN 5034) result in extremely inaccurate irradiance values. Particularly for dynamic system simulations of active or passive solar components, hourly-resolved irradiance values of the direct and diffuse irradiance are necessary; these must indicate the statistical characteristics typical of the location. Since long-term measured or partly synthesised irradiance time sequences are available for only a few locations worldwide (e.g. test reference years for 12 German climate zones), more and more statistical procedures are becoming generally accepted which produce hourly irradiance values from monthly average ones. After a short illustration of the main irradiance absorption and dispersion mechanisms of the atmosphere, what follows deals mainly with the statistical procedures of irradiance calculation.

Extraterrestrial irradiance is weakened in the atmosphere by absorption and reflection and partially converted by dispersion into diffuse irradiance. The relative air mass *m* (also termed *AM*) which solar irradiance has to pass through, gives the relation of atmospheric thickness for a given zenith angle, i.e. $d_{atm}/\cos\theta_z$ to the simple thickness of the atmosphere d_{atm} in the local zenith, and can be calculated for a homogeneous atmosphere by a simple approximation formula:

$$m = \frac{d_{atm}/\cos\theta_z}{d_{atm}} = \frac{1}{\cos\theta_z} \tag{2.20}$$

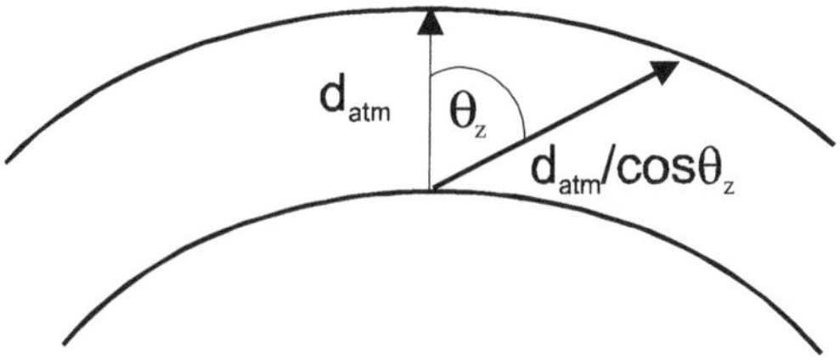

Figure 2.13: Definition of air mass m from atmospheric thickness and zenith angle.

Short-wave irradiance is scattered by air molecules, whose diameter is small in relation to the wavelength of light (around 10^{-10} m), proportionally to $1/\lambda^4$. Above around 0.6 µm the so-called Rayleigh scattering is negligible. Mie dispersion by larger dust particles (aerosols) with diameters of around 10^{-9} m reduces the transmittance according to the Angstroem turbidity formula $\tau = \exp\left(-\beta\lambda^{-\alpha}\mathrm{m}\right)$, the turbidity being characterised by the parameters α and β. Parameter β varies from 0 for very clear to 0.4 for very dull skies, and α depends on the size distribution of the dust particles and is typically about 1.3. Wavelength dependency is thus weaker than for Rayleigh scattering.

Ozone absorbs solar irradiance almost completely under $\lambda = 0.29$ µm and more weakly to around 0.7 µm. Water vapour absorbs in the infrared, with pronounced absorption bands at 1.0, 1.4 and 1.8 µm. Above 2.5 µm almost the entire irradiance is absorbed by CO_2 and H_2O.

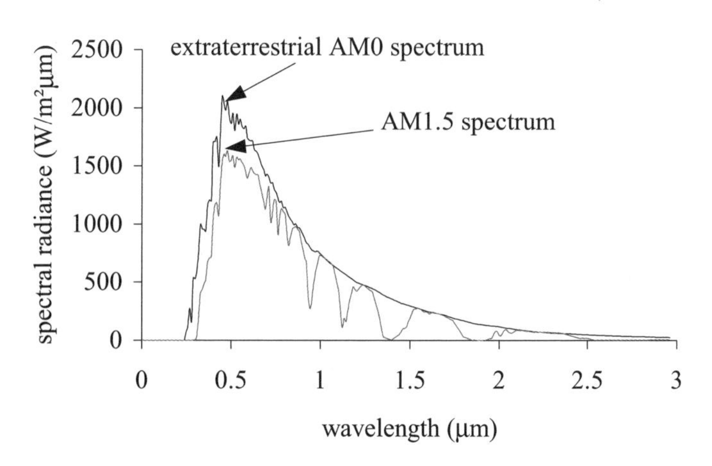

Figure 2.14: An extraterrestrial spectrum with zero air mass (AM0) and a terrestrial spectrum with air mass 1.5 (AM1.5).

The simplest calculation models of irradiance at the earth's surface combine all the above wavelength-dependent effects in a single number, the so-called turbidity factor T, which does not take into account the statistical fluctuations of terrestrial irradiance. T is indicated in some standards as a monthly average value, and varies from a minimum of 3.8 in January to 6.3 in September for German climatic conditions. With a clear sky the direct irradiance on a normal surface is air mass-corrected from the extraterrestrial irradiance by the sun-height angle α_s, and pressure-corrected by the height of the location above sea level H [m]:

$$G_n = G_{en} \exp\left(-\frac{T}{\underbrace{(0.9 + 9.4 \sin \alpha_S)}_{\text{air mass correction}} \underbrace{\exp(H/8000)}_{\text{pressure correction}}} \right) \tag{2.21}$$

The proportion of diffuse irradiance in a clear sky can likewise be calculated as a function of the sun's height and the turbidity factor. The conversion for inclined surfaces takes place by means of tabulated correction factors.

For an overcast sky the model becomes even simpler. One proceeds as usual in daylighting technology on the basis of a symmetrical irradiance or luminous intensity distribution, which has its maximum at the local zenith and decreases towards the horizon. The irradiance intensity in the zenith L_{ez} is determined by

$$L_{eZ} = 1.068 + 74.7 \sin \alpha_s \quad \left[\frac{W}{m^2\, sr} \right] \tag{2.22}$$

If one integrates this irradiance intensity over the entire hemisphere, the following formula is obtained for the horizontal irradiance.

$$G_h = 2.609 + 182.609 \sin \alpha_s \quad \left[\frac{W}{m^2} \right] \tag{2.23}$$

While these simple procedures are sufficient for the analysis of daylight systems, more exact meteorological data records are necessary for the sizing of active solar energy systems.

2.3 Statistical production of hourly irradiance data records

The statistical procedures described below enable the production of a series of hourly irradiance values, a time series, proceeding from a monthly average irradiance value. In order to eliminate the deterministic proportion of the irradiance on the earth's surface, which is determined by extraterrestrial irradiance and the respective position of the sun, the clearness index is used as a statistical variable. The clearness index k_t is defined as the relation of terrestrial to extraterrestrial irradiance on a horizontal surface, calculated for one hour or totalled over the hours of a day or month.

$$k_t = \frac{\sum G_h}{\sum G_{eh}} \tag{2.24}$$

From the given monthly clearness index, as a first step daily values are produced by a autoregressive procedure, and afterwards hourly values are calculated.

2.3.1 *Daily average values from monthly average values*

Two observations of long-term irradiance data form the basis for the statistical production of daily averages from monthly average values:

- Each daily irradiance average value correlates only to the preceding daily value.
- The probability distribution of daily clearness indices around the monthly average value is only determined by the average clearness index. For example, clear months have only small dispersions of the daily values around the average value, and vice-versa.

The time series of daily clearness indices can be calculated either by Markov transition matrices or by autoregressive procedures. Autoregressive procedures are more generally used, since not only correlations with the preceding daily value, but also with values of more than a one-day interval, can be taken into account. For a normally-distributed random variable Z_d with an average value of zero and a standard deviation of $\sigma = 1$ results the new Z_d-values from the correlated preceding values as well as a noise term r_d. The order n of the regression procedure gives the correlations which must be considered with values of more than a one-day interval.

$$Z_d = \rho_1 Z_{d-1} + \rho_2 Z_{d-2} + \cdots + \rho_n Z_{d-n} + r_d \tag{2.25}$$

For the clearness index calculation, a first-order autoregressive procedure is sufficiently exact, since the daily averages correlate particularly with the preceding day, and days which are longer ago have hardly any influence on the clearness index. ρ_1 defines the autocorrelation coefficient for an interval of one day ($n = 1$). For a given time series of N random variables Z_d, the autocorrelation coefficients ρ_n with time interval n can be calculated:

$$\rho_n = \frac{\sum_{i=1}^{N-n} \left(Z_i - \overline{Z}\right)\left(Z_{i+n} - \overline{Z}\right)}{\sum_{i=1}^{N} \left(Z_i - \overline{Z}\right)^2} \tag{2.26}$$

However, since time series are to be synthesised here and therefore their correlation characteristics are unknown, the autocorrelation coefficient of first order ρ_1 has to be given as a parameter. From meteorological data investigations by Gordon and Reddy (1988), it is evident that ρ_1 can vary between 0 and 0.6, depending upon location, but in most cases a value of 0.3 is a good approximation.

The statistical noise r_d is, with an average value of zero, normally distributed, and defined with a standard deviation $\sigma' = \sqrt{1-\rho_1^2}$. The noise is calculated from a random number sequence z from the value area [0,1]:

$$r_d = \sigma'\left(z^{0.135} - (1-z)^{0.135}\right)/0.1975 \quad (2.27)$$

Therefore, for the production of the time series of the normally distributed variables Z_d, the autocorrelation coefficient ρ_1 is given, the random variable Z_d is initialised with $Z_0 = 0$ and for each time-step with the noise term r_d the next value is calculated.

Gordon and Reddy have shown, however, that the daily clearness indices are not normally distributed around the mean monthly value, but that the probability function above the mean clearness index sinks faster than the Gaussian distribution. Therefore the time series of the Gaussian-distributed random variables Z_d, produced in the first step, needs to be converted into the non Gaussian-distributed variable X_d.

For the random variable X_d, the relation of the daily mean clearness index k_{td} to the monthly average value $\overline{k_{tm}}$ is chosen.

$$X_d = \frac{k_{td}}{\overline{k_{tm}}} \quad (2.28)$$

The empirical probability function $P(X_d)$ determined by Gordon and Reddy describes with good accuracy the distribution of the daily clearness indices. The only parameter is the standard deviation of the daily values σ_{Xd} for the respective location, which depends only on the monthly average value $\overline{k_{tm}}$.

$$P(X_d) = AX_d^n\left(1 - \frac{X_d}{X_{\max}}\right) \quad (2.29)$$

where

$$n = -2.5 + 0.5\sqrt{9 + \frac{8}{\sigma_{X_d}^2}} \quad (2.30)$$

$$X_{\max} = \frac{n+3}{n+1} \quad (2.31)$$

$$A = \frac{(n+1)(n+2)}{\left(X_{\max}\right)^{n+1}} \quad (2.32)$$

The dispersion σ_{Xd} of the daily values around the monthly average value decreases with rising monthly clearness index at the location, and can be described by a linear function:

$$\sigma_{X_d}^2 = \begin{Bmatrix} 0.1926 \text{ for } \overline{k_{tm}} \le 0.2 \\ \max\left\{0.01, \left(0.269 - 0.382\overline{k_{tm}}\right)\right\} \text{for } \overline{k_{tm}} > 0.2 \end{Bmatrix} \tag{2.33}$$

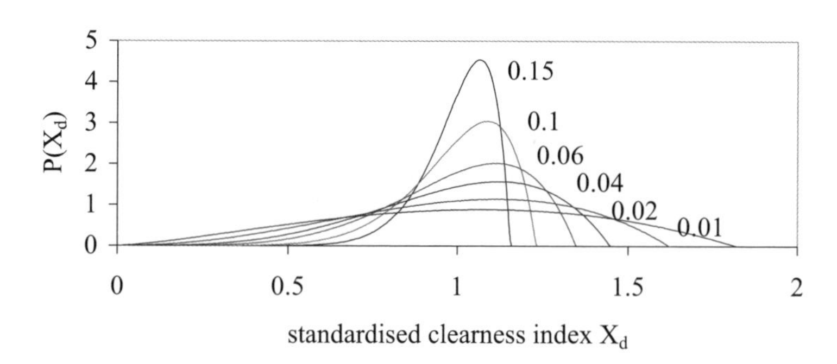

Figure 2.15: Gordon–Reddy distribution function of the standardised daily clearness indices around the monthly average value as a function of the variance σ^2.

In clear months where $\sigma^2 > 0.1$, the daily clearness indices above the monthly average value ($X_d = 1$) drop very sharply.

The initially assumed Gaussian-distributed variable Z_d is converted by a transformation known as Gaussian mapping into the actual probability distribution $P(X_d)$ and so finally the actual time sequence for X_d is obtained. In Gaussian mapping, first the cumulated distribution $F(Z_d)$ of the Gaussian-distributed random variables Z_d is calculated:

$$F(Z_d) = \frac{1}{2}\left(1 \pm \sqrt{1 - \exp\left(\frac{-2Z_d^{\,2}}{\pi}\right)}\right) \tag{2.34}$$

with a positive sign for $Z_d > 0$ and a negative one for $Z_d < 0$.

Following this, the value $F(Z_d)$ is equated with the cumulated value of the non Gaussian-distributed variables $F(X_d)$, and from this the associated X_d-value is unambiguously determined. For X_d there results an implicit equation, which must be solved iteratively:

$$A\,X_d^{\,n+1} = \frac{F(Z_d)}{\dfrac{1}{n+1} - \dfrac{X_d}{(n+2)X_{\max}}} \tag{2.35}$$

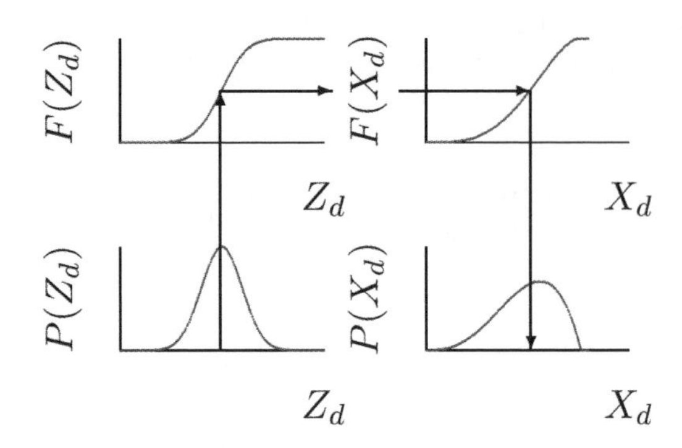

Figure 2.16: Conversion of the normally-distributed random variables Z_d into Gordon–Reddy distributed random variables X_d on the basis of Gaussian mapping.

From the thus calculated X_d-values a series of daily values of the clearness index $k_{td} = X_d \overline{k_{tm}}$ is obtained, with the probability distribution $P(X_d)$ and autocorrelation coefficient ρ_1.

Example 2.5

Calculation of the first six daily k_{td} values for a mean monthly $\overline{k}_{tm}$ value of 0.5 for the month of July (m = 7) with an autocorrelation coefficient ρ_1 of 0.3.

Day number	d = 1	2	3	4	5	6
Random number z [0,1]	0.3	0.1	0.9	0.65	0.2	0.5
Noise term r_d	–0.495	–1.22	1.22	0.365	–0.799	0
Z_d	–0.495	–1.37	0.811	0.608	–0.617	–0.185
$F(Z_d)$	0.31	0.082	0.792	0.729	0.268	0.4266
X_d	0.87	0.57	1.25	1.2	0.83	0.97
k_{td}	0.434	0.285	0.625	0.6	0.415	0.485

with $\sigma^2_{Xd} = 0.078$, $X_{max} = 1.53$, $n = 2.78$ and $A = 3.63$.

A one-year time sequence of daily irradiated energy on a horizontal receptor surface in Stuttgart (48.8° northern latitude) shows the fluctuations of daily irradiance produced by the auto regression method. The yearly total irradiance is about 1190 kWh/m^2.

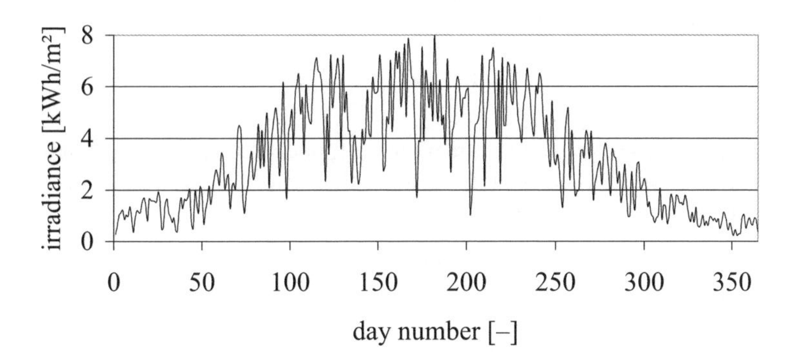

Figure 2.17: Daily irradiance on a horizontal surface in Stuttgart.

2.3.2 *Hourly average values from daily average values*

Hourly values can likewise be calculated from daily averages by an autoregressive model. Analysis of long-term hourly irradiance data shows that significant correlations exist only between directly consecutive hours. Since the hourly average values of the clearness index are normally distributed, the first-order autoregressive procedure can be used directly.

$$y_h = \rho_1 y_{h-1} + r_h \tag{2.36}$$

In order to eliminate the deterministic characteristics of the hourly clearness index, i.e. dependency on sun height, first the expected clearness index $\langle k_{th} \rangle$ of the respective hour is calculated; this depends on the average daily clearness index k_{td} and the zenith angle θ_z. The variable y_h then describes the difference, normalised by the standard deviation, between the hourly, statistically produced clearness index k_{th} and the expected clearness index $\langle k_{th} \rangle$ of the respective hour.

$$y_h = \frac{k_{th} - \langle k_{th}(k_{td}) \rangle}{\sigma_h} \tag{2.37}$$

The noise r_h is calculated according to Equation (2.27) with standard deviation $\sigma' = \sqrt{(1-\rho_1^2)}$. The autocorrelation coefficient ρ_1 depends only slightly on the average daily clearness index and is about 0.38,

$$\rho_1 = 0.38 + 0.06\cos(7.4k_{td} - 2.5) \tag{2.38}$$

where the cosine term has to be calculated in arc measure.

Crucial for the quality of the model is the parameterizing of the expected hourly clearness index, which is then varied by the random number procedure. Aguiar and

Collares-Pereira (1992) have shown that very good results can be obtained with a angle-dependent exponential function.

$$\langle k_{th} \rangle = \lambda + \varepsilon \exp(-\kappa / \sin \alpha_s) \tag{2.39}$$

The parameters of the exponential function are functions of the average daily clearness index. They are empirically derived from a database of 13 European and one African weather stations.

$$\lambda = -0.19 + 1.12 k_{td} + 0.24 \exp(-8 k_{td}) \tag{2.40}$$

$$\varepsilon = 0.32 - 1.6 (k_{td} - 0.5)^2 \tag{2.41}$$

$$\kappa = 0.19 + 2.27 k_{td}^2 - 2.51 k_{td}^3 \tag{2.42}$$

The standard deviation σ_h of the average hourly clearness index k_{th} is likewise empirically derived from the database values, and is represented by a sun-height dependent exponential function.

$$\begin{aligned} \sigma_h &= A \exp\left(B(1 - \sin \alpha_S)\right) \\ A &= 0.14 \exp\left(-20 (k_{td} - 0.35)^2\right) \\ B &= 3 (k_{td} - 0.45)^2 + 16 k_{td}^5 \end{aligned} \tag{2.43}$$

The starting value y_0 for the first hour after sunrise is initialised with zero and the next value is calculated with the autocorrelation coefficient ρ_1 and the noise r_h according to Equation (2.36). The hourly average value of the clearness index results from the definition of the variable y_h.

$$k_{th} = \langle k_{th} \rangle + \sigma_h y_h \tag{2.44}$$

Meaningful k_{th} values must be zero or greater and be below a maximum clearness index for extremely clear sky conditions. A simple approximation for the maximum k_{th} value is

$$k_{th,\max} = 0.88 \cos\left(\pi (h - 12.5) / 30\right) \tag{2.45}$$

For k_{th} values outside the required range of values, new random numbers must be generated.

Example 2.6:

Calculation of hourly k_{th} values (solar time-of-day) for the fourth day of the preceding example (4 July = day 185) with an average daily clearness index of 0.6 in Stuttgart.

The hour angle for sunrise and sunset is ω = 118.8°, i.e. the first hour is between 4 am and 5 am.

First those values which need only be calculated once are determined. This includes:

1. The autocorrelation coefficient ρ_1: $\rho_1 = 0.38 + 0.06\cos(7.4k_{td} - 2.5) = 0.44$

2. The standard deviation σ' of the noise term r_h: $\sigma' = \sqrt{(1-\rho_1^2)} = 0.898$

3. The parameters for the expected hourly $\langle k_{th} \rangle$ value:

$$\lambda = -0.19 + 1.12k_{td} + 0.24\exp(-8k_{td}) = 0.485$$

$$\varepsilon = 0.32 - 1.6(k_{td} - 0.5)^2 = 0.304$$

$$\kappa = 0.19 + 2.27k_{td}^{\ 2} - 2.51k_{td}^3 = 0.465$$

4. The parameters A, B for the standard deviation of the hourly k_{th} -values:

$$A = 0.14\exp\left(-20(k_{td} - 0.35)^2\right) = 0.0397$$

$$B = 3(k_{td} - 0.45)^2 + 16k_{td}^5 = 1.3229$$

With the autocorrelation coefficient and the noise term r_h, which can be calculated via random numbers z, the time series of the variables y_h can be obtained. Then for each hour the expected value $\langle k_{th} \rangle$ and the standard deviation σ_h can be calculated via the sun's height α_s at mid-hour, and thus from y_h the time series of the hourly clearness index k_{th} can also be calculated.

The random variable $y_{h=1}$ is initialised with zero, since no correlation exists with the hourly value before sunrise $y_h = 0$.

h	TLT	z	r_h	y_h	α_s	$\langle k_{th} \rangle$	σ_h	k_{th}
1	4.30h	–	–	0	3.5	0.485	0.138	0.485
2	5.30h	0.3	–0.468	–0.468	12.3	0.519	0.112	0.467
3	6.30h	0.1	–1.15	–1.356	21.9	0.572	0.091	0.529
4	7.30h	0.2	–0.753	–1.349	31.7	0.61	0.074	0.509
5	8.30h	0.9	1.15	0.557	41.5	0.635	0.062	0.552
6	9.30h	0.8	0.753	0.998	50.8	0.652	0.053	0.681

While the expected values of the hourly clearness indices constantly rise as expected with the sun's height, the statistical clearness index k_{th} displays clear fluctuations.

With the thus calculated hourly clearness indices, the global irradiance on a horizontal surface can also be calculated. Extraterrestrial irradiance on a normal surface for day 185 is 1321.9 W/m^2, i.e. on a horizontal surface for the second hour after sunrise with $\alpha_s = 12.32°$, making 282.1 W/m^2 altogether. With the calculated clearness index of 0.467, a total of 131.7 W/m^2 is obtained for global horizontal irradiance.

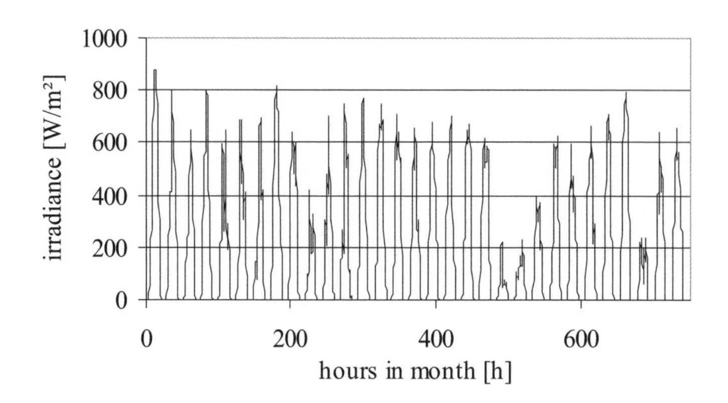

Figure 2.18: Average hourly irradiance on a horizontal surface in Stuttgart in July. The monthly irradiated energy is about 153 kWh/m^2.

2.4 Global irradiance and irradiance on inclined surfaces

2.4.1 Direct and diffuse irradiance

Due to the scattering of extraterrestrial irradiance in the atmosphere, a diffuse irradiance proportion G_d always occurs in addition to direct irradiance G_b (index *b*: beam). The irradiance on the horizontal in W/m^2 is called global irradiance (G_h):

$$G_h = G_{bh} + G_{dh} \tag{2.46}$$

Direct and diffuse irradiance on the horizontal are supplied in many meteorological data records as measured hourly values. If irradiance is determined by statistical procedures, at first only the global irradiance is obtained. The diffuse irradiance proportion correlates directly, however, with the hourly clearness index.

An empirical correlation for 15 locations in North America, Europe and Australia has been determined by Erbs *et al.* (1982).

$$G_{dh} = \begin{Bmatrix} G_h \times (1.0 - 0.09 k_{th}) & \text{for } k_{th} \le 0.22 \\ G_h \times \begin{pmatrix} 0.9511 - 0.1604 k_{th} + 4.388 k_{th}^{\ 2} \\ -16.638 k_{th}^{\ 3} + 12.336 k_{th}^{\ 4} \end{pmatrix} & \text{for } 0.22 < k_{th} < 0.8 \\ G_h \times 0.165 & \text{for } k_{th} \ge 0.8 \end{Bmatrix} \tag{2.47}$$

2.4.2 Conversion of global irradiance to inclined surfaces

The conversion of horizontal global irradiance to inclined surfaces must proceed separately for direct and diffuse irradiance. While the intensity of the direct irradiance depends only on the angle of incidence on the recipient surface, different conversion procedures exist for diffuse irradiance.

The simplest model, by Liu and Jordan (1960), proceeds from even isotropic diffuse irradiance in the sky, which with rising surface inclination results in ever smaller visible proportions. The total irradiance reflected by the ground is likewise assumed to be isotropically distributed. More exact models divide the diffuse irradiance in such a way that apart from the isotropically distributed celestial irradiance, the brightening around the sun (circumsolar irradiance) and also the horizon brightening caused by dispersion are taken into account (Perez *et al.*, 1987). The ray-tracing software programs used in light technology or the standard sky models for daylight calculations calculate the irradiance intensity for each point of the sky hemisphere (Brunger and Hooper, 1993).

2.4.2.1 An isotropic diffuse irradiance model

The total irradiance on the inclined surface G_t (index *t*: tilted) results from the sum of direct irradiance, the isotropic sky diffuse irradiance and also the ground reflection proportion with the reflection coefficient ρ.

$$G_t = \frac{G_{bh}}{\cos\theta_z}\cos\theta + G_{dh}\,F_{surface-sky} + G_h\,F_{surface-ground}\,\rho \tag{2.48}$$

The direct irradiance proportion on a horizontal plane is converted via the cosine of the zenith angle to a surface normal to the sun, and multiplied afterwards by the cosine of the angle of incidence to the inclined surface.

The other diffuse irradiance components are converted with the help of form factors F between the receptor surface and the sky or ground. The form factors depend on the angle of inclination β of the surface to the horizontal.

$$F_{surface-sky} = \frac{1+\cos\beta}{2} \tag{2.49}$$

$$F_{surface-ground} = \frac{1-\cos\beta}{2} \tag{2.50}$$

With these geometrical relations, the total irradiance can be calculated on a randomly inclined surface.

$$G_t = G_{bh}\frac{\cos\theta}{\cos\theta_z} + \underbrace{G_{dh}\frac{1+\cos\beta}{2}}_{G_{d,sky}} + \underbrace{G_h\,\rho\,\frac{1-\cos\beta}{2}}_{G_{d,ground}} \tag{2.51}$$

Example 2.7

Calculation of the terrestrial irradiance with the isotropically diffuse model for a surface normally oriented to the sun for the second hour after sunrise on day 185 (see Example 2.6), i.e. $\theta = 0°$ and $\beta = \theta_z = 77.7°$.

Global horizontal irradiance: $G_h = 131.7\ \mathrm{W/m^2}$

With the clearness index $k_{th} = 0.467$ the result is a value of 95.5 W/m² for horizontal diffuse irradiance, based on the empirical correlation of Erbs *et al.* (1982), and thus a horizontal direct irradiance proportion of 36.2 W/m².

Direct irradiance on the normally-oriented surface: $G_{bt} = 169.6\ \mathrm{W/m^2}$
Diffuse proportion: $G_{d,sky} = 58.3\ \mathrm{W/m^2}$
Diffuse ground reflection at $\rho = 0.2$: $G_{d,ground} = 10.3\ \mathrm{W/m^2}$

The irradiance on the inclined surface is thus: $G_t = 238.5\ \mathrm{W/m^2}$

2.4.2.2 Diffuse irradiance model based on Perez

In the Perez model, the isotropically assumed sky irradiance G_{iso} is overlain by a circumsolar proportion G_{cir} as well as by a term for the horizon brightening G_{hor}.

$$G_{dt} = \underbrace{G_{iso} + G_{cir} + G_{hor}}_{G_{d,sky}} + G_{d,ground} \tag{2.52}$$

On inclined surfaces the total irradiance reflected at the ground increases the diffuse proportion by $G_{d,ground}$.

The circumsolar irradiance is caused by strong forward scattering of the aerosols. The horizon brightening results from scattering at the large air mass of the atmosphere near the horizon, as seen by the observer, and is usually present when the sky is clear.

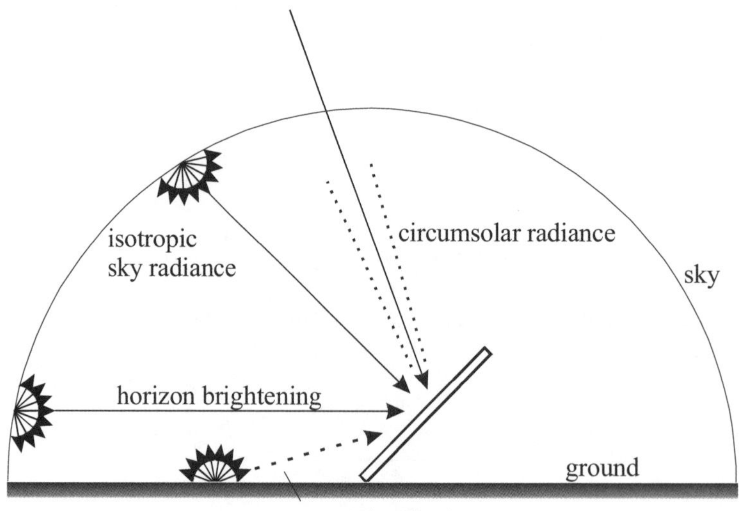

Figure 2.19: Allocation of the irradiance into different components.

$$G_{dt} = \underbrace{G_{dh}(1-F_1)\frac{1+\cos\beta}{2}}_{G_{iso}} + \underbrace{G_{dh}F_1\frac{\cos\theta}{\cos\theta_Z}}_{G_{cir}} + \underbrace{G_{dh}F_2\sin\beta}_{G_{hor}} + \underbrace{G_h\,\rho\,\frac{1-\cos\beta}{2}}_{G_{d,ground}} \qquad (2.53)$$

The isotropic radiance with the form factor $(1+\cos\beta)/2$ is therefore reduced by the proportion of F_1 of circumsolar radiance (first term). The circumsolar radiance with the proportion F_1 is, like the direct radiance, converted to the inclined surface. F_2 describes the horizon radiance, which can be both positive (horizon brightening) and negative (darkening).

The coefficients F_1 and F_2 are parametrised by the zenith angle, a clearness index ε and the brightness coefficient Δ. The clearness index ε essentially shows the ratio of direct to diffuse irradiance (with G_{bn} as direct irradiance on a normal surface), i.e. small clearness indices mean overcast skies and vice versa. Through brightness as a ratio of the air mass-corrected diffuse irradiance G_{dh} to the extraterrestrial irradiance on a normal surface G_{en}, the fact that even very clear skies can be dark (deep blue, small Δ) and overcast skies very bright (large Δ) is taken into account.

$$\varepsilon = \frac{1+\frac{G_{bn}}{G_{dh}}+5.535\times10^{-6}\,\theta_z^3}{1+5.535\times10^{-6}\theta_z^3} \qquad (2.54)$$

$$\Delta = \frac{m\,G_{dh}}{G_{en}} \qquad (2.55)$$

The air mass m is, as before, $m = 1/\cos\theta_z$.

The brightness coefficients F_1 and F_2 are empirically derived coefficients from measurements of different sky conditions. They are calculated using the clearness index ε.

$$F_1 = f_{11}+f_{12}\Delta+\frac{\pi\theta_Z}{180}f_{13} \qquad (2.56)$$

For F_1 only positive values are used (otherwise $F_1 = 0$).

$$F_2 = f_{21}+f_{22}\Delta+\frac{\pi\theta_Z}{180}f_{23} \qquad (2.57)$$

Table 2.1: Perez coefficients for the calculation of the anisotropic diffuse irradiance.

Clearness index ε [–]	f_{11}	f_{12}	f_{13}	f_{21}	f_{22}	f_{23}
≤ 1.056	0.041	0.621	–0.105	–0.040	0.074	–0.031
$1.056 < \varepsilon \leq 1.253$	0.054	0.966	–0.166	–0.016	0.114	–0.045
$1.253 < \varepsilon \leq 1.586$	0.227	0.866	–0.250	0.069	–0.002	–0.062

1.586 < ε ≤ 2.134	0.486	0.670	–0.373	0.148	–0.137	–0.056
2.134 < ε ≤ 3.230	0.819	0.106	–0.465	0.268	–0.497	–0.029
3.230 < ε ≤ 5.980	1.020	–0.260	–0.514	0.306	–0.804	0.046
5.980 < ε ≤ 10.080	1.009	–0.708	–0.433	0.287	–1.286	0.166
>10.080	0.936	–1.121	–0.352	0.226	–2.449	0.383

Thus for the entire irradiance on the inclined surface, one obtains

$$G_t = \frac{G_{bh}}{\cos\theta_z}\cos\theta + G_{dh}\left(1-F_1\right)\frac{1+\cos\beta}{2} + G_{dh}F_1\frac{\cos\theta}{\cos\theta_Z} + G_{dh}F_2\sin\beta + G_h\,\rho\,\frac{1-\cos\beta}{2} \tag{2.58}$$

Example 2.8

Calculation of the irradiance on the 77.7° inclined surface from the last example with the Perez model.

$\varepsilon = 1.5$

$\Delta = 0.34$

$\theta_z = 77.7°$

$$F_1 = f_{11} + f_{12}\Delta + \frac{\pi\theta_Z}{180}f_{13} = 0.227 + 0.866\times0.34 + \pi\times77.7/180\times-0.25 = 0.232$$

$$F_2 = f_{21} + f_{22}\Delta + \frac{\pi\theta_Z}{180}f_{23} = 0.069 + (-0.002)\times0.34 + \pi\times77.7/180\times(-0.062) = -0.0157$$

Isotropic sky component:

$$G_{dh}\left(1-F_1\right)\frac{1+\cos\beta}{2} = 95.5\frac{W}{m^2}(1-0.232)\frac{1+\cos77.7}{2} = 44.5\frac{W}{m^2}$$

Circumsolar proportion: $G_{dh}F_1\dfrac{\cos\theta}{\cos\theta_z} = 95.5\dfrac{W}{m^2}\times0.232\times\dfrac{1}{\cos77.7} = 104\dfrac{W}{m^2}$

Horizon proportion: $G_{dh}F_2\sin\beta = 95.5\dfrac{W}{m^2}\times-0.0157\times\sin77.7 = -1.46\dfrac{W}{m^2}$

G_t = 169.6 W/m^2 + 147 W/m^2 = 316.6 W/m^2, thus 33% more than calculated via the isotropic model.

Using the procedures outlined above, the annual irradiated energy for randomly oriented surfaces can be calculated. This energy, represented as a function of the height and azimuth angle, enables quick determination of the energy supply on randomly oriented surfaces. The optimal annual energy yield is obtained on a south-oriented surface with an angle of inclination β of approximately the geographical latitude Φ minus 10°. With energy losses of only 5%, deviations in the azimuth angle of ±35–40° from south, and in the angle of inclination of ±15–20° of the optimal angle can be tolerated (see Chapter 1).

2.4.3 *Measurement techniques for solar irradiance*

Global irradiance and irradiance on inclined surfaces is measured by pyranometers, with errors under 5%. Pyranometers measure irradiance by means of thermocouples, from the temperature difference of an irradiance-absorbing blackened surface and the housing. Due to the hemispherical glass cover, the angle dependence of the signal is small, and moreover sensitivity is almost independent of wavelength. The voltage level of the thermocouples is low, typically around 5×10^{-6} V per W/m^2 of irradiance. With more exact calibration, measuring accuracies of up to ±1% can be achieved.

Photovoltaic solar cells are cheaper detectors. However, spectral sensitivity and the temperature dependence of the measuring signal, plus reduced efficiency at small irradiances, lead to measuring errors of over 10%. For the measurement of the mean monthly irradiance, PV cells are nonetheless quite suitable.

The diffuse irradiance can be measured with a pyranometer, with shadowing rings to block off direct irradiance. The shadowing ring is adjusted for declination and degree of latitude, and must also be adjusted for seasonal changes in declination (every 2–3 days, depending on the width of the ring). The reduction of diffuse irradiance by the ring is corrected by a variable correction factor between 1.05 and 1.2.

2.5 Shading

When using solar technology in urban areas, it cannot be assumed that solar irradiance strikes receptor surfaces unhindered. Apart from temporary shading of the direct rays by nearby objects, buildings or vegetation, it is seldom the case that a free horizon for diffuse irradiance is available, in particular when facades are used.

For a representation of the shadowing effects of obstructing buildings, firstly, all objects are represented by surface polygons and their corner points. The shadowing caused by a building is then calculated for each corner point by means of the sun vector, and the points of shading are again connected to a polygon.

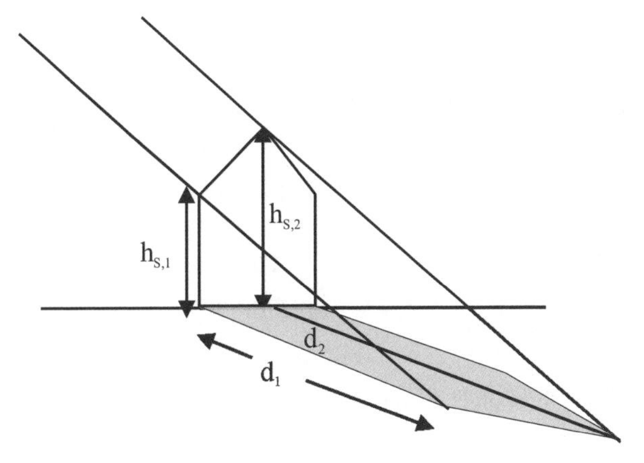

Figure 2.20: Construction of the shadowing from the surface polygon of the blocking building with heights of $h_{S,i}$.

It is simpler, however, to analyse the shading using horizon coordinates (elevation angles and azimuth) from a point receptor. With large receptor surfaces a subdivision of the surfaces is possible.

From the difference between building height h_S and observer height h_O, and also from the building's distance from the point of observation d, the elevation angle α_S for a given azimuth angle γ_S of the shading building can be calculated.

$$\alpha_S = \arctan\left(\frac{h_S - h_O}{d}\right) \tag{2.59}$$

The thus obtained value pairs γ_S and α_S form a polygonal sequence which can be entered in a sun-position diagram, and which enables direct reading of the times when shading of direct irradiance occurs.

For the reduction of diffuse irradiance, the surface between the horizon line of elevation angle $\alpha = 0°$ and the shading contour must be calculated. This shading surface reduces the isotropically assumed diffuse irradiance proportion of the sky by the given angle area.

For this, the shading contour for an azimuth angle area γ_1 to γ_2 is divided into even stages, which can be described by a straight line equation (Quaschning, 1998):

$$\alpha(\gamma) = m\gamma + c \tag{2.60}$$

where m indicates the gradient of the straight lines with $m = \frac{\alpha_2 - \alpha_1}{\gamma_2 - \gamma_1}$ and the constant c results from equating the gradient m and the gradient for $\gamma = 0$:

$$\frac{\alpha_2 - \alpha_1}{\gamma_2 - \gamma_1} = \frac{\alpha_1 - c}{\gamma_1 - 0} \Rightarrow c = \frac{\alpha_1\gamma_2 - \alpha_2\gamma_1}{\gamma_2 - \gamma_1} \tag{2.61}$$

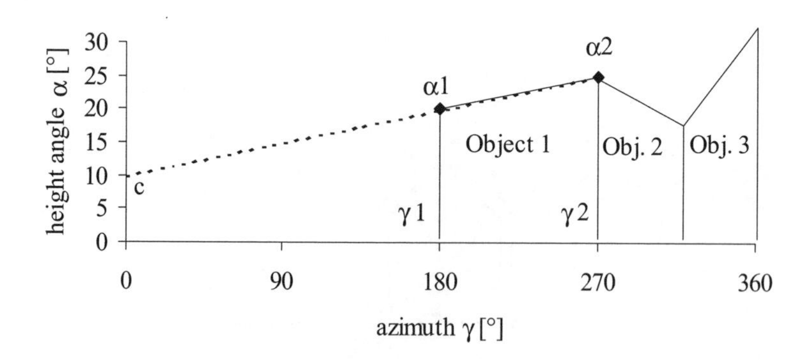

Figure 2.21: Contours of three blocking buildings.

Here, the constant c is determined from the elevation angles of the shading for a linear section between the azimuth angles γ_1 and γ_2 of the first building.

The luminance of each shaded celestial point $L_e(\alpha,\gamma)$ is projected on to the recipient surface with the cosine of the angle of incidence θ, and integrated over the total area. The luminance is defined as the radiant flux emitted by a two-dimensional element dA into a spatial angle $d\Omega$. With an isotropic sky the luminance $L_{e,iso}$ is constant at each celestial point, and can be easily calculated from the horizontal diffuse irradiance (see Chapter 8).

$$L_{e,iso} = \frac{G_{dh}}{\pi} \quad \left[\frac{W}{m^2 sr}\right] \tag{2.62}$$

A small section of the celestial sphere in horizon coordinates can be expressed as $dA = \cos\alpha\, d\alpha\, d\gamma$. The shaded proportion of diffuse celestial radiance, seen from an inclined surface $G_{dt,S}$, results from the luminance $L_{e,iso}$ multiplied by the cosine of the angle of incidence and of the sky section. The limits of integration are, for an azimuth area γ_1 to γ_2 between the horizon ($\alpha = 0°$) and the straight line shading contour ($m\gamma + c$):

$$\begin{aligned} G_{dt,S} &= L_{e,iso} \int_{\gamma_1}^{\gamma_2} \int_{0}^{m\gamma+c} \cos\theta \cos\alpha\, d\alpha\, d\gamma \\ &= L_{e,iso} \int_{\gamma_1}^{\gamma_2} \int_{0}^{m\gamma+c} \left(\sin\alpha \cos\beta + \cos\alpha \sin\beta \cos\left(\gamma - \gamma_F\right)\right)\cos\alpha\, d\alpha\, d\gamma \end{aligned} \tag{2.63}$$

In the case of a horizontal receptor surface (angle of inclination $\beta = 0°$ and surface azimuth $\gamma = 0°$), the projection factor is $\cos\theta = \cos\theta_z = \sin\alpha$, and the integral is simplified to

$$G_{dh,S} = L_{e,iso} \int_{\gamma_1}^{\gamma_2} \int_{0}^{m\gamma+c} \sin\alpha \cos\alpha\, d\alpha\, d\gamma = L_{e,iso} \frac{1}{2} \int_{\gamma_1}^{\gamma_2} \sin^2\left(m\gamma + c\right) d\gamma \tag{2.64}$$

For straight horizontal shading contours $m = 0$, i.e. the elevation angle $\alpha_1 = \alpha_2$. On the basis of Equation (2.61) therefore, $c = \alpha_1$ and for the integral the result is the simple solution:

$$G_{dh,S} = \frac{1}{2} L_{e,iso} \left(\gamma_2 - \gamma_1\right) \sin^2\alpha_1 \quad \text{for } m = 0 \tag{2.65}$$

Also for $m \neq 0$ the integral can be solved for horizontal receivers:

$$G_{dh,S} = \frac{1}{2} L_{e,iso} \left(\gamma_2 - \gamma_1\right) \left(\frac{1}{2} + \frac{1}{4} \frac{\sin 2\alpha_1 - \sin 2\alpha_2}{\alpha_2 - \alpha_1}\right) \quad \text{for } m \neq 0 \tag{2.66}$$

The angles are to be entered here in arc measure.

The solution of the integral for the surface with an angle of inclination β and surface azimuth γ from Equation (2.63) is only possible for gradients of the shading contour $m \neq 0$ on a case-by-case basis, using complex formulae (Quaschning, 1996). Therefore only the solution for $m = 0$ is shown, with which the real contours can be approximated in small sections.

$$G_{dt,S} = \frac{1}{2} L_{e,iso} \left(\cos\beta \left(\gamma_2 - \gamma_1\right) \sin^2\alpha + \sin\beta \left(\alpha + \sin^2\alpha\right) \left(\sin\left(\gamma_2 - \gamma\right) - \sin\left(\gamma_1 - \gamma\right)\right)\right) \quad (2.67)$$

For a given polygonal sequence of the shading contour, the diffuse irradiance proportions of the shaded sections are totalled and an overall shading factor formed, by which the diffuse irradiance is reduced:

$$S_d = \frac{\sum G_{d,S}}{G_d} \quad (2.68)$$

Example 2.9

On an south-oriented roof area, a solar plant is to be erected with the low edge H_o having a height of 8 m. The shading contour of two facing multi-family houses (building 1 is 10 m high and building 2 is 18 m high) with $d = 14$ m distance is to be entered in the sun-position diagram, and the reduction of diffuse and direct irradiance calculated.

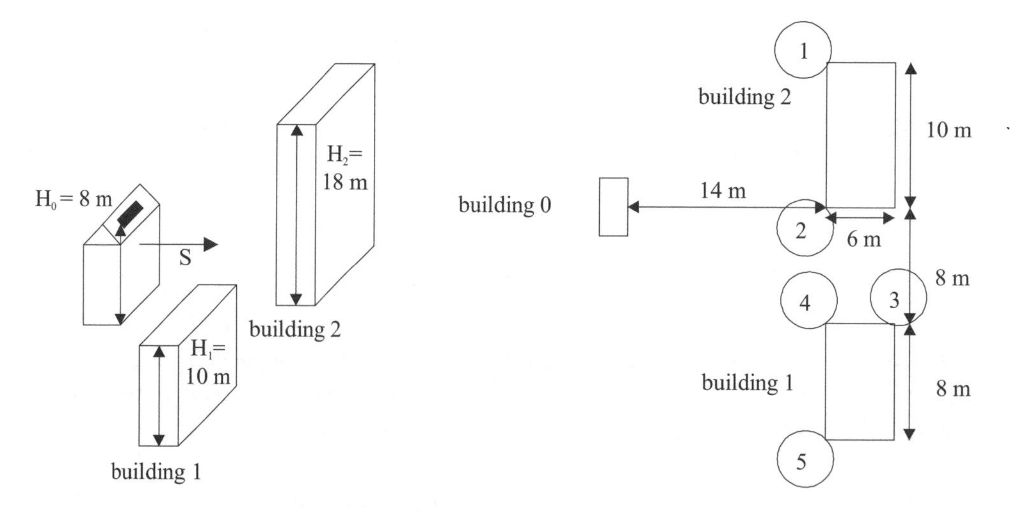

Figure 2.22: Geometry of the shading situation.

For increasing shading azimuth angles, the following value pairs of γ_S, α_S are determined from Figure 2.22:

Point	*Distance d* [m]	*Shading object height* h_S [m]	*Azimuth* γ_S [°]	*Shading height angle* α_S [°]
1	17.2	18	144.5	30.2
2	14	18	180	35.5
3	21.5	10	201	5.3
4	16.1	10	209.7	7.1
5	21.3	10	228.8	5.4

From the values shown in the above table it can be seen that in Stuttgart, for example, shading of the direct irradiance by building 2 occurs during the winter months (October until March) from about 10–12 am. Building 1, on the other hand, with its small elevation angles, does not contribute to the shading. When entering the shading contour into the sun-position diagram, a linear extrapolation between the corner points of the shading is carried out, in order to simplify the diffuse irradiance calculation which follows.

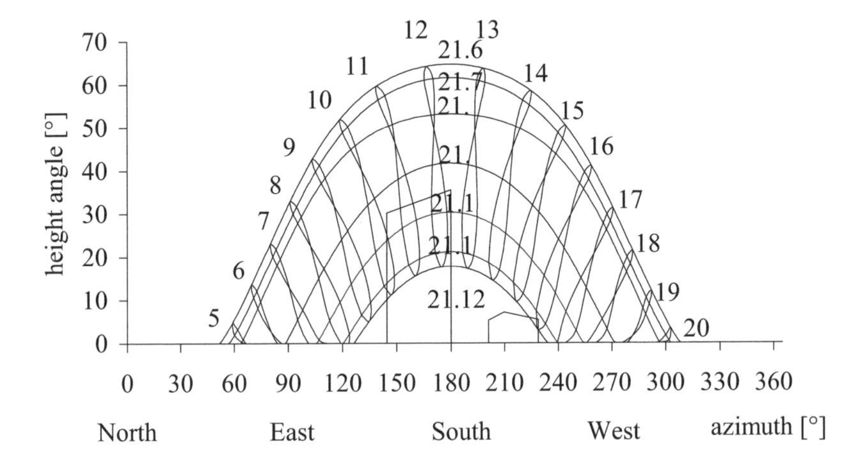

Figure 2.23: Sun-course diagram for Stuttgart with shading buildings drawn in.

For building 2 the shading factor S_d of the diffuse irradiance for a horizontal receptor surface and for a surface with a roof pitch angle of 45° must be calculated in what follows. Although the absolute value of the diffuse irradiance is not necessary for the shading factor, since the radiances shorten in Equation (2.68), a value of 300 W/m^2 is given, in order to be able to calculate the absolute irradiance loss due to shading based on Equation (2.65).

For the horizontal receptor surface with $m \neq 0$, a shaded diffuse proportion of 9 W/m^2 is obtained, and thus a shading factor of 3%. For the 45° inclined surface, the shaded diffuse proportion for $m = 0$ and a mean elevation angle of the shading of $(\alpha_{S1} + \alpha_{S2})/2 = 32.85°$ totals 19.8 W/m^2 and the shading factor is 7%. For the calculation of the shading factor of the inclined surface, the shaded sky proportion was related to the diffuse irradiance to the inclined surface, i.e. to $G_{dh}\left(\frac{1+\cos\beta}{2}\right)$, here 256 W/m^2.

Example 2.10

On a flat roof with a limited surface, the largest possible south-oriented solar plant is to be installed in Stuttgart with a latitude of 48.8°. As a criterion for an acceptable shading situation it is assumed that at solar noon on the winter solstice (21 December) no shading takes place. The relation of the distance of the collector D to the collector length L with a given surface angle of inclination β is to be determined.

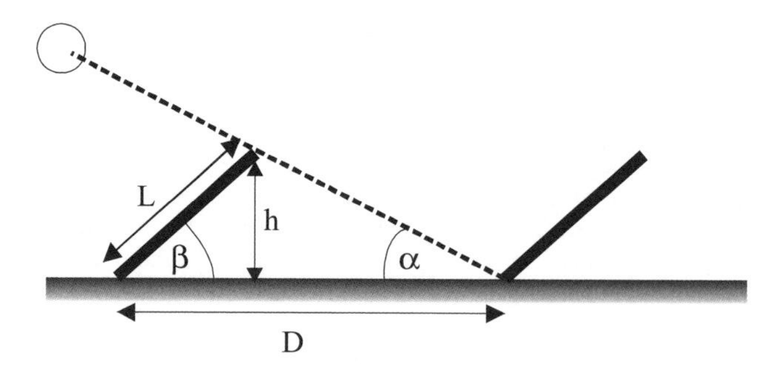

The sun-height angle α at solar noon on 21 December is 17.8°.

The distance D between the collector bases for this position of the sun is calculated as the total of the distances $L\cos\beta$ and h cot α.

$$D = L\cos\beta + h\cot\alpha = L\cos\beta + L\sin\beta\frac{\cos\alpha}{\sin\alpha}$$

For an angle of inclination β, for example 40°, the relation of distance to length should be at least as shown below:

$$\frac{D}{L} = \cos\beta + \sin\beta\cot\alpha = 2.77.$$

3 Solar thermal energy

3.1 Solar-thermal water collectors

The European market for water based solar thermal collectors has been growing by an average of 18% per year over the last decade (Stryi-Hipp, 2001). By the end of 1999, a total surface area of 8.5 million square metres was installed, 75% of which was in Greece, Germany and Austria. In order to reach the European Commission's target of 10 million square metres by 2010, annual growth rates have to double. This goal was nearly reached in 2000, where the European market grew by 30% to a total installed surface area of 1.15 million square metres. For standard solar thermal drinking water systems, the costs have been halved from about 12 000 € in 1984 to 6 000 € in 2002; for the less common heating support systems similar developments are to be expected. The technology is well established on the market and extensive European standardisation work is currently underway to improve the product quality and the customer confidence (Kotsaki, 2001).

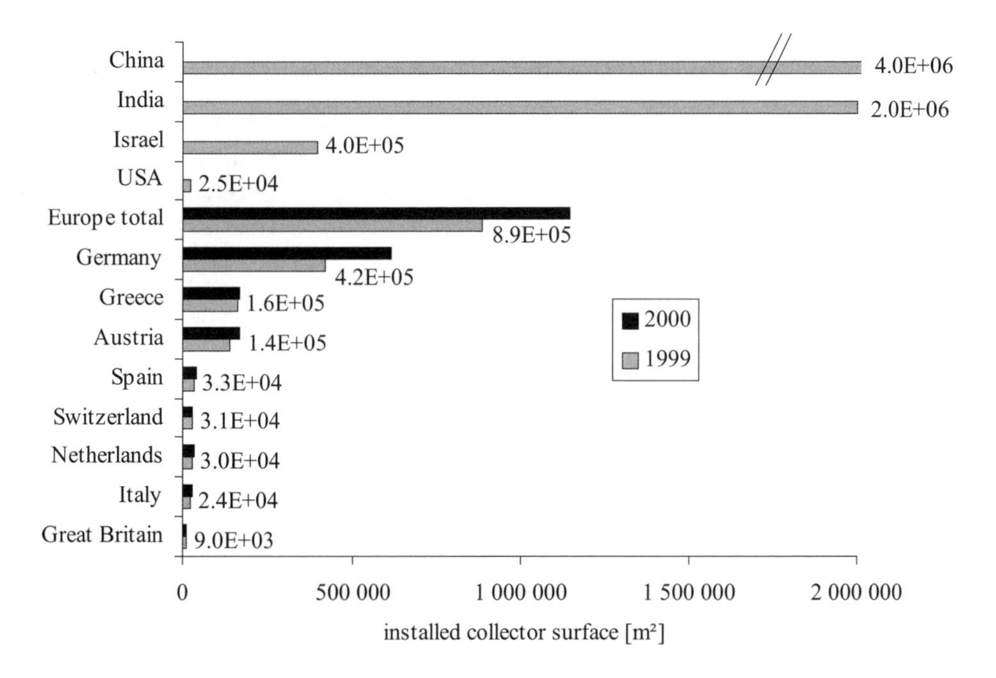

Figure 3.1: Installed solar thermal collector surface area in 1999 and 2000.

3.1.1 Innovations

In the field of collectors, developments are still taking place, aimed at improving selective coatings of absorbers using sputter technology, optimising the heat transfer from the absorber to the tubes through improved welding or soldering technologies (ultrasonic or plasma welding), reducing glazing reflection losses through antireflective layers, optimising vacuum collector geometry and introducing low-level concentrations of the irradiance (CPC collectors).

In system engineering, work is taking place on simplifications of the ducting system (common ducts including sensor cables), on improved heat exchangers, on new strategies for avoiding overheating, on special pumps for solar operation generating high pressures at low volume flows and low electricity consumption, and on new frost-protection fluids, which can handle the high temperature levels generated in vacuum tube collectors.

In storage technology, combined heating/drinking water storage is increasingly being developed, enabling solar support for conventional heating in addition to solar heating of drinking water. The market share of combined heating/drinking water collector systems has reached about 20% of the total collector surface area sold in Europe, and in countries such as Austria, it is up to 50%. It is expected that in Europe about 120 000 solar thermal systems with a total surface area of 1.9 million square metres will be installed as combined heating/drinking water installations (Weiß *et al.*, 2002). New stratified loading systems and various storage tank geometries are currently being developed.

3.1.2 System overview

Solar-thermal systems can be constructed both for decentralised water heating with small collector surfaces of 4–8 m^2 and 300–500 litres of storage capacity, and for central drinking water heating in housing estates, hospitals, sports halls etc. with collector surfaces of over 100 m^2 and storage capacities of around 10 m^3. Moreover, decentralised heating support with system sizes of around 10–20 m^2 and buffer storage of 1–2 m^3 volume can be carried out. A central supply of solar-produced heating energy for housing estates (solar district heating) requires on the other hand seasonal storage with at least ten times more storage capacity (1–2 m^3 of seasonal storage per square metre of collector surface, instead of 0.05–0.1 m^3 for decentralised applications). Solar-thermal systems are modularly structured with collector units of around 2.5–10 m^2 surface, which can replace conventional roofing material and take over the insulation function of the roofing. Completely prefabricated collector roofs including rafters and thermal insulation offer an economical solution, in particular for large-scale installations.

In contrast to photovoltaics, where system expansions are possible without problems, all the components of a solar-thermal system must be coordinated exactly with each other in the planning phase, since the pipework sizes and the pumping power are usually reduced to a low limit for economic reasons, and leave little scope for modular extensions. Storage and heat exchangers are, again with cost in mind, likewise designed for the collector rated output; they can, however, be complemented by additional units.

For buildings, solar-thermal heating of drinking water and heating support are especially relevant. Today's economic swimming pool heating will therefore only be mentioned in passing. High-temperature systems such as parabolic concentrator collectors or solar tower power plants for process heat production require large open spaces and are not relevant for integration in buildings.

3.1.3 *Thermal collector types*

Thermal collectors in the low-temperature range can be classified by the calorific losses between radiation absorbers and their surroundings.

3.1.3.1 Swimming pool absorbers

The simplest type of collector is a non-covered absorber for swimming pool heating, generally made of UV-resistant plastic (polyethylene PE, polypropylene PP or ethyl propylene dien monomers EPDM). The heat transfer coefficient or U-value of the front (U_f) in the direction of the incident irradiance is calculated from the reciprocal value of the thermal resistance $R_{a\text{-}o}$ between the black absorber with temperature T_a and ambient air T_o. The thermal resistance $R_{a\text{-}o}$ consists of the parallel heat transfer resistances for radiation ($1/h_r$) and wind-speed dependent convection ($1/h_c$); it is in the region of 0.04 m²K/W. From it results a heat transfer coefficient U_f of 25 W/m²K.

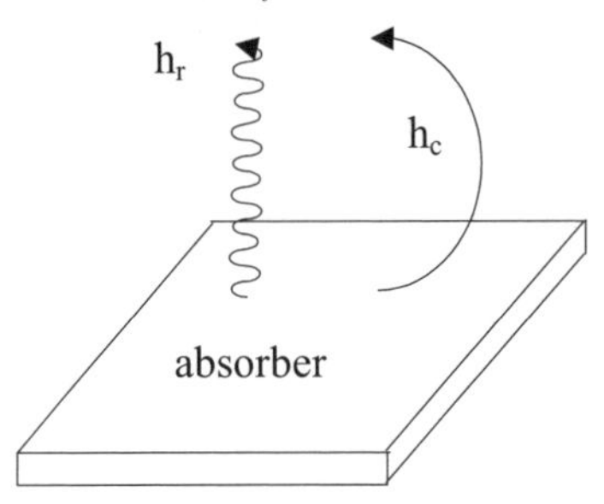

Figure 3.2: Heat losses of a non-covered black absorber.

$$U_f = \frac{1}{R_{a-o}} = \frac{1}{1/(h_r + h_c)} = h_r + h_c \approx 25 \frac{W}{m^2 K} \tag{3.1}$$

3.1.3.2 Flat plate collectors

With most thermal collectors, the calorific loss is reduced by a transparent glass covering of the absorber. The thermal resistance of the standing air layer between the absorber and the glass covering R_{a-g}, around 0.1–0.2 m²K/W, is in addition to the outside thermal resistance R_{g-o} between the cover glass and the ambient air. R_{g-o} consists in turn of a radiation and a convective proportion, and corresponds roughly to the total resistance R_{a-o} of the non-covered collector. Altogether the result is a heat transfer coefficient of around 5–6 W/m²K.

$$U_f = \frac{1}{R_{a-g} + R_{g-o}} \approx \frac{1}{0.15 + 0.04} = 5.3 \frac{W}{m^2 K} \tag{3.2}$$

If the absorber is selectively coated, the radiation exchange between the absorber and the transparent cover is significantly reduced and the U_f value falls to around 3–3.5 W/m²K.

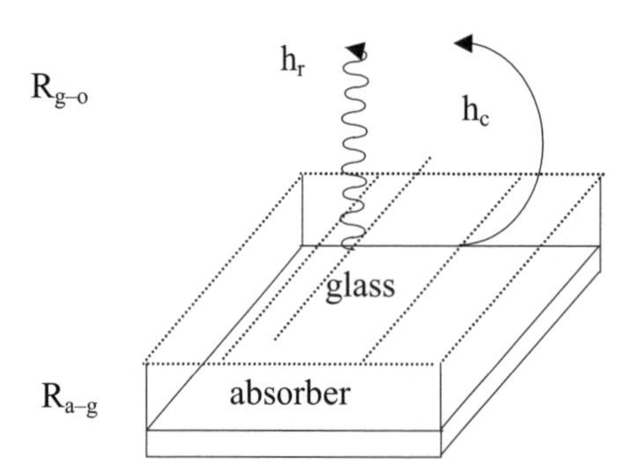

Figure 3.3: Heat loss through the transparent cover of a collector.

3.1.3.3 Vacuum tube collectors

In vacuum collectors the absorbers are also separated thermally from their surroundings by a transparent cover. In addition to the reduction in radiation exchange, convective heat transport is prevented by a very good vacuum of approximately 10^{-3} to 10^{-2} Pa, and the thermal resistance between absorber and cover rises to about 1 m²K/W. The result is U_f values of vacuum collectors of around 1 W/m²K.

A range of different vacuum tube geometries is available on the market. The most compact vacuum tube collector including compound parabolic concentrator mirrors (CPC) is only 45 mm thick. The absorber is either a standard finned tube design within an evacuated glass tube or is coated directly on the inner surface of a double glass tube, with the heat transferred to liquid circulating in separate tubes on the inside of the double glass. Alternatively the mirror coating is placed within an evacuated or gas-filled tube and the irradiance is concentrated on an inner absorber tube.

3.1.3.4 Parabolic concentrating collectors

Parabolic concentrators are especially useful for high temperature applications from 100–200°C, where their efficiency is higher than that of vacuum tube collectors. For some solar cooling systems, temperature levels of 150°C or higher are required, where parabolic concentrators can be efficiently used. Some first designs exist to place parabolic concentrators on building roofs. As they depend on direct irradiance, climates with a high direct irradiance proportion are advantageous. At low investment costs of about 300€/m² the solar heat costs are as low as 0.045 €/kWh on a Turkish site with 1900 kWh/m² direct normal irradiance and 0.11 €/kWh on a southern German site with 890 kWh/m² (Milow, 2002).

Table 3.1: Classification of thermal collectors by the front U_f value.

Collector type	*Heat resistance between absorber and exterior in direction of incident irradiance*	*Front heat transfer coefficient* U_f [W/m²K]
Uncovered swimming pool absorber	Direct radiative and convective heat exchange with environment	> 20
Flat plate collector with uncoated black absorber	Standing air layer between absorber and transparent cover	5–6
Selectively coated flat plate collector	Standing air layer with reduced radiative exchange	3–3.5
Selectively coated vacuum tube collector	Reduced convective and radiative heat exchange between absorber and transparent cover	1–1.5

3.1.4 System engineering for heating drinking water

The system engineering of a standard solar plant for heating drinking water comprises the main components, i.e. the collector and storage, as well as the solar circuit hydraulics and safety system. For small drinking water systems under 10 m² collector surface area, simple sizing rules have proved satisfactory, proceeding from an average warm-water consumption of around 30–60 litres per person and day, and typical collector yields of 350–450 kWh/m²a. Such systems contain the following components:

Table 3.2: Components of a standard solar plant.

Component	*Dimensions*
Flat plate collector Vacuum tube collector	1.25–1.5 m² per person 1.00–1.2 m² per person
Storage tank (pressurised)	40–70 l per m² collector surface area
Heat exchanger in collector circuit (integrated in storage tank)	30–40 W/K power per m² collector surface (corresponds to 0.15–0.20 m² surface area)
Pipework	15 mm external diameter up to 8 m² collector surface and 50 m pipe length
Solar station with pump, manometre, security valve, expansion vessel	25–80 W pumping power (frequency controlled special solar pump is best, otherwise the smallest available heating system pump)
Controller (temperature difference measurement between collector and storage tank, maximum temperature limitation storage tank)	Relay for pump control

With a system of these dimensions, annual solar cover of 40–60% can be obtained, about 70–100% in the summer and 10–20% in the winter.

3.1.4.1 The solar circuit and hydraulics

Besides the selection of the collector and storage, the hydraulics of the collector circuit with pipework, valves and pump have to be dimensioned, and also the safety engineering with the pressure level, expansion vessel and excess-pressure protection has to be worked out. With a standard solar plant for heating drinking water, all hydraulic and safety-related functions are brought together in a pre-mounted solar station.

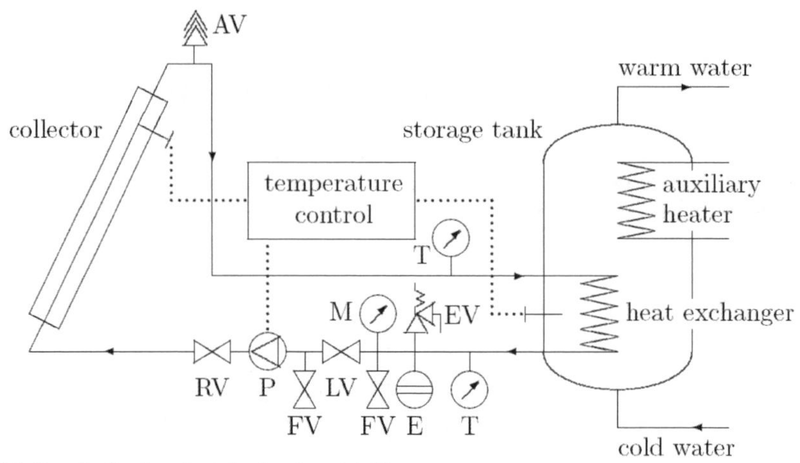

Figure 3.4: Standard solar plant for heating drinking-water.

The system consists of a collector, storage, temperature difference regulation and the following components; air vent *AV*, temperature sensor *T*, expansion vessel *E*, excess pressure valve *EV*, filling/emptying valves *FV*, manometer *M*, lockable valve *LV*, pump *P* and a lockable non-return valve *RV*.

Pumps

For the collector circuit, conventional circulating heating pumps with a low-power annular gap motor (< 100 W electrically) have mainly been used so far. These are optimally dimensioned for one-family or two-family houses, depending on the amount of heat needed, with flow rates between 1 and 4 m³/h. Thermal solar plants for heating drinking water are, however, usually driven with small flow rates, between 0.1–0.5 m³/h, i.e. with only 10% of the delivery of standard heating pumps. With such small deliveries the efficiency of a heating pump is between 2% and 7%, while with optimal sizing, efficiency of up to 20% can be achieved. The small efficiencies lead to a not insignificant electricity consumption of approximately 100 kWh/a in a small system with 2000 kWh yield per year, i.e. in terms of primary energy, 15% (300 kWh) is used for pump energy. Regulating the pump speed reduces annual power input by 50%.

The pumping height of a standard heating pump is relatively small, a maximum of 4–5 m. When using such pumps, de-aeration of the collector circuit at its highest point is important for the operation of the system. High supply pressures with small volume flow deliveries are obtained, on the other hand, with gear pumps with wet or dry rotating motors, which provide a pumping height of 35 m at a flow rate of 0.1 m³/h, so the breather, which is usually difficult to access, is unnecessary, and the initial air cushion can be fully squeezed out of the collector circuit.

Heat transfer liquid

The heat transfer liquid in the solar circuit must be provided with antifreeze in Central European climates, to avoid freezing of the external pipework system. Mixtures of water and antifreeze agents (glycols, e.g. polypropylene glycol) are used; they are provided with different inhibitor salts for corrosion protection. Mixing proportions of approximately 40% antifreeze to 60% water are usual, causing the freezing temperature to sink to –20°C. With an increasing proportion of antifreeze the viscosity of the heat transfer liquid rises by a factor of 3–5, which must be taken into account in pressure-loss calculations, and which lowers the thermal capacity by 10–20%.

With solar plants with high standstill temperatures (vacuum tube collectors), great attention must be paid to the temperature reliability of the antifreeze.

Overheating protection and expansion vessels

The expansion vessel is an essential component of overheating protection since when the system is at a standstill, with the usual system pressure at 3 to 6 × 10^5 Pa, the solar circuit liquid boils in the collector. The expansion vessel in small systems has to take up not only the volume increase of the solar circuit liquid of approximately 10%, but also the complete evaporating collector fluid content V_c of 0.5–2 litres per square metre of collector surface. The maximum volume increase ΔV of the solar liquid at system standstill is calculated from the maximum occurring temperature difference between the standstill temperature T_s and the temperature during initial filling T_o, from the volume expansion coefficient of the fluid β', from the volume of the collector circuit contents V_{cc} and from the evaporated collector contents V_c.

$$\Delta V = \beta'(T_{\mathrm{s}} - T_{\mathrm{o}})V_{\mathrm{cc}} + V_{\mathrm{c}} \tag{3.3}$$

The volume expansion coefficient β' rises with temperature and for water in the temperature range 60–80°C is about 5.87 × 10^{-4} K^{-1}, and for antifreeze mixtures around 10 × $10^{-4} K^{-1}$. The volume of the collector circuit V_{cc} is calculated for the pipes and heat exchanger without the fluid contents of the collectors.

The gas cushion of the diaphragm expansion vessel with content V_g is compressed from the selected pre-pressure p_{min} when filling the system, to the maximum operating pressure p_{max} at system standstill. To avoid air entry, the pre-pressure is selected to be about 0.2–0.5 × 10^5 Pa higher than the static pressure. This results from the difference in height between the collector field and expansion vessel Δh [m] with a static increase in pressure of 10^4 Pa per metre.

$$p_{\mathrm{min}} = (\Delta h + 5) \times 10^4 \quad [Pa] \tag{3.4}$$

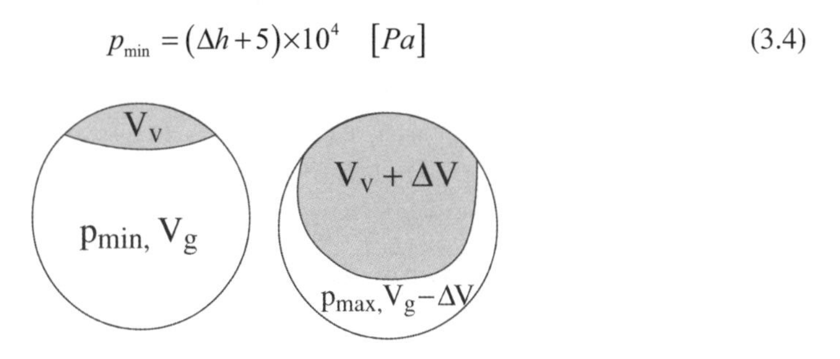

Figure 3.5: Expansion vessel at minimum and maximum operating pressures.

The maximum operating pressure p_{max} is given by the design pressure of the safety respectively overflow valves. The volume of the expansion tank V_a must be designed in such a way that the volume compression of the gas volume V_g is sufficient, due to the difference between minimum and maximum operating pressure, to absorb the volume increase ΔV on the liquid side of the diaphragm. Based on the ideal gas equation, the result is, assuming a constant temperature,

$$p_{\min} V_g = p_{\max}\left(V_g - \Delta V\right)$$
$$V_g = \frac{p_{\max}}{p_{\max} - p_{\min}} \Delta V \tag{3.5}$$

In addition to the necessary gas volume to absorb the volume increase, the liquid side of the expansion vessel always includes some safety water volume V_V, so that at very low temperatures the system pressure does not drop. This safety volume should be about 1–2% of the entire system volume ($V_{cc}+V_c$):

$$V_a = V_g + V_V \tag{3.6}$$

In large systems (> 100 m²) closed pressure expansion vessels are designed only for the volume increase of the liquid. Steam produced in the collector is blown off through safety or overflow valves and the heat distribution medium is collected in a pressure-free receptacle. After the collectors have cooled, a filling pump takes over the automatic feedback of the liquid to the solar circuit. The non-return valve *RV* is necessary for the avoidance of nocturnal cooling of the storage tank through a thermosyphonic rise of the solar circuit fluid.

Example 3.1

Calculation of the volume of an expansion tank for a 15 m² solar plant with a height difference between the expansion tank and collector field of 20 m, and a safety valve with a maximum operating pressure of 4.5×10^5 Pa. For the system the following collector circuit volume is assumed:

Heat exchanger liquid content with 4.6 m² of heat transfer surface: 3.6 litres
Pipe volume at a total of 40 m pipe length and 22 mm external pipe diamter (DN22): 12.6 litres
The volume expansion of the collector circuit contents V_{cc} for $\beta' = 10 \times 10^{-4}$ K^{-1} and a standstill temperature T_s of flat plate collectors of about 180°C (T_o = 15°C) is about $10\times10^{-4} K^{-1} \times (180-15) K \times 16.2 l = 2.67 l$.
The evaporating collector contents V_c are about15 litres, so that ΔV = 17.67 litres. The safety water volume V_V is $V_V = 0.01\times(3.6+12.6+15) l = 0.31 l$.
The minimal operating pressure given by the height difference is 2.5×10^5 Pa, so the expansion tank volume is

$$V_a = \frac{4.5\times10^5}{4.5\times10^5 - 2.5\times10^5} 17.67 l + 0.31 = 40.1 l \ .$$

Heat exchangers
For heat transfer between the collector primary circuit and the storage tank, internal heat exchangers are usually chosen for small systems for reasons of cost and space. Gilled pipe or plain-end pipe heat exchangers have a specific heat transfer rate of 200–500 W/m²K.

With maximum surface areas of 1–2 m² within 300–500 litre storage tanks and a mean temperature difference of 5 K, internal heat exchangers are limited in their performance to about 5 kW. Thus in larger systems, external heat exchangers requiring a second pumping circuit have to be used. Through the forced convection on both sides of the heat-transferring surface, the transfer rates of counter-current plate heat exchangers rise to 1000–4000 W/m², so at plate distances of a few millimetres per m³ construction volume, very high power density can be transferred.

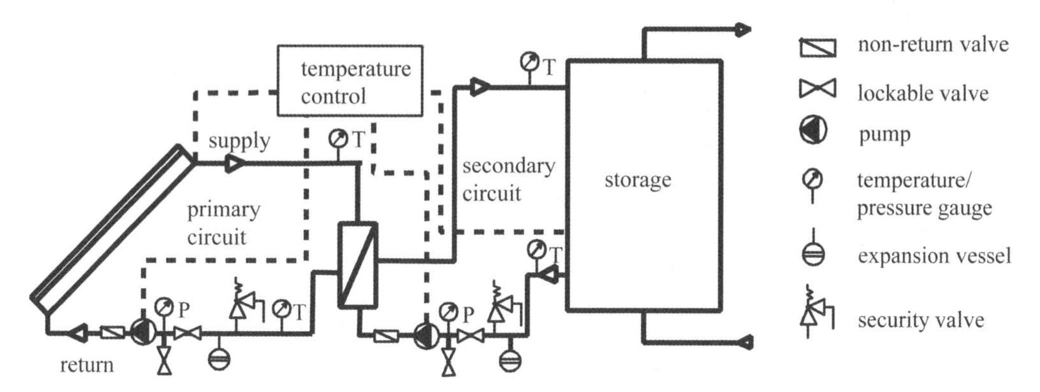

Figure 3.6: Circuit with external heat exchanger.

For the secondary circuit, a further pump is necessary, as well as an expansion tank and safety valves.

Collector interconnecting and pressure losses
For an even flow through the collectors at high flow rates, and thus good heat transfer between absorbers and fluid, serial interconnecting of collectors is favourable. A limiting factor is the pressure loss $\Delta p = \sum \xi \rho / 2 v^2$, which rises as a square of the flow velocity $v = \dot{V} / A_q$; this determines the electrical power P_{el} of the solar circuit pump. The flow velocity is determined by the mass flow $\dot{m} = \rho \dot{V}$ in the collector circuit, which is about 10–15 kg per m² collector surface and hour in so-called low flow systems. In standard systems it is between 30–60 kg/(m²h) and flows through the tubing cross-section A_q [m²].

The pressure drops are determined with the help of pressure loss factors ξ, which are calculated from friction factors in pipes and from tabulated factors for various fittings. For bends and T-fittings, an overall addition is set of a factor of 1.5 on the pipe pressure losses in practice. The pressure losses of pumps, heat counters etc. are determined from data sheets.

$$P_{el} = \frac{\dot{V}\Delta p}{\eta} = \frac{\dot{V}\sum \xi \frac{\rho}{2}\left(\frac{\dot{V}}{A_q}\right)^2}{\eta} \tag{3.7}$$

The efficiency η of small solar pumps at 100 W electrical output is around 2–7% and rises to 70–80% for high power dry rotor pumps. With small systems of up to 10 m² collector surface and 50 m pipe length, the pressure losses in the collector are less than 0.2×10^5 Pa and in the rest of the solar circuit about 0.3×10^5 Pa, so without a pressure loss calculation, small, three-stage heating pumps ($P_{el} < 100$ W) can be used.

With medium system sizes of approximately 15–40 m², typical total pressure losses are between $0.3–0.8 \times 10^5$ Pa, in each case a third of which is caused by the collector field, the pipes and the heat exchanger respectively. In the collector field the pressure loss should be limited to a maximum of 0.3×10^5 Pa by combined parallel/series connection. In systems between 40–100 m², a collector field pressure loss of 0.4×10^5 Pa is acceptable; the rest of the solar circuit should remain limited to 0.6×10^5 Pa. In large-scale installations over 100 m², typical pressure losses of 0.7×10^5 Pa in the collector field and 0.9×10^5 Pa in the solar circuit can be expected.

Example 3.2

Estimate of the flow rate of a small system of 10 m², with a heating pump of 80 W maximum power, and determination of the pump power of a large-scale installation with 100 m² of collector surface and a specific flow of 30 kg/m²h.

With a pump efficiency of 7%, a typical increase in pressure of 5×10^4 Pa and an electrical power of 80 W, a flow rate of

$$\dot{V} = \frac{\eta P_{el}}{\Delta p} = \frac{0.07 \times 80W}{5 \times 10^4 Pa} = 1.1 \times 10^{-4} \frac{m^3}{s} = 0.4 \frac{m^3}{h}$$

can be transported. With a 10 m² system, that is about 40 l/m²h, which corresponds to a typical flow of a standard system. For a large-scale installation with an assumed pump efficiency of 40%, the electrical power is

$$P_{el} = \frac{\dot{V}\Delta p}{\eta} = \frac{\frac{3m^3}{3600s} \times 1.6 \times 10^5 Pa}{0.4} = 333.3\ W$$

Controllers

The usual temperature-difference controllers for warm-water systems control the solar circuit pump via a relay switch, as a function of the temperature difference between the collector exit and the store at the height of the heat exchanger. Temperature-difference regulation is usually provided by hysteresis, to avoid frequent switching of the pump during flow-begin and subsequent temperature reduction (switching-on temperature difference around 5 K, switching-off temperature difference 2–3 K).

3.1.4.2 Heat storage

Short-term stores in solar thermal systems, for heating drinking water and supporting heaters, are predominantly steel tanks with pressure levels between 2–6 × 10^5 Pa. For direct drinking water storage, these are either enamelled on the inside or made of high-grade steel. The stored drinking or heating water has a thermal capacity of 4190 J/kgK = 1.16 Wh/kgK, so at a usable temperature difference of about 40°C, an energy quantity of

$$Q_{st} = mc_p\Delta T = 1\,kg \times 1.16\,Wh/kgK \times 40K = 46.4\,Wh \tag{3.8}$$

per litre of storage volume can be stored. If a heater buffer store is loaded to 80°C and a low-temperature heating/inlet temperature of 40°C can still be used, 46 kWh of usable energy can be stored with a 1000 litre store. In a low-energy building with a heating power requirement of 5 kW, about 10 h of the heating energy requirement can be covered; a very short time store!

Drinking water stores for solar plants are equipped, in contrast to conventional stores, with two heat exchangers, one of which must be situated in the lower, cool storage area, in order to utilise even small rises in temperature at the collector. The second internal heat exchanger is necessary for auxiliary heating in the upper third of the store (the standby section). From the solar heat transfer in the lower storage area alone, and the subsequent free convection of the heated liquid within the drinking water store, a high system inertia is the result, so usable temperatures of > 40°C are only achieved after approximately 4 hours of irradiance.

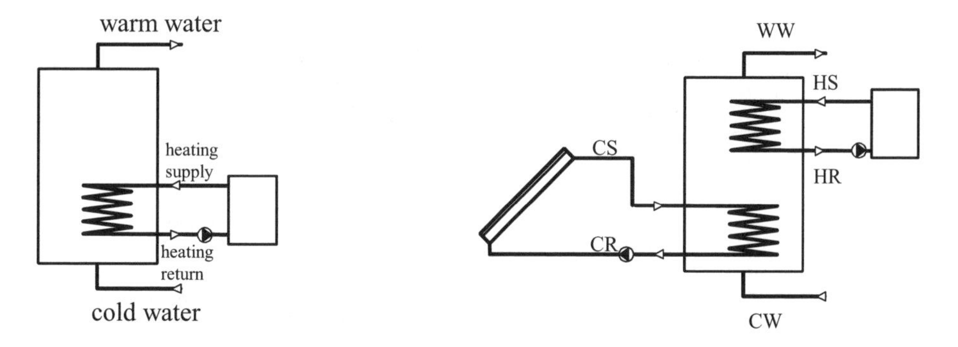

Figure 3.7: Drinking-water storage with one or two heat exchangers.

In new storage concepts, to reduce system inertia the solar-produced heat is led directly into the upper storage area. This is possible, for example, via a riser situated in the store. The warm collector water is led by the riser from above to below; the storage water warms up first in the riser and flows up into the upper storage area with the drinking water heat exchanger. The cooled storage water at the heat exchanger sinks in a second conduit pipe back into the lower storage area. The drinking water is warmed up via an internal heat exchanger when flowing through the upper storage area.

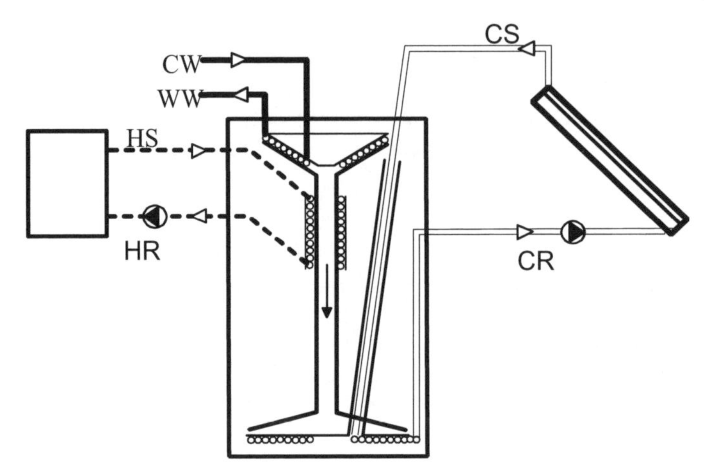

Figure 3.8: Drinking-water storage with a riser to bring solar heat quickly into the upper storage area.

Alternatively, the warm collector water can be switched at high temperatures directly to the upper storage area, via external switching with three-way valves. At low collector temperatures the solar heat is brought into the colder lower storage area.

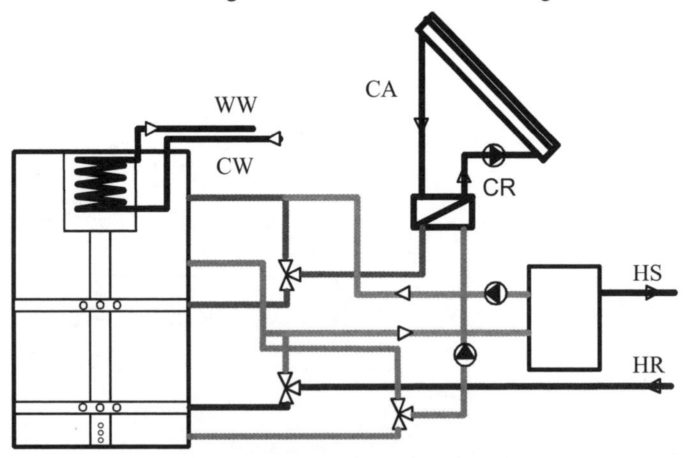

Figure 3.9: Combined drinking water and heating storage with external switching of the collector supply into the upper or middle storage area.

Apart from heating drinking water on the throughflow principle, such heating-water filled buffer stores enable a simple solar heating support, since from the middle storage area heat can be drawn off directly for space heating. The hydraulic connection of the buffer store to the conventional heating system must, however, be carefully planned for the preservation of the temperature stratification and for negative consequences on the efficiency of condensation boilers, if return temperatures are too high.

Heat losses of stores

The heat losses of a store consist of the heat transfer losses of the thermal insulation and also the convective losses by fluid circulation through connections and armatures.

The product of that effective heat transfer coefficient U_{eff} and cladding surface A of the store ($U_{eff}A$) can be determined from a simple energy balance and also measured: after homogeneous heating of the store, the temperature of the store T_{st} drops without warm-water withdrawal simply by calorific losses of the covering to the ambient air at temperature T_o and by free convection in the tank connections.

$$mc\frac{dT_{st}}{dt} = -\left(U_{eff}A\right)\left(T_{st}(t) - T_o\right) \tag{3.9}$$

As a boundary condition at the point in time $t = 0$, the storage temperature is given after heating up the store: $T_{st}\big|_{t=0} = T_{st,0}$. From the exponential drop in storage temperature, the effective calorific loss of the store can be determined in W/K.

$$T_{st}(t) = \left(T_{st,0} - T_o\right)\exp\left(-\frac{\left(U_{eff}A\right)}{mc}t\right) + T_o \tag{3.10}$$

Example 3.3

Determination of the effective calorific losses of a 750-litre heater buffer store from measured storage temperature values on a night without direct heat withdrawal. The room temperature is constant at 13°C.

The measured values are plotted as a temperature proportion of time-dependent storage temperature minus ambient temperature for the initial test temperature difference between memory and environment against time, here in minutes. The exponential involution of the condition of temperature over time results in the function:

$$\frac{T_{st}(t) - T_o}{T_{st,0} - T_o} = \exp\left(-\frac{\left(U_{eff}A\right)}{mc}t\right) = \exp\left(-0.0001 \times t\right)$$

For a store with contents m = 750 kg and a thermal capacity of the heating water of 1.16 Wh × 60 min/h = 69.9 Wmin, an effective calorific loss of $U_{eff}A = 10^{-4} \times mc = 5.22$ W/K results from the exponential coefficient of 10^{-4} min^{-1}.

Time [min]	Storage temperature [°C]
0	49.20
30	49.15
60	49.00
90	48.85
120	48.75
150	48.60
180	48.50
210	48.40
240	48.25
270	48.15
300	48.05
330	47.95
360	47.80
390	47.70
420	47.55
450	47.45
480	47.30
510	47.20
540	47.10
570	46.95

Typical heat loss values for a 400-litre drinking water store are between 1.7 and 3W/K, depending upon the standard of insulation and type of pipe connections, and for a 1000-litre store between 3.7 and 5.5W/K. Substantially higher loss coefficients point to missing non-return valves and strong convection of the storage water through the connection pipes. The effective calorific loss of the heater buffer store in the example is thus in the upper area of the loss values.

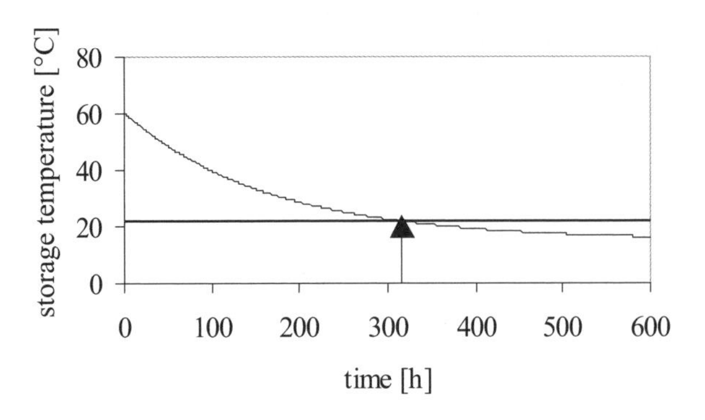

Figure 3.10: Exponential drop in storage temperature. After 308 h, i.e. some 13 days, the temperature has fallen to 1/e of the initial value.

Solar stores are often cylinder geometries with flat ball caps as covers and bases. For simplification, the ball caps can be approximated in small short-term stores by flat surfaces. The U-values of the store top cover U_t and bottom U_b are calculated from the layer thickness s and heat conductivity λ of the thermal insulation (if present) as well as from the thermal resistance between the thermal insulation and room air $1/h_i$ (standard value 0.13 m²K/W).

$$U_t = \left(\frac{s}{\lambda} + \frac{1}{h_i} \right)^{-1} \qquad U_b = \left(\frac{1}{h_i} \right)^{-1} \tag{3.11}$$

The length-related U_l-value of the standing cylinder results from the solution of the stationary thermal heat conduction equation in cylindrical coordinates and depends on the outside diameters of the store d_s and the thermal insulation d_i. The thermal resistance between the fluid and the store wall can be neglected when calculating the U_l-value.

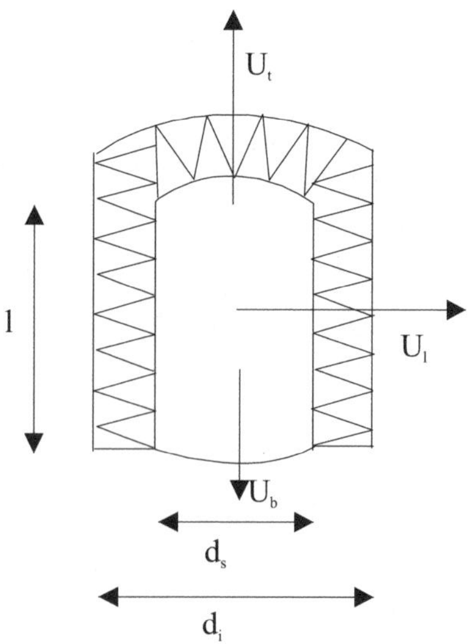

Figure 3.11: Dimensions and heat transfer coefficients of a heat store.

$$U_l = \frac{\pi}{\frac{1}{2\lambda}\ln\frac{d_i}{d_s} + \frac{1}{h_i d_i}} \tag{3.12}$$

The entire calorific loss of the store per Kelvin temperature difference to the ambient air thus results in:

$$UA = U_l l + U_t A_t + U_b A_b \tag{3.13}$$

Example 3.4

Calculation of the calorific losses of the cladding surface of the above 750-litre store with the following geometry:

Store height l:	2.03 m
Store diameter d_S:	0.69 m
120 mm PU foam insulation with heat conductivity λ:	0.04 W/mK
Heat transfer coefficient inside h_i:	8 W/m²K

The bottom of the store is not insulated.

From the above values the total diameter d_i = 0.93 m results, thus a length-related U_l-value of 0.81 W/mK and due to the store height a calorific loss U_l of 1.65 W/K. In addition there are calorific losses at the top and the bottom, both with a surface of 0.37 m². The heat transfer coefficient of the insulated cover is $U_t = \frac{1}{s/\lambda + 1/h_i} = 0.32\frac{W}{m^2K}$ and of the uninsulated bottom $U_b = \frac{1}{1/h_i} = 8\frac{W}{m^2K}$. The entire calorific loss of the cladding surface is thus:

$$UA = U_l l + U_t A_t + U_b A_b = 1.65W/ + 0.12W/K + 2.96W/K = 4.73W/K$$

Although the bottom constitutes only 7% of the total cladding surface of 5.14 m², its calorific loss coefficient dominates. With good temperature stratification of the store, however, the temperature at the store's lower surface, typically 20°C, is clearly lower than the top cover area temperature of about 60°C, so the total calorific loss $\dot{Q}_l$ at a mean storage temperature of 40°C is distributed roughly thus:

$$\dot{Q}_l = \underbrace{1.65W/K \times (40-15)K}_{41.25W} + \underbrace{0.12W/K \times (60-15)K}_{5.4} + \underbrace{2.96W/K \times (20-15)K}_{14.8} = 61.45W$$

3.1.4.3 Piping and circulation losses

Piping calorific losses arise in the form of reheating power from the temporary operation of the collector and warm-water withdrawal, as well as from constant heat losses in circulation pipes. The heating up power Q_h can easily be calculated, at a given temperature difference to the surroundings, via the mass and thermal capacity of the pipe (m_p, c_p) and of the heat distribution medium (m_f, c_f):

$$Q_h = (m_p c_p + m_f c_f)(T - T_o) \tag{3.14}$$

The output losses of a circulation pipe are calculated via the length-related U_l-value [W/mK] of the insulated pipe:

$$Q_c = U_l l (T - T_o) t_c \tag{3.15}$$

where t_c are the operating hours of the circulation pump and l is the length of the circulation pipe.

Example 3.5

Calculation of the heating up losses of a DN15 (15 mm exterior diameter) collector pipe filled with antifreeze, from 10°C to 50°C.

Thermal capacity of copper c_p:	0.39 kJ/kgK
Density of copper ρ_p:	8867 kg/m³
Thermal capacity of fluid c_f:	3.5 kJ/kgK
Density of fluid ρ_f:	1060 kg/m³
Pipe length l:	30 m
Outer diameter d_{po}:	0.015 m

Inner diameter d_{pi}: 0.013 m

$$m_p = \rho_p V_p = \rho \frac{\pi\left(d_{po}^2 - d_{pi}^2\right)}{4} l = 11.7\text{kg} \qquad m_f = \rho_f V_f = \rho \frac{\pi d_{pi}^2}{4} l = 4.22\text{kg}$$

$$Q_h = \left(m_p c_p + m_f c_f\right)\left(T - T_0\right) = \left(4.56\frac{\text{kJ}}{\text{K}} + 14.77\frac{\text{kJ}}{\text{K}}\right)(40\text{K}) = 773.2\,\text{kJ} = 215\,\text{Wh}$$

Example 3.6

Calculation of the circulation losses of a 30-m long, DN15 pipe with 50°C warm-water, to room air at 20°C, with a daily working time of 10 hours. The pipe has 30 mm of insulation ($\lambda = 0.04$ W/(mK)).

$$U_l = \frac{\pi}{\frac{1}{2\lambda}\ln\frac{d_i}{d_{lo}} + \frac{1}{h_i d_i}} = \frac{\pi}{\frac{1}{2\times 0.04}\ln\frac{0.075}{0.015} + \frac{1}{8\times 0.075}} = 0.144\frac{W}{mK}$$

3.1.5 System technology for heating support

Between the monthly solar irradiance or collector yield maxima and the heating requirement of buildings, a half-year phase shift exists. Sensible heat stores have, however, only a small heat storage capacity of a few days. This means that decentralised heating support systems should be designed to cover no more than 15–30% of the heating requirement, as otherwise summer overheating, longer standstill periods and falling surface-specific collector yields are inevitable. Collector surfaces between 10–20 m² per housing unit can be used, however, apart from the year-round drinking water supply, for heating support with no significant yield reduction. At collector areas below 30 m², only short-term storage is necessary and system yield barely improves with storage size, as long as a minimum storage volume of about 0.8–1.5 m³ is available. More important for system yield is the correct hydraulic connection of the conventional heater to the storage volume: if, for example, the return flow to the boiler is connected at the bottom of the buffer store, the useful temperature stratification will be destroyed due to the high boiler volume flows and yield will drop by about 5–6 % (Kerskes *et al.*, 2002).

As system concepts, either two-store systems with separate drinking water and heater buffer stores are available, or more economical combined heating/drinking water stores. The temperature stratification of the store is important for a high solar yield, as long operation times of the collector circuit pump are only possible at low temperature levels in the lower storage area. A stratified transfer of the solar heat is possible through internal risers with diaphragm flaps, which enable a release of warmed fluid of low density only at a height at which the surrounding store fluid also has a low density and thus a high temperature.

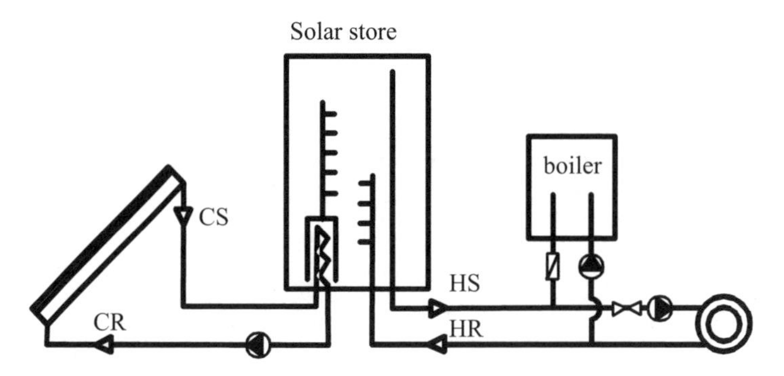

Figure 3.12: Internal step-by-step loading of a heat buffer store.

The riser with diaphragm flaps is heated through an internal heat exchanger by the solar collector warm-water supply flow CA. The heating circuit supply is taken from the upper storage area (heating supply HS) and step-fed in again depending on the heating return temperature (heating return HR). Alternatively the feed can be controlled externally via temperature-dependent three-way valves. Decisive for the temperature stratification is, apart from the feeding of the solar heat, above all the heat extraction or supply of the heating system, which is always connected with high mass flows. The heating supply is connected to the upper storage area, the colder heating return should if possible be led into the store by a step-loading installation, so that at low temperature differences of the space heating and relatively high return temperatures, there is no direct feed into the lower storage area. Through the high heating water mass flows, the lower storage area without a step-loader is heated very quickly to the heating return temperature, i.e. at least 30°C, and often to 40–70°C. A heating return temperature rise through the store is not recommended if a condensing boiler is used whose condensation potential depends largely on low return temperatures which fall below the condensation point.

The heating of drinking water by internal heat exchangers is no problem as long as the heat exchangers contact the whole store from bottom to top. Integrated small drinking water stores, if placed only in the upper standby section, can destroy the temperature stratification if provision is not made for a flow of the cooled storage water into the lower area, by special diverting pipes in the store. As an alternative to internal high-grade steel heat exchangers, use is also made of external plate-type heat exchangers with such a high transfer power that cold water passing through can be warmed up to the required temperature. In order to achieve as even an outlet temperature as possible, the external heating water pump is revolution-adjusted as a function of the warm-water flow. In addition, a mixing valve provides an even outlet temperature and an upper temperature limit.

A theoretical investigation of the solar yield of five storage concepts for the heating support of a low-energy building with 15 m² of collector surface and 1050 litres of storage volume has shown that only the concept of an integrated drinking water store in the upper storage area leads to a clear yield reduction, 13% compared with the best system (Pauschinger, 1997 Kerskes *et al.*, 2002). All other concepts – a two-store system, a system with external through-flow heating of drinking water, a system with an integrated drinking water store reaching from bottom to top, and a step-loading store with external heating of drinking water – produce yields of approximately 300 kWh/m²a, deviating by less than 3% from each other. If no storage is used, the system yield drops by about 20%.

3.1.6 Large solar plants for heating drinking water with short-term stores

Large central solar plants for heating drinking water, with collector surfaces of over 100 m², are particularly suitable for buildings with a continuously high warm-water requirement of over 5 m³/day, such as hospitals, nursing homes, old people's homes and large housing estates. All large-scale installations have central buffer storage with storage contents > 5 m³; they differ mainly in the concept of the heat transfer to the drinking water.

Concepts are common to pre-heat drinking water on the throughflow principle via an external heat exchanger with conventional auxiliary heating in a drinking water store.

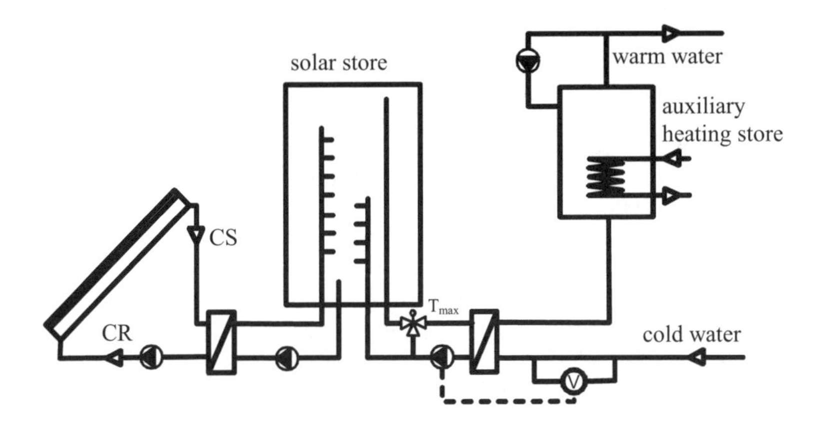

Figure 3.13: Drinking water pre-heating via external heat exchangers regulated by the measured drinking water flow.

Since cold water temperature around 10°C is always offered at the consumption-side entry to the external heat exchanger, the return temperature into the buffer store is low in this concept, and the solar plant yield is up to 10% higher than with classical load storage concepts. Apart from control problems caused by strongly varying tapping rates, the circulation losses cannot be covered via the solar buffer with this version.

The classical load storage concept is based on the transport of heat from the buffer store into a drinking water store, as soon as the temperature level in the buffer is above the drinking water temperature in the store. In retrofit systems, existing stores can be used as auxiliary heating stores and the additional drinking water store for the solar plant is purely a pre-heating store. Here, too, the lack of cover of circulation losses is unfavourable.

If a large drinking water store is used for pre-heating and auxiliary heating, the circulation losses can also be covered by solar energy.

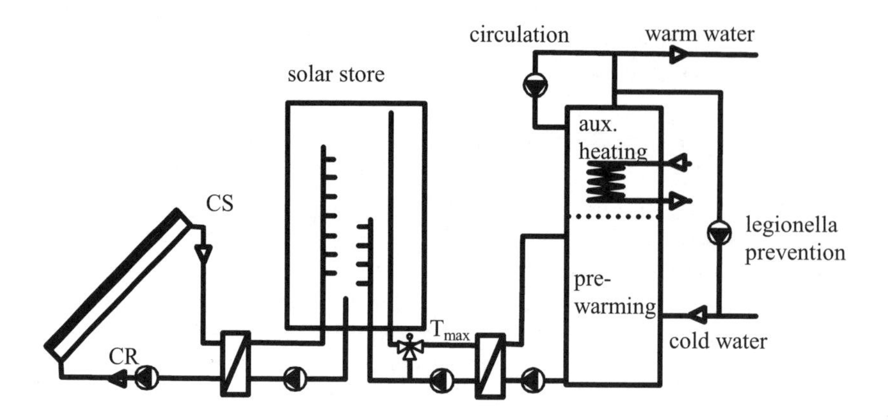

Figure 3.14: Transfer from a central buffer store to a drinking-water store.

The pre-heating area is either designed as a separate store or as a large vertical store combined with the auxiliary heating store (the broken line in Figure 3.14). In order to be able to through-heat the pre-heating area to avoid the formation of legionellae, heat can be taken out of the upper storage area.

Load regulation

The load regulation of the buffer store is set either by means of a temperature difference control between the collector and lower storage area, as in small systems, or frequently by combined irradiance/temperature difference control. Since in large solar plants the heat transfer to the buffer takes place via external heat exchangers and often large pipe lengths are available, not only the collector outlet temperature T_1 but also the collector circuit temperature T_2 at the heat exchanger entrance must be measured, since it is relevant for the heat transfer to the buffer.

Either a temperature signal of the collector or a radiation sensor is used to switch on the collector circuit pump, (Peuser *et al.*, 2000). The control strategy reads: P_1 on, if $T_1 - T_3 >$ 8 K; P_2 on, if $T_2 - T_3 > 7$ K and if P_1 is on.

In pure temperature-difference control, the collector circuit pump P_1 is turned on if the collector temperature T_1 is about 8 K above the lower store temperature T_3. With long external pipes, a minimum pump operating time of some minutes is necessary, so that the warmed collector liquid reaches the heat exchanger entrance T_2. To be sure of avoiding freezing of the heat exchanger on the heating-circuit side, a bypass must be added before the heat exchanger, if external pipes are long. The storage pump P_2 does not operate till T_2 is also about 7 K above the lower store temperature T_3 and the collector pump P_1 is operating at the same time. Thus at high room temperatures the storage pump will not be activated by a high temperature T_2, at night, for example. As in small systems, the power-off temperature difference should clearly be lower than the switching-on temperature difference, at about 2–3 K.

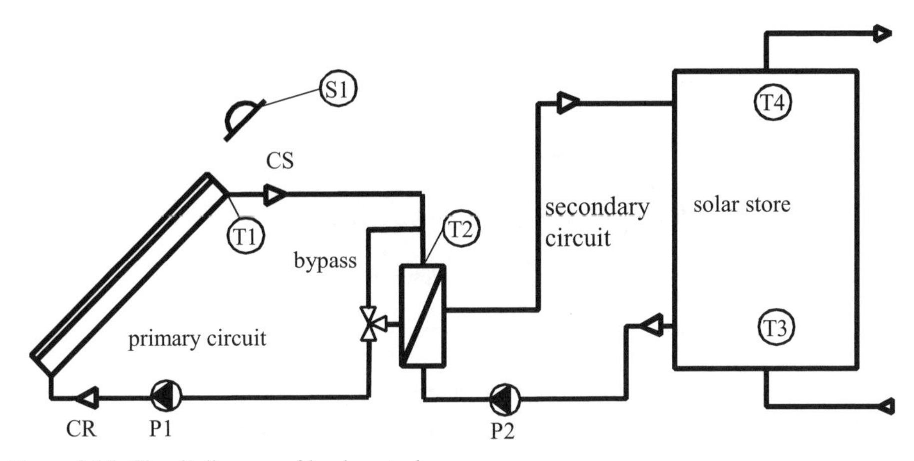

Figure 3.15: Circuit diagram of load control.

Alternatively to the collector temperature T_1, a radiation sensor S_1 with a typical switching-on irradiance of 200 W/m² can be used for switching the collector circuit pump. Again, the storage pump P_2 only operates if the heat exchanger temperature is around 7 K above the lower store temperature. If the storage pump does not connect during the start-up minimum operating time of the collector pump, switch-off occurs when a lower irradiance value of about 150W/m² is reached. With a running storage pump, however, only the temperature difference between the heat exchanger entrance and the store temperature should be used as a power-off criterion, and not the irradiance signal.

When the maximum temperature T_4 of the store is attained, not only the storage pump but also the collector pump must be switched off. Thus the collector goes into standstill, and the collector liquid evaporates and is pressed into the expansion tank, or blown off under control. Since the components near the collector (aeration valve, sensors, thermal insulation) are designed for high standstill temperatures, no problems arise. If, however, the collector pump is not switched off, high-temperature heat-bearing fluid circulates in the entire collector circuit, leading to a very strong thermal stress of all components.

Discharge control

The highest control demands exist for heating drinking water on the throughflow principle. Since the best heat transfer efficiencies result when there is the same flow on the buffer store side and the tapping side, the buffer store discharge pump P_3 must be directly revolution-adjusted as a function of the tapping flow rate. The regulation signal is received either by a direct volume flow measurement, or indirectly by a dynamic temperature difference measurement.

For example, a dipping sensor in the cold water intake (T_5) reacts far faster to changes in fluid temperature than an external pipe temperature sensor: when withdrawing water, this different reaction rate leads to an initially high temperature difference signal, which can be used for volume flow measurement. In practice, this regulation concept has so far proved problematic.

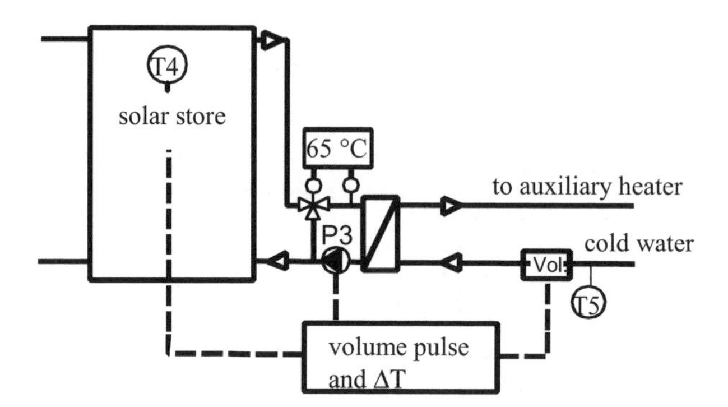

Figure 3.16: Discharge control of the buffer store on pre-heating drinking water flowing through directly.

Transfer from the solar buffer into a drinking water store is simpler and more robust via an external heat exchanger regulated by a temperature-difference measurement between the upper buffer store temperature T_4 and the lower drinking water store temperature T_5.

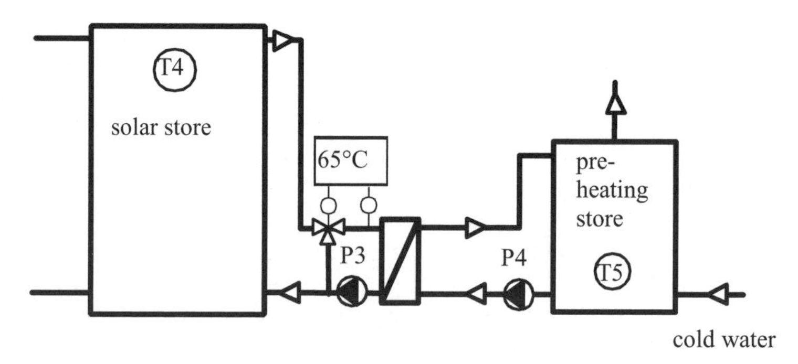

Figure 3.17: Discharge control on transfer to a drinking-water store.

The buffer store discharge pump P_3 and the drinking water reloading pump P_4 operate if the buffer store temperature T_4 is about 7 K above the lower temperature in the pre-heating store T_5. At a temperature difference of less than 3 K, both pumps are switched off.

3.1.6.1 Design of large solar plants

In large solar plants, strict sizing of the collector surface to the minimum summer requirement has worked satisfactorily, to avoid standstill times of the collector field if at all possible, and to offer economical solutions at high solar yields. With such low collector surfaces, typical yearly degrees of cover of around 20% are achieved, while with small systems with a clearly higher specific system cost reduction per added square metre of collector surface, a slight oversizing in the summer with degrees of cover of 60% is useful.

Consumption measurements in a number of building types show that during the summer, low demand periods have to be expected due to vacations; these periods cause a consumption reduction of up to 50% of the average value. Though consumption in a hospital is around 50–60 litres daily, per full occupancy person, in a vacation month such as July only 35 litres are measured, related to the same planned occupancy number. Taking sizing for these low demand periods into account, design can proceed on the basis of the following values.

Table 3.3: Specific warm-water consumption in litres per full occupancy person (*p*) per day (*d*) at 60°C, determined from summer low demand periods.

	Residential building	*Hospital*	*Old peoples' home*	*Student halls of residence*
Specific warm-water consumption [*l/p d*]	20–25	30–35	30–35	20–25

On the basis of daily total drinking water throughput thus calculated, the collector surface can be estimated. Annual yield calculations result in minimum solar heating costs of 0.13 €/kWh at a utilisation of 70 litres of warm water requirement per square metre of collector surface. Larger collector surfaces mean lower utilisation of the solar plant with higher standstill times, a smaller yield and higher system costs. In contrast, higher utilisation, i.e. under-sizing of the collector surface, is no problem. The specific solar buffer volume per square metre of collector surface depends both on the uniformity of the tapping profile and on the degree of utilisation of the system. With an even tapping profile in multi-family houses, hospitals, old peoples' homes etc. and a utilisation of 70 litres of warm-water per square metre of collector surface, buffer sizing of 40–50 litres per m² of collector is sufficient. With larger collector surfaces, i.e. a lower utilisation of 40–50 l/m², the buffer volume should also increase to 60–70 l/m².

In buildings with greatly reduced consumption, e.g. at weekends (industrial premises), buffer sizing of 70–100 l/m² is recommended, depending on the degree of utilisation.

Table 3.4: System efficiency and buffer store volume depending on the degree of utilisation and the evenness of the tapping profile.

Degree of utilisation (litres/m² of collector surface)	*Evenness of the tapping profile*	*System efficiency* [%]	*Specific buffer volume (l/m² of collector surface)*
70	even tapping profile of a multi-family house	47	40–50
40	even tapping profile of a multi-family house	37	60–70
70	tapping profile of a0 workshop with no weekend consumption	36–38	60–80
40	tapping profile of a workshop with no weekend consumption	26–28	70–100

3.1.7 Solar district heating

Solar district heating means a central heating supply to large residential or industrial estates. A solar district heating network consists of a central heating installation with a buffer store and a boiler for auxiliary heating, a heat distribution network, building transfer stations for heating and warm-water, plus the collector field, often distributed decentrally on buildings. In addition to the short-term buffer store in the central heating installation, the summer excess heat is deferred seasonally till the heating period by a long-term heat store, and the degree of solar cover can be increased.

Energy can be supplied both for heating drinking water and for heating support via the solar district heating network. For an approximate design, both the heating requirement and that for heating drinking water must be known. The following reference values can be used for approximate sizing.

Table 3.5: Approximate sizing of solar district heating systems (Hahne *et al.*, 1998).

	Solar district heating with short-term store	*Solar district heating with long-term store*
Minimum plant size	From 30–40 housing units or 60 persons respectively	from 100–150 housing units (70 m^2 each)
Flat plate collector surface	0.8–1.2 m^2 per person	1.4–2.4 m^2 per MWh yearly heating demand 0.14–0.2 m^2 per m^2 heated surface
Store volume (water equivalent)	0.05–0.1 m^3/m^2	1.5–4 m^3/MWh 1.4–2.1 m^3/m^2 flat plate collector
Usable solar energy	350–500 kWh/m^2a	230–350 kWh/m^2a
Degree of solar cover	drinking water: 50% total: 10–20%	total: 40–70%

District heating systems differ in their types of heat distribution networks. In central drinking water heating systems, the supply and return pipe for heating as well as a warm-water pipe and a circulation pipe for drinking water supply to each building are led (4–conductor network) from the central heating installation. Supply and return of the collector field needs two further pipes for the central heating installation (4 + 2 conductor network). Due to high circulation losses of the drinking water pipes, such a concept is sensible only for small systems with 20–30 housing units.

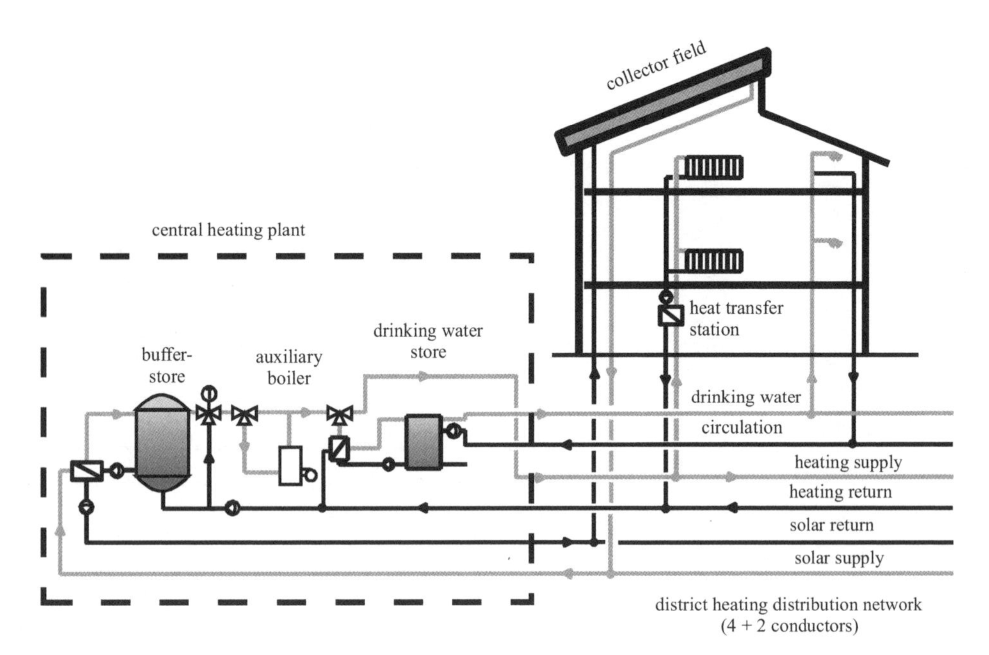

Figure 3.18: 4 + 2 conductor network: supply and return heating with house transfer station, warm water and circulation pipe for direct heating of drinking water, plus two pipes for the collector field.

With larger district heating systems, the use of a second house transfer station for drinking water heating is preferable, due to high circulation losses. The heat distribution network is reduced to two conductors, which must maintain all year the necessary temperature of 60–70°C for heating drinking water. Furthermore, as above, two pipes are necessary for the collector field (2 + 2 conductor network). The heat exchanger of the warm water transfer station can either be designed for throughflow heating of warm water (with small consumption in one- or two-family houses) or else for transfer to drinking water stores.

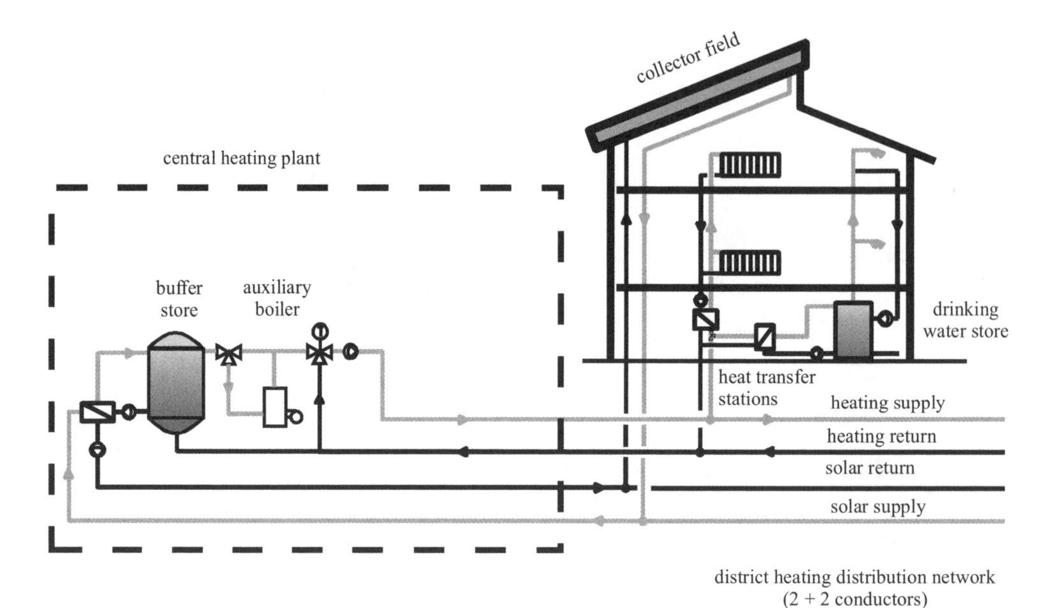

Figure 3.19: 2 + 2 conductor network: supply and return pipes for the house transfer stations heating and warm water.

A further pipe can be saved if only the hot collector supply (on the secondary circuit side) is led into the buffer store of the central heating installation, while the solar return pipe coincides with the heat return pipe. Via a house transfer station on the solar side, the heating or warm-water return can be warmed up directly in the solar transfer station (return temperature rise) and flow only then via the solar supply pipe to the central heating installation, if heating is required in the building. In this case the heat return pipe is not used. When heat is required without solar energy, hot buffer water is drawn via the heat supply pipe from the upper storage area, and fed without return rise by the solar plant via the heat return pipe back into the buffer. If no heating requirement exists and if the solar plant is supplying energy, the direction of flow is reversed in the heat return pipe, and stored water is drawn from the lower buffer store area to be heated in the solar transfer station, and returned via the solar supply pipe to the buffer.

If the degree of solar cover for drinking water and heating support is to be between 40–70% of the total energy requirement, long-term heat stores with storage volumes of 1–10 m³ per m² of collector surface are necessary.

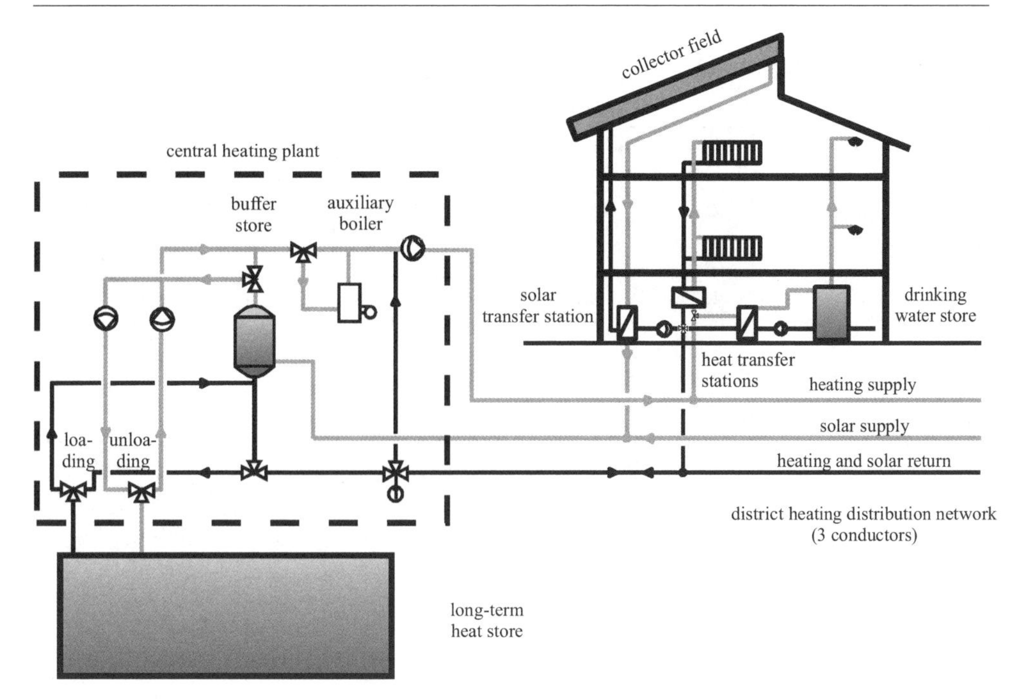

Figure 3.20: Three-conductor network with heating supply, secondary solar supply and a shared solar and heater return pipe. With the long-term heat store, solar heat can be stored seasonally.

A short characterisation of the most important available storage concepts as well as approximate ratios of storage volume to collector surface at 50% solar cover are taken from Kübler and Fisch (1998) and Hahne *et al.* (1998).

3.1.8 Costs and economy

The costs of a small standard system (< 6m²) can be divided into three; collector costs, costs of the store and solar station, and of assembly and pipework. At average equipment prices of some 3500–5000 € per housing unit, the result is solar heat prices of about 0.15–0.25 €/kWh, without taking subsidies into consideration. Related to a square metre of effective area, capital outlays of 35–70 €/m² in addition to the auxiliary heat supply must be expected. The best cost–benefit ratio with additional investments of 15–20 € per m² of effective area is obtained for large-scale solar installations with a short-term store. In the context of the German funding programme "Solarthermie 2000", the determined real solar heat costs are between 0.1–0.13 €/kWh. Here too, the collector costs plus assembly and pipework each account for one-third of the total costs; the storage and regulation costs clearly fall, however.

Table 3.6: Characteristics of long-term heat stores.

Store type	*Hot water store*	*Gravel/water store*	*Earth pipe store*
Storage concept	container store or earth-basin store	earth-basin with gravel/water filling without a load bearing cover construction	heat transfer pipes in the ground, to a maximum depth of 150 m
Construction	reinforced concrete, steel- or glassfibre-reinforced plastic, or a pit with a cover and lid, stainless steel- or sheeting cover	cover with plastic sheeting	U-shaped coaxial plastic pipes with 1.5–3 m separation
Maximum/ minimum volumes	max. 100 000 m³, the largest store designed so far being 28 000 m³	-	>100 000 m³ due to high lateral heat losses
Heat insulation	15–30 cm at the lid and the store walls, and also under the store if the pressure can be withstood	as with water stores	only in the covering layer 5–10 m from the surface
Store volume/flat plate collector surface	1.5–2.5 m³/m²	2.5–4 m³/m²	8–10 m³/m²
Approximate costs (€/m³) at 20 000 m³ storage volume	70–80	65–85	25
Other characteristics	container store costly	with a gravel proportion of 60–70 vol. %, around 50% larger building volume as a water store	easily constructed

In the most economical large-scale installations with collector surfaces of around 1000 m², the system costs, including planning and value-added tax, are around 375–465 € per m² of collector surface (Stuttgart-Burgholzhof, Goettingen, Neckarsulm). Solar district heating systems with long-term heat stores require the highest floor space-related additional investments, around 75–140 €/m². Here, however, the solar-covered proportion of the total heat requirement lies between 40 and 70%, while solar heated drinking water supply in small or large-scale installations covers at most 20% of the total heat requirement.

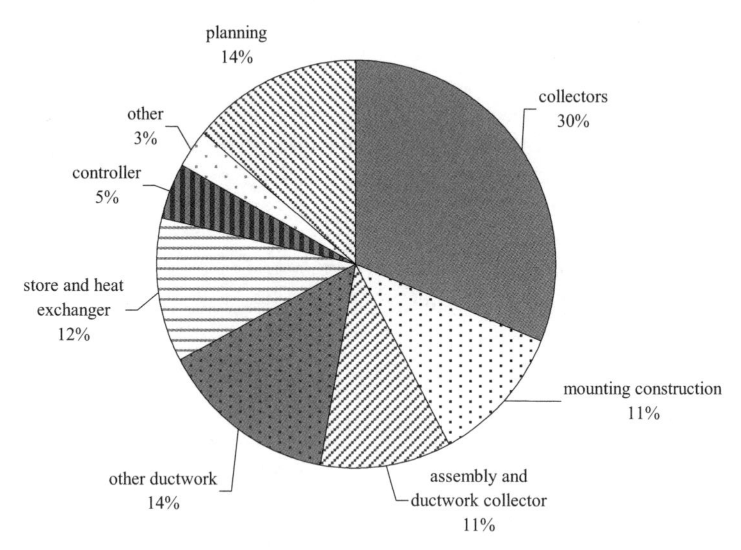

Figure 3.21: Cost allocation of 35 large-scale solar thermal installations with short-term stores from the Solarthermie 2000 programme.

The total costs were about 620 €/m² including planning and value-added tax.

3.1.9 Operational experiences and relevant standards

Like conventional heating systems, solar plants require regular maintenance. The most frequently occurring defects in practice are problems during installation and initial operation (leakages in pipes, insufficient aeration) as well as unsatisfactory control strategies and hydraulic problems.

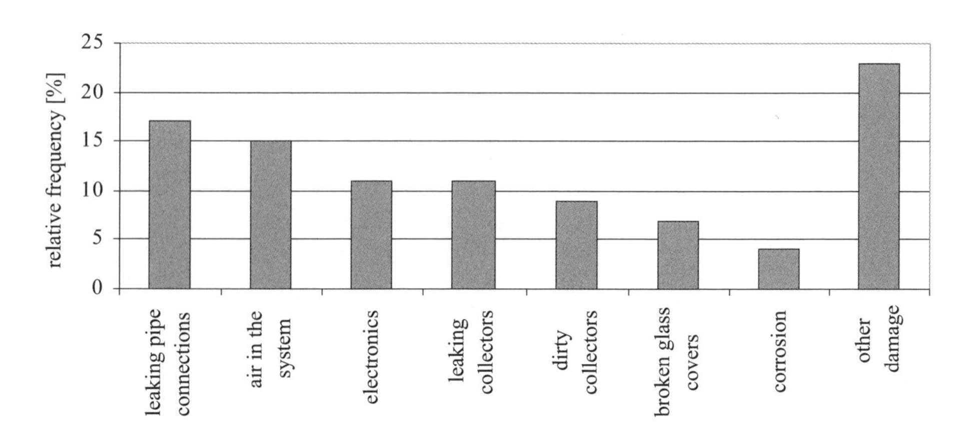

Figure 3.22: Relative frequency (in %) of defects in thermal solar plants.

The occurrence of leaky fittings, especially on the roof, can be reduced by the use of larger collector units. Problems with air in the system are reduced by having as high flow rates as possible, i.e. series connection of the collectors. Ventilation valves must be fitted at the highest places, and air cushions avoided; alternatively, pumps with a high increase in pressure can be installed to force air through the entire cycle to the pump aeration valve.

For the authorisation and performance evaluation of thermal collectors, the following standards and guidelines are relevant:

Collectors
ISO 9806: performance evaluation and quality testing
EN 12975: thermal testing of collectors

Storage tests
ISO 9459, part 4a
EN 12975, part 3
EN 12977, part 3

Systems:
Solar drinking water systems
ISO 9459, part 5:
- ⇨ performance evaluation of systems with forced circulation
- ⇨ performance evaluation of compact units
- ⇨ investigation of systems with forced circulation, using the component test procedure

Solar drinking water systems with heating support
ISO 9459, part 4: investigation of systems with forced circulation using the component test procedure
In-situ test for acceptance test (short-time procedure)
Small systems: ISO 9459, part 5 – DST (dynamic system testing)
Large-scale installations from 100 m² up: ISO 9459, part 4 – CTSS (component testing – system simulation),
ISTT (in situ term testing)

Furthermore, for large-scale installations the steam boiler regulation is relevant, as it regulates the safety engineering and commissioning procedures of the system.

3.1.10 Efficiency calculation of thermal collectors

The available power and efficiency of thermal collectors is determined by the optical characteristics of the transparent cover and the absorber, and by the calorific losses between absorber and environment. If the mean absorber temperature T_a is known, the surface-related available power is obtained simply by looking at an energy balance. The irradiance transmitted through the cover with a transmission coefficient τ and absorbed at factor α minus the calorific losses results in the available useful power $\dot{Q}_u$ per square metre of collector surface A.

$$\frac{\dot{Q}_u}{A} = G\tau\alpha - U_t\left(T_a - T_o\right) \tag{3.16}$$

The efficiency of the solar-thermal collector results from the proportion of the surface-related available power divided by the irradiance in the collector level.

$$\eta = \frac{\dot{Q}_u}{AG} = \underbrace{\tau\alpha}_{\eta_0} - U_t \frac{\left(T_a - T_o\right)}{G} \tag{3.17}$$

η_0 represents the optical efficiency of the collector. The heat transfer coefficient of the collector U_t consists of the losses at the collector front, back and sides. The ratio of the temperature difference between absorber and environment $T_a - T_o$ and irradiance G is called a reduced parameter.

The temperature on the absorber sheet metal T_a is, however, a complicated function of the distance from the heat-removing fluid tubes and the flow length, so an average value can only be determined very laboriously from a measured temperature distribution. What is measurable, however, is the fluid inlet temperature into the collector or the mean fluid temperature, which at not too low flow rates is given by the arithmetical average value between entry and exit temperatures. Above all the representation of the available energy as a function of the fluid inlet temperature is very useful for system simulations, since the fluid inlet temperature is given by the storage return temperature.

To be able to determine analytically the available energy at a given fluid inlet temperature or mean fluid temperature, the temperature distribution on the absorber sheet metal must first be calculated as the solution of a thermal conduction problem. Subsequently the local fluid temperature is calculated by the heat transfer to the fluid. At a given mass flow, the entire rise in temperature and available energy can then be calculated by integration over the flow length, and represented as a function of the fluid inlet temperature.

3.1.10.1 Temperature distribution of the absorber

To determine the conversion of the absorbed irradiance to available power, the heat flow caused by the temperature gradient on the sheet metal towards the fluid tube must be calculated.

The temperature distribution on the absorber sheet metal transverse to the direction of flow is calculated by the thermal conduction characteristics of the sheet metal and by the convective heat transfer coefficient between sheet metal and fluid. The calculation of the temperature distribution takes place via the solution of the energy balance equation for a point on the absorber sheet metal. The absorbed irradiance $G\tau\alpha$ is transferred on the one hand into calorific losses to the environment, and on the other hand into a heat flow led by the absorber sheet metal towards the fluid tube. Following Fourier's law of thermal conduction, this heat flow $\dot{Q}\big|_x$ at point x on the sheet metal is proportional to the temperature gradient and to the cross-section area, which is calculated from the plate thickness δ and the unit length $l = 1$ in the direction of flow. Due to the small absorber plate thickness, the thermal conduction problem is regarded as only one-dimensional, i.e. along the sheet metal.

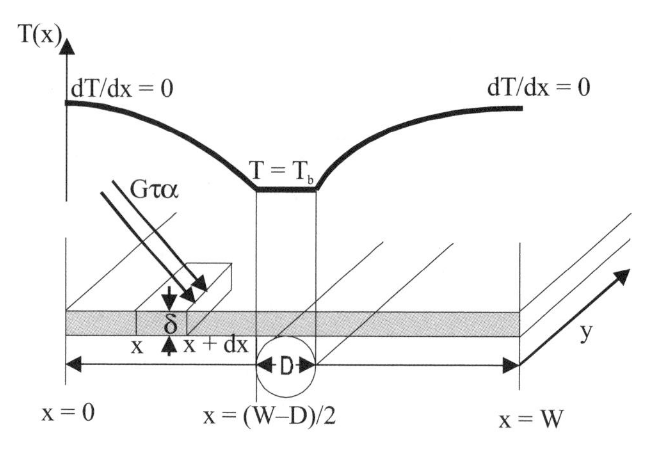

Figure 3.23: Geometry of the absorber sheet metal with fluid pipe and surface element for energy balance.

The energy surplus between the absorbed radiation and heat losses leads to a change in the temperature gradient within a surface element $l\ dx$ and is led off laterally to the fluid pipes over the cross-sectional area $l\delta$. Using a Taylor sequence development, the difference of the heat flows results in:

$$\begin{aligned}\dot{Q}\big|_x-\dot{Q}\big|_{x+dx} &= \left(-\lambda(l\delta)\frac{dT}{dx}\right)\bigg|_{x_0}-\left(-\lambda(l\delta)\frac{dT}{dx}\right)\bigg|_{x_0+\Delta x}\\ &=-\lambda(l\delta)\left(\frac{dT}{dx}\bigg|_{x_0}-\left(\frac{dT}{dx}\bigg|_{x_0}+\frac{d^2T}{dx^2}\bigg|_{x_0}dx+..\right)\right)=\lambda(l\delta)\frac{d^2T}{dx^2}\bigg|_{x_0}dx\end{aligned} \tag{3.18}$$

The power balance for the surface element $l\ dx$ (with unitary length $l = 1$ in the direction of flow y) of the absorber sheet metal is thus:

$$\begin{aligned}&G\tau\alpha(ldx)-U_t(T-T_o)(ldx)+\lambda(l\delta)\frac{d^2T}{dx^2}dx=0\\ &\Leftrightarrow G\tau\alpha-U_t(T-T_o)+\lambda\delta\frac{d^2T}{dx^2}=0 \quad \left[\frac{W}{m^2}\right]\end{aligned} \tag{3.19}$$

From this a second-order differential equation for the temperature field is obtained.

$$\frac{d^2T}{dx^2}=\frac{U_t}{\lambda\delta}\left(T-T_o-\frac{G\tau\alpha}{U_t}\right) \tag{3.20}$$

As the zero point of the x-axis, the midpoint between two fluid pipes is chosen, at which the temperature is maximum and the temperature gradient is zero (first boundary condition):

$$\left.\frac{dT}{dx}\right|_{x=0} = 0 \tag{3.21}$$

As the second boundary condition, a temperature of the sheet metal over the fluid pipe is assumed. The geometry of the fluid pipe with lateral sheet metal corresponds to a classical cooling fin in which the sheet metal, in contrast to the collector, serves not to absorb heat but to lead it off. The temperature over the fluid pipe is, as with a cooling fin problem, characterised with base temperature T_b. This temperature can be eliminated in a second step, after determination of the useful energy led to the pipe and the convective heat transfer to the fluid.

$$T\big|_{x=(W-D)/2} = T_b \tag{3.22}$$

The differential equation can be solved by substitution of the variables $\Psi = T - T_o - G\tau\alpha / U_t$ and by using the constants $m = \sqrt{U_t/(\lambda\delta)}$ $\left[m^{-1}\right]$

$$\frac{d^2\Psi}{dx^2} - m^2\Psi = 0 \tag{3.23}$$

with the boundary conditions

$$\left.\frac{d\Psi}{dx}\right|_{x=0} = 0 \tag{3.24}$$

$$\Psi\big|_{x=(W-D)/2} = T_b - T_o - \frac{G\tau\alpha}{U_t} \tag{3.25}$$

and the general solution

$$\Psi = C_1 \sinh\ (mx) + C_2 \cosh\ (mx) \tag{3.26}$$

From the boundary conditions the constants C_1 and C_2 are obtained.

$$\left.\frac{d\Psi}{dx}\right|_{x=0} = C_1 m \underbrace{\cosh(0)}_{1} + C_2 m \underbrace{\sinh(0)}_{0} = 0 \Rightarrow C_1 = 0 \tag{3.27}$$

$$\Psi\big|_{x=(W-D)/2} = \underbrace{C_1}_{0} \sinh\left(m(W-D)/2\right) + C_2 \cosh\left(m(W-D)/2\right) = T_b - T_o - \frac{G\tau\alpha}{U_t}$$

$$\Rightarrow C_2 = \frac{T_b - T_o - \dfrac{G\tau a}{U_t}}{\cosh\left(m(W-D)/2\right)} \tag{3.28}$$

After back substitution, the result for the temperature on the absorber sheet metal is

$$T(x) = T_o + \frac{G\tau\alpha}{U_t} + \left(T_b - T_o - \frac{G\tau\alpha}{U_t}\right)\frac{\cosh(mx)}{\cosh(m(W-D)/2)} \tag{3.29}$$

From the temperature distribution, and using Fourier's law, the amount of heat at the point $x = (W–D)/2$ led from both sheet metal sides to the fluid pipe can now be determined. As the temperature distribution right and left of the fluid pipe is symmetric, the amount of heat calculated on one side can simply be doubled. The heat flow is characterised by $\dot{Q}_{fin}$.

$$\begin{aligned}\dot{Q}_{fin} &= 2\times\left(-\lambda(l\delta)\left.\frac{dT}{dx}\right|_{x=(W-D)/2)}\right)\\ &= -2\lambda(l\delta)\left(T_b - T_o - \frac{G\tau\alpha}{U_t}\right)\frac{\sinh(m(W-D)/2)}{\cosh(m(W-D)/2)}\times m\\ &= 2\frac{\lambda(l\delta)m}{U_t}\left(G\tau\alpha - U_t(T_b - T_o)\right)\tanh(m(W-D)/2)\\ &= (W-D)l\left(G\tau\alpha - U_t(T_b - T_o)\right)\frac{\tanh(m(W-D)/2)}{m(W-D)/2}\end{aligned} \tag{3.30}$$

with $\frac{\lambda\delta m}{U_t} = \frac{1}{m}$ used in accordance with the definition.

The fin efficiency F shows the ratio of the actual heat flow $\dot{Q}_{fin}$, based on Equation (3.30), to the ideal heat flow $\dot{Q}_{ideal} = (W-D)l\left(G\tau\alpha - U_t(T_b - T_o)\right)$ which would result if all the sheet metal were at the lower base temperature with correspondingly low heat losses.

$$F = \frac{\tanh(m(W-D)/2)}{m(W-D)/2} \tag{3.31}$$

$$\dot{Q}_{fin} = (W-D)lF\left(G\tau\alpha - U_t(T_b - T_o)\right) \quad [W] \tag{3.32}$$

In addition to the heat flow $\dot{Q}_{fin}$ led to the pipe, the irradiance directly absorbed via the external diameter D of the pipe must also be taken into account.

$$\dot{Q}_{tube} = Dl\left(G\tau\alpha - U_t(T_b - T_o)\right) \quad [W] \tag{3.33}$$

The sum of these two heat amounts is transferred convectively to the fluid and finally produces the useful power per pipe in the direction of flow.

$$\dot{Q}_{fin} + \dot{Q}_{tube} = \left((W-D)F + D \right) l \left(G\tau\alpha - U_t \left(T_b - T_o \right) \right) = \dot{Q}_{n(N=1)} \quad [W] \tag{3.34}$$

The thermal resistance between the sheet metal over the fluid tube with a base temperature T_b and the fluid consists of a convective part between the tube sheet and fluid, with thermal resistance $1/h_{fi}$ at a tubing inside diameter of D_i, and of the contact resistance between the sheet metal and tube. The convective heat transfer coefficient h_{fi} is about 100 W/m^2K with a laminar flow and reaches values of 300–1000 W/m^2K with a turbulent flow. The effective contact conductivity is determined from the conductivity of the contact material, the width of the contact b_{con} and the thickness d_{con}, and is usually negligible in today's absorber constructions: $\lambda_{con,eff} = \frac{\lambda_{con} b_{con}}{d_{con}} \left[\frac{W}{mK} \right]$.

$$\dot{Q}_{u(N=1)} = \frac{1}{\left(\frac{1}{h_{fi} \pi D_i l} + \frac{1}{\lambda_{con,eff} l} \right)} \left(T_b - T_f \right) \quad [W] \tag{3.35}$$

The base temperature can now be eliminated by equating the available power from Equation (3.34). The available power [W] of an absorber pipe (number $N = 1$) for unit flow length l can be represented as a function of the local fluid temperature.

$$\dot{Q}_{u(N=1)} = \frac{1/U_t}{\frac{1}{h_{fi} \pi D_i l} + \frac{1}{\lambda_{con,eff} l} + \frac{1}{\left((W-D)F + D \right) U_t l}} \left(G\tau\alpha - U_t \left(T_f - T_o \right) \right) \tag{3.36}$$

3.1.10.2 Collector efficiency factor *F'*

The collector efficiency factor F' introduced by Duffie–Beckmann is the ratio of thermal resistances from Equation (3.36), which arises as a result of normalisation to the surface area of an absorber strip of width W and unit length $l = 1$:

$$F' = \frac{1/U_t}{W \left[\frac{1}{h_{fi} \pi D_i} + \frac{1}{\lambda_{con,eff}} + \frac{1}{\left((W-D)F + D \right) U_t} \right]} \tag{3.37}$$

The efficiency factor indicates the ratio of the actual available power to the higher available power, which would result for an absorber sheet metal at the low fluid temperature (with correspondingly smaller calorific losses).

$$\dot{Q}_{u(N=1)} = Wl\,F' \left(G\tau\alpha - U_t \left(T_f - T_o \right) \right) \quad [W] \tag{3.38}$$

3.1.10.3 Heat dissipation factor F_R

The available energy calculated so far describes the heat supply from the absorber sheet metal to the fluid at a point y on the collector. This heat supply leads to a local rise in temperature of the fluid, which depends on the mass flow through the fluid pipe.

The calculation of the rise in temperature over the complete collector length then enables the calculation of the available useful power from the collector. For an absorber pipe the following energy balance results for a total mass flow through the collector $\dot{m}$, i.e. a mass flow per pipe of $\dot{m}/N$:

$$\left(\frac{\dot{m}}{N}\right)c_p\frac{dT_f}{dy}-WF'\left(G\tau\alpha-U_t\left(T_f-T_o\right)\right)=0 \tag{3.39}$$

with the boundary condition

$$T_f\big|_{y=0}=T_{f,in} \tag{3.40}$$

The differential equation is solved by substitution

$$\Psi=G\tau\alpha-U_t\left(T_f-T_o\right) \tag{3.41}$$

and separation of the variables:

$$\frac{d\Psi}{\Psi}=-U_t\frac{NWF'}{\dot{m}c_p}dy \tag{3.42}$$

$$\Psi=C_1\exp\left(-\frac{U_tNWF'}{\dot{m}c_p}y\right) \tag{3.43}$$

The boundary condition results in

$$\Psi\big|_{y=0}=C_1=G\tau\alpha-U_t\left(T_{f,in}-T_o\right) \tag{3.44}$$

and thus for the fluid temperature at any point y in the direction of flow:

$$T_f\left(y\right)=T_o+\frac{G\tau\alpha}{U_t}+\left(\left(T_{f,in}-T_o\right)-\frac{G\tau\alpha}{U_t}\right)\exp\left(-\frac{U_tNWF'}{\dot{m}c_p}y\right) \tag{3.45}$$

To determine the fluid output temperature, the collector length L is used for y. The product of the number of tubes N, absorber strip width W and collector length L thereby corresponds to the collector surface A.

$$T_{f,out}=T_o+\frac{G\tau\alpha}{U_t}+\left(\left(T_{f,in}-T_o\right)-\frac{G\tau\alpha}{U_t}\right)\exp\left(-\frac{U_tF'A}{\dot{m}c_p}\right) \tag{3.46}$$

The available power of the collector $\dot{Q}_u$ can thus be represented analytically as a function of the fluid inlet temperature, the ambient temperature and the irradiance.

$$\begin{aligned}
\dot{Q}_u &= \dot{m}c_p\left(T_{f,out}-T_{f,in}\right)\\
&= \dot{m}c_p\left(T_o+\frac{G\tau\alpha}{U_t}+\left(\left(T_{f,in}-T_o\right)-\frac{G\tau\alpha}{U_t}\right)\exp\left(-\frac{U_tF'A}{\dot{m}c_p}\right)-T_{f,in}\right)\\
&= \left(\frac{\dot{m}c_p}{U_t}+\dot{m}c_p\frac{\left(\left(T_{f,in}-T_o\right)-\frac{G\tau\alpha}{U_t}\right)\exp\left(-\frac{U_tF'A}{\dot{m}c_p}\right)}{G\tau\alpha-U_t\left(T_{f,in}-T_o\right)}\right)\left(G\tau\alpha-U_t\left(T_{f,in}-T_o\right)\right)\\
&= \left(\frac{\dot{m}c_p}{U_t}\left(1-\exp\left(-\frac{U_tF'A}{\dot{m}c_p}\right)\right)\right)\left(G\tau\alpha-U_t\left(T_{f,in}-T_o\right)\right)
\end{aligned} \tag{3.47}$$

The first term of the equation, normalised to the collector surface A, is defined after Duffie–Beckmann as the heat dissipation factor F_R. It indicates the ratio of the actual available power to the attainable available power, if the complete absorber were at the cold fluid inlet temperature.

$$F_R=\frac{\dot{m}c_p}{AU_t}\left(1-\exp\left(-\frac{U_tF'A}{\dot{m}c_p}\right)\right) \tag{3.48}$$

The heat dissipation factor, which depends on the mass flow and geometrically on the collector efficiency factor, leads to a simple available-power equation:

$$\dot{Q}_u=AF_R\left(G\tau\alpha-U_t\left(T_{f,in}-T_o\right)\right) \tag{3.49}$$

The thermal efficiency is determined by the ratio of available power per square metre collector surface and irradiance.

$$\eta=\frac{\dot{Q}_u/A}{G} \tag{3.50}$$

The mean fluid temperature of the collector is obtained by integrating the fluid temperature using Equation (3.45) over the collector length.

$$\bar{T}_f=\frac{1}{L}\int_0^L T_f(y)\,dy=T_{f,in}+\frac{\dot{Q}_u\left(1-\frac{F_R}{F'}\right)}{AF_RU_t} \tag{3.51}$$

The mean absorber temperature is obtained by equating the available power equation as a function of the input temperature and as a function of the mean absorber temperature.

$$\bar{T}_a = T_{f,in} + \frac{\dot{Q}_u / A}{U_t F_R}(1 - F_R) \tag{3.52}$$

Example 3.7

Calculation of the available power and temperatures of a flat plate collector at G = 800 W/m² irradiance, ambient temperature T_o = 10°C and an inlet temperature into the collector $T_{f,in}$ from the lower part of the storage tank at 30°C. To represent the influence of the mass flow on the available power and the temperature conditions, the calculation for a low-flow system with $\dot{m}$ = 10 kg/m²h is to be carried out, and also for a standard system with $\dot{m}$ = 50 kg/m²h.

The optical efficiency $\eta_0 = \tau\alpha = 0.9 \times 0.9 = 0.81$ and the entire calorific loss U_t with 4 W/m²K are given. The data of the collector are as follows:

Width of the absorber strip W	15 cm
Length of the absorber strip L	2.5 m
External pipe diametre D (DN8, 1mm wall strength)	8 mm
Heat conductivity of the absorber sheet metal λ_{copper}	385 W/mK
Sheet thickness δ	0.5 mm
Contact resistance $1/\lambda_{con,eff}$	0
Heat transfer coefficient h_{fi}	1000 W/m²K

Calculations:

Fin efficiency $\quad F = 0.966$ with $m = \sqrt{\dfrac{U_t}{\lambda\delta}} = 4.56 \ \left[m^{-1}\right]$

Efficiency factor F': $\quad F' = \dfrac{1/4}{0.15(0.053 + 1.722)} = 0.94$

Heat dissipation factor F_R:

$$F_R\big|_{\dot{m}=10\frac{kg}{m^2h}} = 2.9(1 - \exp(-0.94/2.9)) = 0.8$$

$$F_R\big|_{\dot{m}=50\frac{kg}{m^2h}} = 14.5(1 - \exp(-0.94/14.5)) = 0.91$$

The heat dissipation factor F_R is mass-flow dependent.

Available power:

$$\left.\frac{\dot{Q}_u}{A}\right|_{\dot{m}=10\frac{kg}{m^2h}} = 0.8\left(800 \times 0.81 - 4\left(30-10\right)\right) = 454.4\frac{W}{m^2}$$

$$\left.\frac{\dot{Q}_u}{A}\right|_{\dot{m}=50\frac{kg}{m^2h}} = 516.9\frac{W}{m^2}$$

The available power and efficiency improve with rising mass flow. At typical flow conditions of collectors between 10 and 50 kg/m^{2}h, the efficiency varies by 12%.

Efficiency:

$$\left.\eta\right|_{\dot{m}=10\frac{kg}{m^2h}} = 0.57$$

$$\left.\eta\right|_{\dot{m}=50\frac{kg}{m^2h}} = 0.64$$

Outlet temperature: $T_{f,out} = 69°C \quad (39°C)$

The advantage of the smaller mass flows shows up particularly in the outlet temperature. During a single flow-through with the low-flow system, a rise in temperature of 39 K is achieved, with the higher through flows only 9 K.

Mean fluid temperature: $\overline{T}_f = 51.1°C \quad (31.5°C)$

Mean absorber temperature: $\overline{T}_a = 58.4°C \quad (42.8°C)$

Between the mean absorber temperature and mean fluid temperature there is a difference of 7.3 or 8.3 K, respectively.

3.1.10.4 Heat losses of thermal collectors

The thermal collectors available on the market are today produced almost exclusively with a single transparent cover, or even no cover. The heat transfer coefficient between the absorber and ambient air via the collector front (U_f) cannot be set as constant, since the temperatures of the absorber cover a substantially larger range than usual temperatures occurring in buildings. For each given absorber temperature, therefore, the convective and radiant heat transfer coefficients should be calculated iteratively and the U_f-value of the front determined. The heat transfer coefficients through the insulated collector rear U_b and side panels U_s can, however, be regarded as constant and calculated from the layer thickness s [m], heat conductivity λ [W/mK] of the insulating material, and from the outside thermal resistance $1/h_a$ [m²K/W] between the insulating material and the environment.

$$U_b = U_s = \left(\frac{s}{\lambda} + \frac{1}{h_a}\right)^{-1} \qquad (3.53)$$

The outside thermal resistance consists of a radiation-dependent and a convective, wind velocity-dependent proportion ($1/h_a=1/(h_c+h_r)$). It is quite sufficient to calculate using the standard value of outside thermal resistance of $1/h_a$ = 0.04 m²K/W which is usual in construction. If all heat transfer coefficients are calculated from the absorber against the ambient temperature, then the parallel heat transfer coefficients can be added to the total loss coefficient U_t. The losses through the collector sides with the small side panel surface A_s are related here to the aperture surface of the collector A.

$$U_t = U_f + U_b + U_s \frac{A_s}{A} \tag{3.54}$$

Heat transfer coefficient of the transparent cover U_f

The heat transfer coefficient of the transparent cover consists of the total of the thermal resistances between the absorber and the environment. The thermal resistances between absorber and glazing R_{a-g}, and between glazing and environment R_{g-o}, are calculated from the parallel heat transfers for convection h_c and radiation h_r.

$$R_{a-g} = \frac{1}{h_{c,a-g} + h_{r,a-g}} \qquad R_{g-o} = \frac{1}{h_{c,g-o} + h_{r,g-o}} \tag{3.55}$$

$$U_f = \frac{1}{R_{a-g} + R_{g-o}} \tag{3.56}$$

The temperature-dependent heat transfer coefficients are first calculated assuming a cover plate temperature T_g. As the heat flow $\dot{Q}_f$ over the collector front between absorber and environment,

$$\frac{\dot{Q}_f}{A} = U_f (T_a - T_o) \tag{3.57}$$

equals the heat flow between absorber and cover,

$$\frac{\dot{Q}_{a\to g}}{A} = \left(h_{c,a-g} + h_{r,a-g}\right)\left(T_a - T_g\right) \tag{3.58}$$

a new cover glass temperature can be calculated by equating the heat flows.

$$\frac{\dot{Q}_f}{A} = U_f (T_a - T_o) = \left(h_{c,a-g} + h_{r,a-g}\right)\left(T_a - T_g\right) = \frac{\dot{Q}_{a\to g}}{A}$$
$$\Rightarrow T_g = T_a - \frac{U_f (T_a - T_o)}{\left(h_{c,a-g} + h_{r,a-g}\right)} \tag{3.59}$$

In the following, the equations for the necessary heat transfer coefficients are discussed.

Heat transfer coefficient for radiation h_r

Using the Stefan–Boltzmann law, from a diffusely radiating black surface A_1 [m²] with an emission coefficient $\varepsilon = 1$, the radiating power

$$\dot{Q}_1 = A_1 \sigma T_1^4 \tag{3.60}$$

is emitted into the room, the Stefan–Boltzmann constant being $\sigma = 5.67 \times 10^{-8}$ W/m²K⁴. A part Φ_{12} of this radiation power is absorbed by a second black surface A_2 placed at random in the room. A_2 for its part also emits radiation as a function of the temperature T_2, with $\dot{Q}_2 = A_2 \sigma T_2^4$ of which the first surface absorbs a part Φ_{21}. The net radiation exchange $\dot{Q}$ between surfaces A_1 and A_2 results from the difference between the radiation to the respective other surface and the returning radiation.

$$\dot{Q} = \dot{Q}_{1\to 2} - \dot{Q}_{2\to 1} = \underbrace{\Phi_{12} A_1 \sigma T_1^4}_{\text{radiative flux } A_1 \text{ to } A_2} - \underbrace{\Phi_{21} A_2 \sigma T_2^4}_{\text{radiative flux } A_2 \text{ to } A_1} \tag{3.61}$$

As both surfaces have been defined as black emitters with absorption and emission coefficients of 1, no reflection of the radiation at the surfaces takes place.

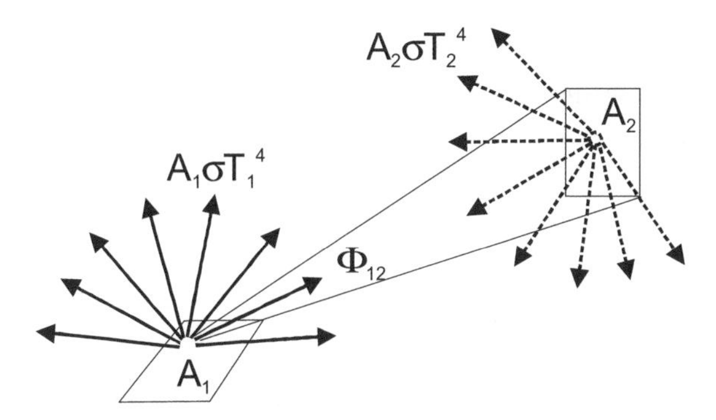

Figure 3.24: Radiant flux of a small (differential) two-dimensional element A_1 at temperature T_1 to the surface element A_2, which absorbs the proportion Φ_{12} and in turn emits radiation.

If the two surfaces are at the same temperature, the net radiant heat exchange equals zero and the reciprocity condition for the form factors Φ_{12} und Φ_{21} is obtained:

$$\Phi_{12} A_1 = \Phi_{21} A_2 \tag{3.62}$$

The heat flow by radiation exchange between two black emitters is thus given by:

$$\dot{Q} = \Phi_{12} A_1 \sigma \left(T_1^4 - T_2^4\right) = \Phi_{21} A_2 \sigma \left(T_1^4 - T_2^4\right) \tag{3.63}$$

The form factors Φ_{ij} as a geometrical ratio of the radiation-receiving surface A_j to the entire hemisphere, into which power is irradiated, are generally complicated functions of the surface geometry. Only in those cases where the complete emitted radiation from the second surface can be absorbed are the irradiating numbers easy to determine.

For the calculation of the heat transfer coefficients in solar thermal collectors, two cases are of special importance; the radiation exchange between two flat-parallel plates (absorber and transparent cover) and the radiation exchange between the flat cover and the sky hemisphere. If with the parallel plates it is approximately assumed that these are infinitely expanded, the entire radiant flux of surface A_1 is absorbed by surface A_2 and vice versa, i.e. $\Phi_{12} = \Phi_{21} = 1$.

In the radiation exchange of a flat surface with the sky hemisphere, the radiation emitted by surface A_1 is completely received by the sky hemisphere, i.e. $\Phi_{12} = 1$. Conversely, the proportion which the small surface A_1 sees of the total radiation of the sky hemisphere is very small. Based on the reciprocity condition from Equation (3.62), here $\Phi_{21} = A_1 / A_2 \approx 0 \quad for\ A_1 \ll A_2$.

Normally the emission or absorption coefficient of surfaces is less than one (grey emitters), so the radiation received from the second surface with $(1 - \varepsilon_2) = (1 - \alpha_2)$ is partly reflected back, and in turn can be absorbed by surface A_1 with ε_1 and reflected back with $(1 - \varepsilon_1)$ etc.

To create a radiation energy balance for surface A_1, the radiation $\dot{Q}_{1\to2\to1}$ reflected by surface A_2 and absorbed again by A_2 must be deducted from the radiated energy quantity $\dot{Q}_1$. Furthermore, the radiation $\dot{Q}_{2\to1}$ emitted by surface A_2 and absorbed by A_1 must be treated as an energy gain. This balance should be regarded for the simple case of flat-parallel plates with a form factor of 1. For radiation with intensity $\dot{Q}_1$ emitted by surface A_1, the following amount of heat returns:

$$\begin{aligned}\frac{\dot{Q}_{1\to2\to1}}{\dot{Q}_1} &= \underbrace{(1-\varepsilon_2)}_{\text{first reflection from } A_2} \times \underbrace{\varepsilon_1}_{\text{absorption at } A_1} + \underbrace{(1-\varepsilon_2)(1-\varepsilon_1)}_{\text{first back reflection from } A_1} \times \underbrace{(1-\varepsilon_2)\varepsilon_1}_{\text{second reflection from } A_2 \text{ absorbed by } A_1} \\ &+ (1-\varepsilon_2)(1-\varepsilon_1)(1-\varepsilon_2) \underbrace{(1-\varepsilon_1)}_{\text{second reflection from } A_1} \times \underbrace{(1-\varepsilon_2)\varepsilon_1}_{\text{third reflection from } A_2 \text{ absorbed by } A_1} + \ldots \\ &= (1-\varepsilon_2)\varepsilon_1\left(1+(1-\varepsilon_2)(1-\varepsilon_1)+((1-\varepsilon_2)(1-\varepsilon_1))^2+..\right) = \frac{(1-\varepsilon_2)\varepsilon_1}{1-(1-\varepsilon_2)(1-\varepsilon_1)}\end{aligned} \tag{3.64}$$

In addition, surface A_2 emits the intensity $\dot{Q}_2$, of which surface A_1 absorbs a proportion ε_1 and reflects $(1 - \varepsilon_1)$. After multiple reflections the infinite sequence of the radiation absorbed by A_1 results in:

$$\frac{\dot{Q}_{2\to 1}}{\dot{Q}_2} = \varepsilon_1 + (1-\varepsilon_1)(1-\varepsilon_2)\varepsilon_1 + \ldots = \frac{\varepsilon_1}{1-(1-\varepsilon_1)(1-\varepsilon_2)} \quad (3.65)$$

The energy balance of surface A_1 thus results for the radiation exchange between flat-parallel plates with $A_1 = A_2$ and $\Phi_{12} = \Phi_{21} = 1$:

$$\begin{aligned}\dot{Q}_{1,net} &= \dot{Q}_1 - \dot{Q}_{1\to 2\to 1} - \dot{Q}_{2\to 1} \\ &= \dot{Q}_1 - \dot{Q}_1 \frac{(1-\varepsilon_2)\varepsilon_1}{1-(1-\varepsilon_2)(1-\varepsilon_1)} - \dot{Q}_2 \frac{\varepsilon_1}{1-(1-\varepsilon_2)(1-\varepsilon_1)} \\ &= \frac{\dot{Q}_1\varepsilon_2 - \dot{Q}_2\varepsilon_1}{\varepsilon_1 + \varepsilon_2 - \varepsilon_1\varepsilon_2} \\ &= \frac{\varepsilon_1\sigma A_1 T_1^4 \varepsilon_2 - \varepsilon_1\varepsilon_2\sigma A_2 T_2^4}{\varepsilon_1 + \varepsilon_2 - \varepsilon_1\varepsilon_2} = \frac{1}{\frac{1}{\varepsilon_2} + \frac{1}{\varepsilon_1} - 1}\sigma A\left(T_1^4 - T_2^4\right)\end{aligned} \quad (3.66)$$

This heat flow equation is linearised by extracting the temperature difference $T_1 - T_2$ and thus the heat transfer coefficient for radiation h_r is determined.

$$\dot{Q} = h_r A\left(T_1 - T_2\right) \quad (3.67)$$

$$h_r = \frac{\sigma}{1/\varepsilon_1 + 1/\varepsilon_2 - 1}\left(T_1^2 + T_2^2\right)\left(T_1 + T_2\right) \quad (3.68)$$

For the calculation of the radiation exchange between the absorber at temperature T_a and glazing at temperature T_g, $T_1 = T_a$ and $T_2 = T_g$ are set.

Selective coating

The long-wave emission coefficients of black absorber colour are typically 95%, those of glass are 88%. The emissivity can be reduced if a solar radiation-absorbing coating is applied on a substrate with a small emission coefficient, for example a metal. Galvanisation of metals with black chrome is increasingly replaced by environmentally friendly cathodic sputtering methods, which also consume only one-tenth of the energy required for galvanisation. Emissivities of 5% can be reached, while shortwave irradiance absorption stays at 95%.

If one or more form factors Φ_{ij} are not equal to one, the energy balances must be modified accordingly. This can be analysed by the example of the radiation exchange of a surface against the sky hemisphere.

The radiation $\dot{Q}_1 = \varepsilon_1\sigma A_1 T_1^4$ emitted by surface A_1 meets the sky hemisphere with form factor $\Phi_{12} = 1$ and is reflected with $(1 - \varepsilon_2)$. Now, however, only a fraction of the irradiance reflected from the sky reaches surface A_1, i.e. $\Phi_{21} = A_1/A_2$. The infinite sequence of the radiation reabsorbed by A_1 is thus modified as follows:

$$\frac{\dot{Q}_{1\to2\to1}}{\dot{Q}_1} = \underbrace{(1-\varepsilon_2)}_{\text{1. refl. from } A_2} \underbrace{\frac{A_1}{A_2}\varepsilon_1}_{\text{1. abs. from } A_1} + \underbrace{(1-\varepsilon_2)}_{\text{1. refl. from } A_2} \underbrace{\left(1-\frac{A_1}{A_2}\varepsilon_1\right)}_{\text{1. refl. from } A_1} \underbrace{(1-\varepsilon_2)}_{\text{2. refl. from } A_2} \underbrace{\frac{A_1}{A_2}\varepsilon_1}_{\text{2. abs. from } A_1} +.. $$

$$= \frac{(1-\varepsilon_2)\varepsilon_1 \frac{A_1}{A_2}}{1-(1-\varepsilon_2)\left(1-\frac{A_1}{A_2}\varepsilon_1\right)} \tag{3.69}$$

In a similar way, the radiation emitted from A_2 at A_1 is modified by the factor A_1/A_2.

$$\frac{\dot{Q}_{2\to1}}{\dot{Q}_2} = \frac{\varepsilon_1 \frac{A_1}{A_2}}{1-\left(1-\frac{A_1}{A_2}\varepsilon_1\right)(1-\varepsilon_2)} \tag{3.70}$$

The net radiant flux of surface A_1 is thus given by:

$$\dot{Q}_{1,net} = \frac{1}{\frac{A_1}{A_2}\frac{1}{\varepsilon_2}+\frac{1}{\varepsilon_1}-\frac{A_1}{A_2}} \sigma A_1 \left(T_1^4 - T_2^4\right) \tag{3.71}$$

from which Equation (3.66) results as a special case for equal surfaces $A_1 = A_2$. If surface A_1 is very much smaller than surface A_2, then Equation (3.71) is simplified to:

$$\dot{Q}_{1,net} \approx \varepsilon_1 \sigma A_1 \left(T_1^4 - T_2^4\right) \quad for\ A_1 \ll A_2 \tag{3.72}$$

If the sky temperature is defined with T_{sky} ($T_2 = T_{sky}$), the result for the heat transfer coefficient between cover ($T_1 = T_g$) and sky is:

$$h_r = \sigma\varepsilon_1 \left(T_g^2 + T_{sky}^2\right) \left(T_g + T_{sky}\right) \tag{3.73}$$

The sky temperature takes into account the transparency of the atmosphere, i.e. missing back radiation in the wavelength range between approximately 8 to 14 μm. The sky temperature can be calculated from the dew point temperature T_{dp} [K]:

$$T_{sky} = T_o \left(0.8 + \frac{T_{dp} - 273}{250}\right)^{0.25} \tag{3.74}$$

As a simpler equation without calculation of the water vapour partial pressure and the dew point temperature, the modified Swinbank equation [K] can be used (Fuentes, 1987):

$$T_{sky} = 0.037536\,T_o^{1.5} + 0.32\,T_o \tag{3.75}$$

To be able to calculate the heat flows as usual by the temperature difference between cover T_g and ambient temperature T_o, the heat transfer coefficient for radiation is normalised to this temperature difference.

$$h_r = \sigma\varepsilon_1 \left(T_g^2 + T_{sky}^2\right)\left(T_g + T_{sky}\right)\frac{\left(T_g - T_{sky}\right)}{\left(T_g - T_o\right)} \tag{3.76}$$

Convective heat transfer coefficients h_c

Convective heat transfer occurs in the form of natural convection between the absorber and transparent cover, and as forced convection by wind forces between the cover and the environment. The characteristic nominal value for the convective heat transfer is always the Nußelt number *Nu*, from which the convective heat transfer coefficient can be calculated as a function of a characteristic length *L* and of the heat conductivity λ of the air.

$$Nu = \frac{h_c L}{\lambda} \tag{3.77}$$

Free convection in a standing air layer

In a standing air layer between flat-parallel plates, the plate distance *d* is used as a characteristic length *L* for the calculation of h_c. For the calculation of the Nußelt number, there exists a set of empirical correlations for flat plate collector geometries with a temperature gradient ΔT between absorber and cover and different collector angles of inclination β. Based on Duffie and Beckmann (1980) the following equation can be used up to collector angles of inclination of 75°; for angles of inclination over 75° the function value of 75° is retained.

$$Nu = 1 + 1.44\left[1 - \frac{1708(\sin 1.8\beta)^{1.6}}{Ra\cos\beta}\right]\left[1 - \frac{1708}{Ra\cos\beta}\right]^{+} + \left[\left(\frac{Ra\cos\beta}{5830}\right)^{1/3} - 1\right]^{+} \tag{3.78}$$

The plus sign of the bracketed term means that only positive results are to be used; with negative bracketed terms the term is set at zero. The Rayleigh number describes the lift by the thermally caused density variations and is given by the product of the Grashof (*Gr*) and Prandtl (*Pr*) numbers:

$$Ra = Gr\,Pr = \frac{g\beta'\Delta T L^3}{\nu^2} \times \frac{\nu\rho c_p}{\lambda} = \frac{g\beta'\Delta T L^3}{\nu a} \tag{3.79}$$

where

g: Gravitation constant [m/s²]

$\beta' = 1/T$: Volume expansion coefficient of ideal gases [K^{-1}]

ΔT: Temperature difference between the panels [K]

L:	Characteristic length, here the panel separation d [m]
ν:	Kinematic viscosity [m²/s]
$a = \lambda/\rho\, c_p$:	Temperature conductivity [m²/s]

Forced convection by wind

The convective heat flow at the glass cover of a collector is mainly caused by wind, i.e. forced convection. A smaller proportion results from free convection between the pane and the ambient temperature. A good approximation is given by the following overlay of the two heat transfer coefficients:

$$h_c = \sqrt[3]{h_{c,w}^3 + h_{c,free}^3} \tag{3.80}$$

The forced convection transition coefficient $h_{c,w}$ is calculated from a Nußelt correlation for a plate with flow parallel to the plate with a turbulent boundary layer (VDI, 1994).

$$Nu_{turb} = \frac{0.037\,\mathrm{Re}^{0.8}\,\mathrm{Pr}}{1 + 2.443\,\mathrm{Re}^{-0.1}\left(\mathrm{Pr}^{\frac{2}{3}} - 1\right)} \Rightarrow h_{c,w} = \frac{Nu_{turb}\lambda}{L} \tag{3.81}$$

The Reynolds number Re results from the wind velocity v_w and the plate length L under parallel flow conditions from $\mathrm{Re} = v_w L/\nu$. For the material values ν, ρ, λ, c_p the temperature of the ambient air is used. The heat transfer coefficient $h_{c,w}$ is likewise calculated with the panel length L as a characteristic length. For the calculation of the free convection proportion, the glass temperature T_g and the ambient temperature T_o must be known.

$$h_{c,free} = 1.78\left(T_g - T_o\right)^{1/3} \quad \left[Wm^{-2}K^{-1}\right] \tag{3.82}$$

A simplified approach takes only the wind velocity into account, so

$$h_{c,w} = 4.214 + 3.575\,v_w \tag{3.83}$$

Example 3.8

Calculation of of the front U-value U_f of collectors with ($\varepsilon = 0.1$) and without selective coating ($\varepsilon = 0.9$) of the absorber under the following boundary conditions:

Absorber temperature	$T_a = 70°C$
Ambient temperature	$T_o = 10°C$
Wind velocity	$v_w = 3$ m/s

Emission coefficient of transparent cover	$\varepsilon_g = 0.88$
Plate distance absorber-glass	$d = 2.5$ cm
Collector angle of inclination	$\beta = 45°$
Assumed pane temperature for first iteration	$T_g = 40°C$

1st iteration:

The heat transfer coefficient for radiation $h_{r,a\text{-}g}$ between absorber and glass is 0.79 W/m²K for the selectively coated absorber, and 6.44 W/m²K for the black absorber. The convective heat transfer coefficient of the standing air layer is determined by the material values of the mean air temperature of (70 + 40)/2 = 55°C (λ = 0.0286 W/m K, $\nu = 1.85 \times 10^{-5}$ m²/s, ρ = 1.045 kg/m³, c_p = 1.009 kJ/kg K). The Prandtl number is 0.71 and the Grashof number 43787 at a mean temperature difference of 30K between absorber and assumed pane temperature, so the Rayleigh number is 31089. From this results a Nußelt number of 2.78 and a convective heat transfer coefficient $h_{c,a-g}$ = 2.9 W/m²K.

Between the glass cover and the sky, according to Swinbank a radiant heat transfer coefficient of $h_{r,g-o}$ = 7.23 W/m²K is obtained at a sky temperature of 269 K, and for the wind-velocity dependent convective heat transfer based on simplified Equation (3.83) $h_{c,w}$ = 14.9 W/m²K.

From this results the heat transfer coefficient of the transparent cover after the first iteration of $U_f = \frac{1}{\frac{1}{2.9+0.79}+\frac{1}{14.9+7.23}} = 3.17 \quad Wm^{-2}K^{-1}$ for the selectively coated absorber, and 6.57 W/m²K for the black absorber.

With these values the new pane temperature is calculated:

$$T_{g,new} = 70°C - \frac{3.17\,Wm^{-2}K^{-1} \times 60°C}{2.9\,Wm^{-2}K^{-1} + 0.79\,Wm^{-2}K^{-1}} = 18.6°C$$

or 27.8°C for the non-selective absorber. The temperature is clearly lower than the originally set temperature of 40°C. With this temperature the heat transfer coefficients are again calculated. After the end of iteration the following values are obtained:

Parameter	*Selective coating*	*Non-selective*
$h_{r,a-g}$	0.72 W m^{-2}K^{-1}	6.08 W m^{-2}K^{-1}
$h_{c,a-g}$	3.4 W m^{-2}K^{-1}	3.2 W m^{-2}K^{-1}
$h_{r,g-o}$	11.88 W m^{-2}K^{-1}	8.3 W m^{-2}K^{-1}
$h_{c,w}$	14.9 W m^{-2}K^{-1}	14.9 W m^{-2}K^{-1}
T_g	18.0°C	27.1°C
U_f	3.56 W m^{-2}K^{-1}	6.64 W m^{-2}K^{-1}

3.1.10.5 Optical characteristics of transparent covers and absorber materials

The transmittance of a transparent cover results from the reflection losses at the boundary surfaces and the absorption losses in the cover material itself. The reflection losses depend on the refractive index of the cover material and on the mechanical structuring of the cover. The lowest reflection losses are achieved if an antireflective coating with a refractive index is used, which has a value between the materials "air" and the cover material and if the glass is very plain. The mechanical structuring, which is still often used to avoid glare from direct irradiance reflections and to optically hide the interior of the collector, increases reflection losses, especially at higher incidence angles. Compared to a plane, non-structured glass the annual additional reflection losses of a structured glass are up to 6% higher.

Absorption of the cover material can be reduced by a lower iron content of the glass, which increases transmission from typically 84% for standard float glass to 89–91% for "solar glass". A low iron glass with antireflection coating can achieve 96% transmission at zero incidence angle.

At the boundary surface of two media with different refractive indices (here, for example, air and glass), the reflection characteristics of the surface can be calculated from constancy conditions for the electrical and magnetic field. For this, the almost unpolarised natural light is divided into two components, which strike the boundary surface parallel or perpendicularly to the plane of incidence. The square of the field strengths results in the radiating power with the associated coefficient of reflection r.

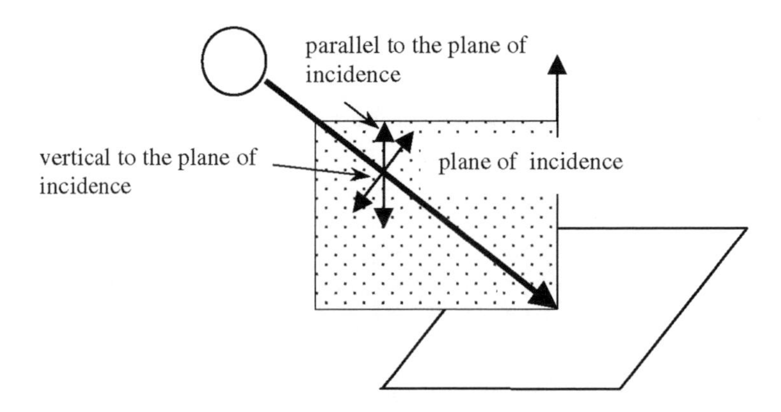

Figure 3.25: Polarisation directions parallel and perpendicular to the plane of incidence.

The plane of incidence is defined by the surface normal and the irradiance vector. The reflected radiating power G_r in relation to the incident power G_i results from the average value between parallel and perpendicularly polarised components, which must first be calculated separately.

$$r = \frac{G_r}{G_i} = \frac{1}{2}\left(r_\perp + r_\parallel\right) \tag{3.84}$$

The reflection factors are calculated according to the Fresnel formulae from the angle of incidence θ_1 and the angle of refraction in the material θ_2.

$$r_\perp = \frac{\sin^2(\theta_2 - \theta_1)}{\sin^2(\theta_2 + \theta_1)} \tag{3.85}$$

$$r_\parallel = \frac{\tan^2(\theta_2 - \theta_1)}{\tan^2(\theta_2 + \theta_1)} \tag{3.86}$$

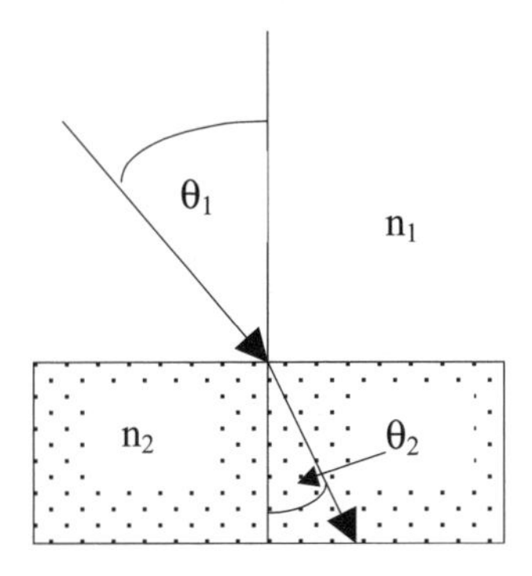

Figure 3.26: Illustration of the angles and refractive indices.

The angles are, according to Snell's law, a function of the refractive indices n_1 and n_2:

$$\frac{n_1}{n_2} = \frac{\sin\theta_2}{\sin\theta_1} \tag{3.87}$$

With perpendicular irradiance, both angles are zero and the reflection coefficient of the boundary surface becomes

$$r(0) = \left(\frac{n_1 - n_2}{n_1 + n_2}\right)^2 \tag{3.88}$$

The reflection factors at the boundary surface are equal for perpendicular irradiance, then decrease to zero as a function of the angle of incidence θ_1 with parallel polarised light, and finally rise to one at parallel incidence. With perpendicular polarisation the reflection factor constantly rises.

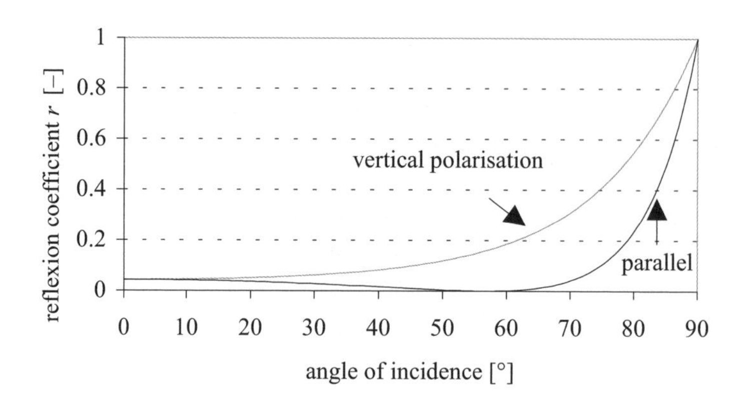

Figure 3.27: Reflection coefficient r for parallel and perpendicularly polarised incoming radiation after Fresnel.

Table 3.7: Common refractive indices of transparent materials.

Material	Refractive index
Air	1.0
Glass	1.526
Polycarbonate	1.6
Polymethyl Methacrylate (Plexi-glass)	1.49

Absorption in the material

The reduction in radiation intensity in the cover material itself (dG) is proportional to the absolute intensity G of the radiation and to the extinction coefficient K of the material:

$$\begin{aligned} dG &= -G \times K\,dx \\ G &= C\exp\left(-K\left(x - x_0\right)\right) \end{aligned} \tag{3.89}$$

The boundary condition is given by the incident intensity G_0 at the point $x = x_0$. The distance travelled in the material results from the thickness of the material L and the cosine of the angle in the material θ_2 to $L/\cos\theta_2$, so the transmittance τ_a of the material (without surface reflections) results from the intensity of the irradiance G_t after a single ray passage to the entering intensity G_0 at x_0.

$$\tau_a = \frac{G_t}{G_0} = \exp\left(-K\frac{L}{\cos\theta_2}\right) \tag{3.90}$$

Table 3.8: Extinction coefficients of transparent materials.

Material	*Extinction coefficient K (m^{-1})*
Solar glass	4
Typical window glass	30
Absorbing sun protection glass	130–270

Transmission and reflection coefficients of the transparent cover

If the passage of an incoming ray is followed, taking into account the reflection losses at the entrance boundary surface in the material, then the intensity of the first outgoing ray is obtained including further reflection losses at the exit boundary surface between material and air. The rays reflected at the exit boundary surface are followed further and lead finally, after further reflections, to further exiting rays of lower intensity. The overall transmittance results as the ratio of an infinite series of the exiting radiation intensity to the incident radiation.

Similarly the entire reflection coefficient is calculated from the infinite series of the reflected rays.

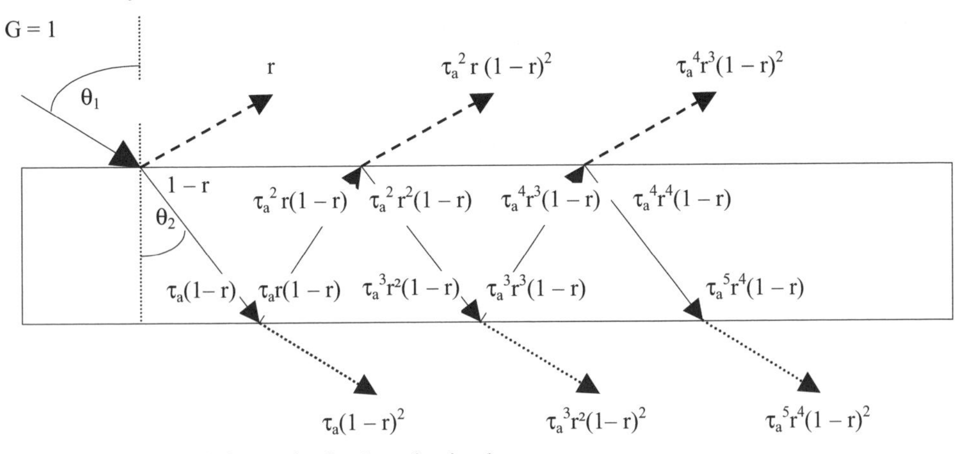

Figure 3.28: Transmission and reflection of a simple cover.

Here the transmission or reflection coefficients must be calculated separately for both polarisation directions, and afterwards can be averaged arithmetically for unpolarised natural light.

$$\tau_\perp = (1-r_\perp)^2 \tau_a \sum_{n=0}^{\infty} r_\perp^{\,2n} \tau_a^{\,2n} = \frac{\tau_a (1-r_\perp)^2}{1-(r_\perp \tau_a)^2}$$

$$\tau_\parallel = \frac{\tau_a (1-r_\parallel)^2}{1-(r_\parallel \tau_a)^2} \tag{3.91}$$

$$\tau = \frac{\tau_\perp + \tau_\parallel}{2}$$

$$\rho_\perp = r_\perp + r_\perp (1-r_\perp)^2 \tau_a{}^2 \sum_{n=0}^{\infty} r_\perp{}^{2n} \tau_a{}^{2n} = r_\perp + \frac{r_\perp (1-r_\perp)^2 \tau_a^2}{1-(r_\perp \tau_a)^2}$$

$$\rho_\| = r_\| + \frac{r_\| (1-r_\|)^2 \tau_a^2}{1-(r_\| \tau_a)^2} \tag{3.92}$$

$$\rho = \frac{\rho_\perp + \rho_\|}{2}$$

The absorption coefficient of the material can likewise be calculated via an infinite series, or directly from the transmission and reflection degrees.

$$\alpha_\perp = 1 - \tau_\perp - \rho_\perp = (1-r_\perp)\tau_a \sum_{n=0}^{\infty} (1 + r_\perp{}^n \tau_a{}^n) \tag{3.93}$$

Example 3.9

Calculation of the transmission and reflection degrees of a single-glazed, 4 mm thick solar glass cover (with $K = 4\ \mathrm{m}^{-1}$ and $n = 1.526$) for angles of incidence of 0° and 60°.

With perpendicular incidence the reflection coefficient for perpendicular and parallel polarised light is $r(0) = 0.043$. With $\tau_a = 0.984$:

$$\tau(0) = \frac{0.98(1-0.043)}{1-(0.98 \times 0.043)^2} = 0.9$$

$$\rho(0) = 0.043 + \frac{0.043(1-0.043)^2 0.98^2}{1-(0.043 \times 0.98)^2} = 0.081$$

For non-perpendicular incidence both polarisation directions must be regarded separately. For θ_1 = 60° is θ_2 = 35°. Thus $r_\perp = 0.18$, $r_\| = 0.0017$ and $\tau_a = 0.98$.

The transmission coefficients are $\tau_\perp = 0.673$, $\tau_\| = 0.98$ and $\tau = 0.83$ and the reflection factors $\rho_\perp = 0.31$, $\rho_\| = 0.0028$ and $\rho = 0.16$, i.e. twice as high as with perpendicular incidence.

Absorption of absorber substances and transmission–absorption product

The absorption factor of absorber substances can be regarded with good accuracy as angle-independent.

Table 3.9: Absorption coefficients and emissivities of typical absorber substances.

Material	*Absorption coefficient α*	*Emissivity ε*
Black paint	0.95	0.95
Black chrome on nickel (galvanically applied selective coating)	0.95	0.1
TiNOx (vacuum sputter coating)	0.94	0.038

Taking into account multiple reflections between the transparent cover and absorber, the result is, from the infinite series, an effective transmission absorption product:

$$(\tau\alpha) = \tau\alpha \sum_{n=0}^{\infty} \left((1-\alpha)\rho\right)^n = \frac{\tau\alpha}{1-(1-\alpha)\rho} \tag{3.94}$$

where ρ is the reflection factor for light reflected diffusely at the underside of the glass (this can be calculated simply for an average angle of incidence of 60°).

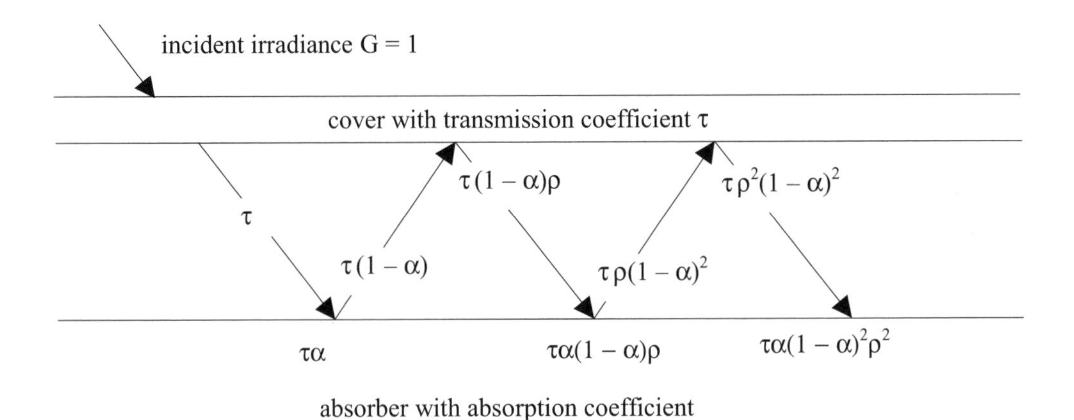

Figure 3.29: Effective transmission–absorption product of a single-glazed collector.

3.1.11 Storage modelling

For the calculation of the calorific losses of stores, a simple energy balance (Equation (3.9)) has already been created for the temperature decrease of the stored water as a function of time. A prerequisite for the analytic solution of the differential equation was the assumption of a homogeneously mixed store, with no heat supply or withdrawal from the store during cooling. In normal operation, however, apart from the transmission heat losses from the store $\dot{Q}_t$, the solar collector and the auxiliary heat source supply the amount of heat $\dot{Q}_c$ and $\dot{Q}_h$, and a heat load $\dot{Q}_l$ is withdrawn by consumers for warm-water or heating.

The temporal development of the store temperature T_s results from the energy balance:

$$mc\frac{dT_s}{dt} = \dot{Q}_c + \dot{Q}_h - \dot{Q}_l - \dot{Q}_t \qquad (3.95)$$

If the parameter δ_c or δ_l is used for the operation of the collector circuit or the load circuit of consumers, with the values 1 in operation and 0 at pump standstill or zero load, the energy balance can be represented as a function of the temperatures. The return temperature into the collector and the withdrawal temperature for consumers (consumer supply) equals the store temperature T_s for a homogeneously mixed storage tank. The return temperature of the consumers $T_{l,return}$ is (in warm-water systems) the cold water temperature, in heating support the return temperature of the heating. The outlet temperature $T_{f,out}$ from the collector (with mass flow $\dot{m}_c$) is above the store temperature during operation of the collector circuit. For homogeneously mixed stores with temperature T_s, the following energy balance applies:

$$(mc)_s \frac{dT_s}{dt} = \delta_c (\dot{m}c)_c (T_{f,out} - T_s) + \dot{Q}_h - \delta_l (\dot{m}c)_l (T_s - T_{l,return}) - U_{eff} A (T_s - T_o) \qquad (3.96)$$

Since neither the load profile with the operating conditions δ_l nor the auxiliary heating $\dot{Q}_h$ are analytical functions of time, the above differential equation cannot be solved analytically. A simple forward difference method enables the calculation of the store temperature at the time-step $n + 1$ from the values of the preceding time-step n.

$$T_{s,n+1} = T_{s,n} + \frac{\Delta t}{(mc)_s} \begin{pmatrix} \delta_c (\dot{m}c)_c (T_{f,out,n} - T_{s,n}) + \dot{Q}_{h,n} - \delta_l (\dot{m}c)_l (T_{s,n} - T_{l,return,n}) \\ -U_{eff} A (T_{s,n} - T_{o,n}) \end{pmatrix} \qquad (3.97)$$

Far more favourable for solar operation is, however, a thermally non-mixed store. A realistic storage model must take into account a temperature stratification. For this, the store is divided across its height L into several layers. For each layer an energy balance is created, which as above contains the supply of solar heat and auxiliary energy, plus calorific losses and possible heat dissipation by consumers. In addition to these terms, the amount of heat $\dot{Q}_{free}$ is exchanged in layer i with the surrounding layers $i - 1$ and $i + 1$ by natural convection and thermal conduction. Exact mathematical modelling of the convection current is complex. In the simplest approximation, thermal conduction and convection are summarised in an effective vertical heat conductivity λ_{eff}, and the heat flow between layer $i - 1$ and i, or from i to $i + 1$, is calculated using the Fourier equation. The net heat flow for layer i of height z and cross section A_q results from the difference of the two heat flows.

$$\dot{Q}_{free} = \dot{Q}_{free,i-1\to i} - \dot{Q}_{free,i\to i+1} = -A_q \frac{\lambda_{eff}}{z}\left(T_{s,i} - T_{s,i-1}\right) - \left(-A_q \frac{\lambda_{eff}}{z}\left(T_{s,i+1} - T_{s,i}\right)\right)$$
$$= A_q \frac{\lambda_{eff}}{z}\left(T_{s,i+1} - 2T_{s,i} + T_{s,i-1}\right) \quad (3.98)$$

The vertical temperature stratification in the store is diminished by a high effective heat conductivity. The cooling of the upper standby section leads to an increased heat requirement for auxiliary heating. Low collector temperatures cannot be used due to heating of the lower storage area. The effective heat conductivity with good stores without internal installations according to measurements taken by the University of Stuttgart (ITW, 1995) is in the area of the heat conductivity of water (λ = 0.644 W/mK at 50°C). With good stores with internal heat exchangers, λ_{eff} can be set at 1–1.5 W/mK. Furthermore, amounts of heat $\dot{Q}_{for}$ are exchanged between the layers by forced convection, depending on the mass flow balance of the store. Since the external storage tank fittings are connected with only a few layers, a separate balance has to be created for each layer. In the simplest case without external links in a layer, the thermal capacity flows are $(\dot{m}c)_{i-1} = (\dot{m}c)_i$ and the heat flow by forced convection is given by:

$$\dot{Q}_{for} = (\dot{m}c)_{i-1}(T_{i-1} - T_i) + (\dot{m}c)_i (T_i - T_{i+1})$$
$$= (\dot{m}c)_i (T_{i-1} - T_{i+1}) \quad (3.99)$$

For layers with external connections, the external heat flows must also be taken into account, and the total of the mass flows for each layer must be zero. The total energy balance for a layer is then:

$$mc\frac{dT_{s,i}}{dt} = \dot{Q}_{c,i} + \dot{Q}_{h,i} - \dot{Q}_{l,i} - \dot{Q}_{t,i} + \dot{Q}_{free,i-1,i+1} + \dot{Q}_{for,i-1,i+1} \quad (3.100)$$

With the terms of forced and free convection, a coupling of the equations for layer i with the two layers i – 1 and i + 1 occurs. The solution to the set of equations is substantially simplified if in each time-step only the energy entry from the preceding layer i – 1 is considered, which leads to a change of temperature in the node i, so the store temperatures $T_{s,i}$ can be calculated downward successively.

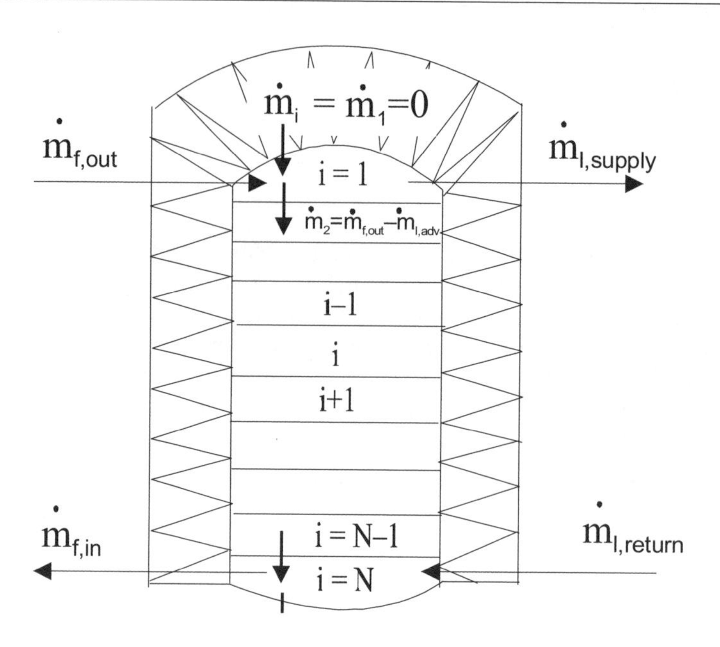

Figure 3.30: Stratified store with *N* layers.

For the calculation of the energy exchange between the layers from top ($i = 1$) to bottom ($i = N$), first the effective mass flows are determined. If, for example, two connections for the collector circuit with mass flow $\dot{m}_c$ and two links for the load circuit with mass flow $\dot{m}_l$ are selected, with flow in opposite directions, the effective mass flow between the layers $\dot{m}_i$ is formed from the difference of the two mass flows. For the first and last layer the effective mass flow is zero.

$$\dot{m}_i = \dot{m}_c - \dot{m}_l \quad \text{for} \quad i = 2, N \tag{3.101}$$

$$\dot{m}_i = 0 \quad \text{for} \quad i = 1 \quad \text{and} \quad i = N + 1 \tag{3.102}$$

A positive effective mass flow $\dot{m}_i$ with energy entry from layer $i - 1$ in layer i is taken into account by parameter $\delta_i^+ = 1$ (otherwise $\delta_i^+ = 0$). A negative effective mass flow from layer $i + 1$, i.e. dominance of the load mass flow and thus cooling of layer i, is taken into account by parameter δ_i^-. Thus the energy balance for the temperature node i reads:

$$\begin{aligned}(m_i c)_s \frac{dT_{s,i}}{dt} = {} & \delta_i^c (\dot{m}c)_c \left(T_{f,out} - T_{s,i}\right) + \dot{Q}_{h,i} - \delta_i^l (\dot{m}c)_l \left(T_{s,i} - T_{l,return}\right) \\ & - U_{eff} A_i \left(T_{s,i} - T_o\right) + \delta_i^+ \dot{m}_i c \left(T_{s,i-1} - T_{s,i}\right) + \delta_i^- \dot{m}_{i+1} c \left(T_{s,i} - T_{s,i+1}\right) \\ & - A_q \frac{\lambda_{eff}}{z} \left(T_{s,i} - T_{s,i-1}\right)\end{aligned} \tag{3.103}$$

with A_i as the exterior surface and A_q as the cross-section area of the respective node, and:

$$\delta_i^c = \begin{cases} 1 & for \quad i = 1 \\ 0 & for \quad i \neq 1 \end{cases}$$ i.e. collector supply into the top layer,

$$\delta_i^l = \begin{cases} 1 & for \quad i = N \\ 0 & for \quad i \neq N \end{cases}$$ i.e. load return into the bottom layer,

$$\delta_i^+ = \begin{cases} 1 & for \quad \dot{m}_i > 0 \\ 0 & for \quad \dot{m}_i \leq 0 \end{cases}$$ i.e. energy input from layer $I - 1$ to layer i

$$\delta_i^- = \begin{cases} 1 & for \quad \dot{m}_{i+1} < 0 \\ 0 & for \quad \dot{m}_{i+1} \geq 0 \end{cases}$$ i.e. energy input from layer $i + 1$ to layer i

Example 3.10

Calculation of the temperature distribution for a 750-litre buffer store, loaded with a mass flow from the collector field of 150 kg/h and a constant collector fluid outlet temperature $T_{f,out}$ of 60°C . The heat transfer coefficient of the store walls is to be 0.5 W/m²K , and the effective heat conductivity 1 W/mK. The store diameter $d_{s,a}$ is 0.69 m, the height 2.0 m. The ambient temperature T_o and also the initial temperature of the store is 12°C. Auxiliary heating is switched off and no heat is withdrawn from the store, i.e. all load terms are zero.

For the highest temperature node, the term of the effective heat conductivity is omitted. The collector mass flow at temperature $T_{f,out}$ is brought into the highest node, so $\delta_l{}^c = 1$. In the first time-step the calorific losses to the environment, at the same initial store temperature as ambient temperature, are still zero. With a time-step of 60 seconds, there is a rise in temperature in the highest node of:

$$\Delta T_{s,1} = \underbrace{\Delta t}_{60s} / \left(\underbrace{m_i}_{150l} \underbrace{c}_{4190J/kgK} \right)_s \left(\begin{array}{l} \underbrace{\delta_1^c}_{1} \left(\underbrace{\dot{m}}_{0.04167kg/s} c \right)_k \left(\underbrace{T_{f,out}}_{60°C} - \underbrace{T_{s,1}}_{12°C} \right) + \underbrace{\dot{Q}_{h,i}}_{0} \\ - \underbrace{\delta_i^l}_{0} (\dot{m}c)_l \left(T_{s,1} - T_{l,return} \right) - \underbrace{U}_{0.5W/m^2K} \underbrace{A_1}_{1.24m^2} \left(\underbrace{T_{s,1}}_{12°C} - \underbrace{T_o}_{12°C} \right) \\ + \underbrace{\delta_i^+}_{0} \dot{m}_i c \left(T_{s,i-1} - T_s \right) + \underbrace{\delta_i^-}_{0} \dot{m}_{i+1} c \left(T_{s,i} - T_{s,i+1} \right) \end{array} \right) = 0.8°C$$

The second node is warmed by effective thermal conduction and forced convection from the first layer, with the forced convection clearly more significant.

$$\Delta T_{s,2} = \underbrace{\Delta t}_{60s} / \left(\underbrace{m_i}_{150l} \underbrace{c}_{4190J/kgK} \right)_s \left(\begin{array}{l} \underbrace{\delta_2^c}_{0} \left(\underbrace{\dot{m}}_{0.04167kg/s} c \right)_k \left(\underbrace{T_{f,out}}_{60°C} - \underbrace{T_{s,2}}_{12°C} \right) + \underbrace{\dot{Q}_{h,i}}_{0} - \underbrace{\delta_2^l}_{0} (\dot{m}c)_l \left(T_{s,2} - T_{l,return} \right) \\ - \underbrace{U}_{0.5W/m^2K} \underbrace{A_1}_{0.8696m^2} \left(\underbrace{T_{s,2}}_{12°C} - \underbrace{T_o}_{12°C} \right) + \underbrace{\delta_2^+}_{1} \underbrace{\dot{m}_2}_{0.04167kg/s} c \left(\underbrace{T_{s,i-1}}_{12.8°C} - \underbrace{T_{s,i}}_{12°C} \right) \\ + \underbrace{\delta_i^-}_{0} \dot{m}_{i+1} c \left(T_{s,i} - T_{s,i+1} \right) - \underbrace{A_q}_{0.3739m^2} \underbrace{\lambda_{eff}}_{1W/mK} / \underbrace{z}_{0.4m} \left(\underbrace{T_{s,2}}_{12.0°C} - \underbrace{T_{s,1}}_{12.8°C} \right) \end{array} \right) = 0.0134°C$$

The temperatures represented as a function of time clearly show the temperature stratification of the store on loading from above by the collector:

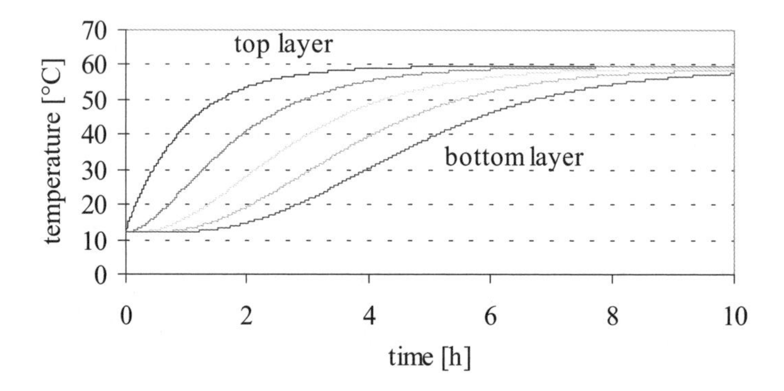

Figure 3.31: Temperature gradient in the store at different heights.

If at the same time a load mass flow (e.g. the half mass flow of the collector of 75 kg/h at a load return temperature of 12°C) is withdrawn from the store, the temperature gradients flatten accordingly.

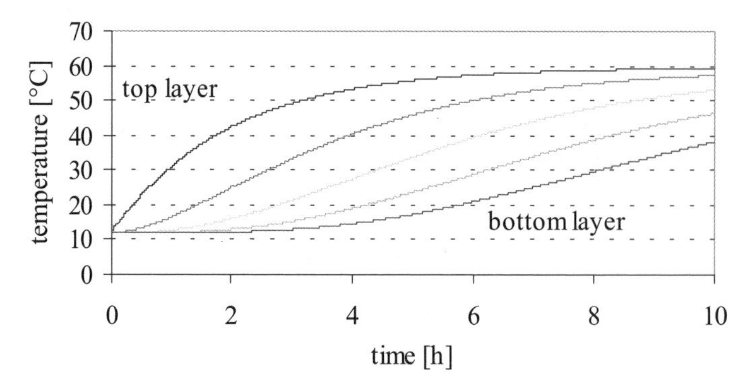

Figure 3.32: Temperature gradient of a layer store with a load mass flow of 75 kg/h and a collector mass flow of 150 kg/h.

3.2 Solar air collectors

Solar air collectors are thermal collectors which use air as a heat distribution medium and enable much simpler safety and system engineering by avoiding freeze-protection and overheating problems. In low-energy and passive house development with low heating requirements and air distribution systems for controlled ventilation, a certain renaissance of air heating has taken place, since with small air volume flows covering the heating requirement, thermal comfort can now be ensured.

Solar air collectors are particularly suitable for integration into building shells, since the air collector, being a well insulated element at the rear, fulfils the thermal demands on external construction components, and at the front is in keeping with usual building materials by a glass cover or by trapezoidal sheet metal constructions. Due to the use of air, possible collector leakages are not problematic, even for buildings with integrated warm facade constructions.

However, the large dimensions of the air ducts (due to the low thermal heat capacity of air and the lack of direct storage possibilities for the heat produced) are a negative factor. For reasons of cost and efficiency, special stone stores have not gained acceptance in buildings. More suitable is the activation of storage masses in the building itself (hypocaust systems) or heat storage in conventional hot water tanks by means of an air–water heat exchanger. At present, only a few commercial air collector systems are available, but they are all examples of interesting building-integration solutions.

Collector types

Solar air collectors differ mainly in the type of absorber cover and the air circulation along the absorber. Simple and economical air collector systems for pre-heating outside air do without a transparent cover and suck air in through fine perforations in the absorber sheet metal. At the same time, the trapezoidal sheet metal construction of the absorber sheet metal serves as a weather shell for the building. An example of such a product with a perforated absorber and a thermally-insulated bearing construction on the back is the SOLARWALL® air collector system.

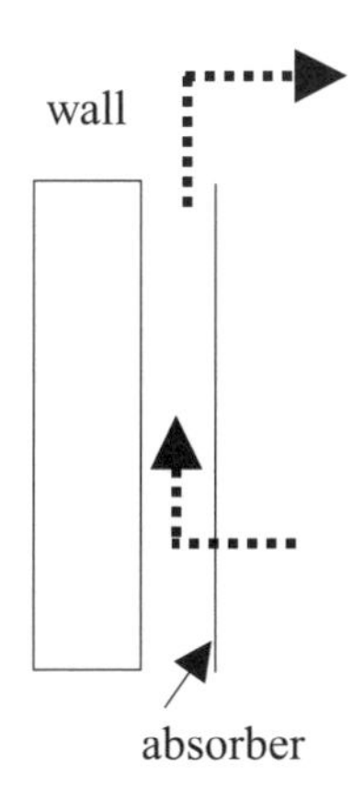

Figure 3.33: Solar air heater with a perforated external trapezoidal sheet metal as the absorber.

In transparently-covered absorber systems, the air is either led between the cover and absorber (so called Trombe-wall) or, to improve the thermal characteristics, led under the absorber sheet metal (under-flowed absorber). Absorbers flowed – around on both sides, or porous ones, have not gained market acceptance.

The calorific losses of the absorber to the environment depend, as with water collectors, on the cover system. Unglazed absorbers are comparable to swimming pool collectors and are used for applications with small rises in temperature. Calorific losses of under-flowed absorbers with a standing air layer between the absorber and cover are calculated like the thermal water collectors. With overflowed absorbers, the flow rate dependent convective heat transfer coefficients in the air gap enter the calculation. Calculation of the useful thermal energy is completely different, since a thermal conduction problem between the absorber sheet metal and the fluid tube no longer has to be solved; the entire absorber sheet metal transfers heat convectively to the air.

Due to the substantially poorer heat transfer characteristics of air compared to water, absorber ribs are used in most collector systems for surface enlargement. Here the design with continuous, even ribs is the technically simplest and most economical.

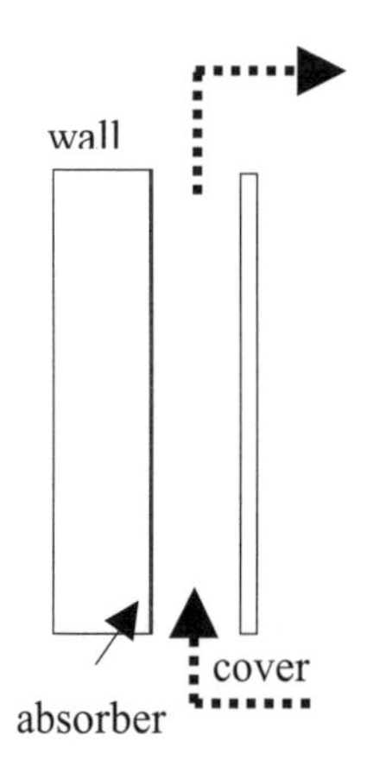

Figure 3.34: Transparently covered air collector with a netted cover.

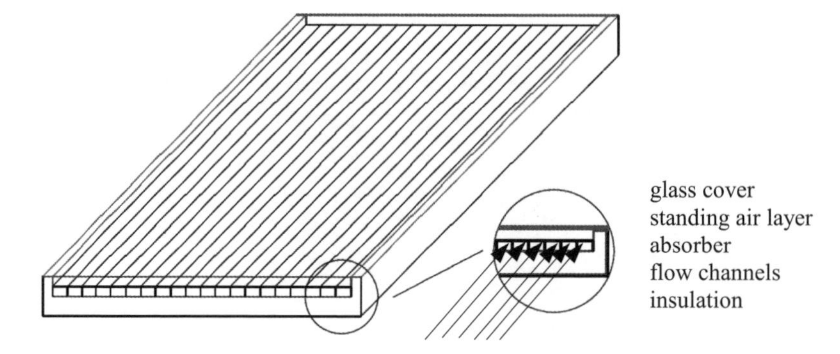

Figure 3.35: Transparently covered air collector with an underflowed absorber.

3.2.1 System engineering

Pre-heating of fresh air

The most favourable use of air collectors is, as with water-based systems, the pre-heating of outside air with altogether low absorber and air temperatures and thus small calorific losses. Systems for air pre-heating are used for heating fresh air in buildings without heat recovery systems. With decentralised air collector fields within the parapet area of a facade with an air intake into the space behind, the entry and exit air distribution system always necessary for heat recovery can be omitted. Even in an energy-optimised low-energy building, a commercial air collector can produce heating energy savings of between 150 kWh/m² in a lightweight construction and 210 kWh/m² in a massive construction for german climatic conditions (Eicker, 1998). Flow rates recommended by the manufacturers are typically 60 m³/m²h.

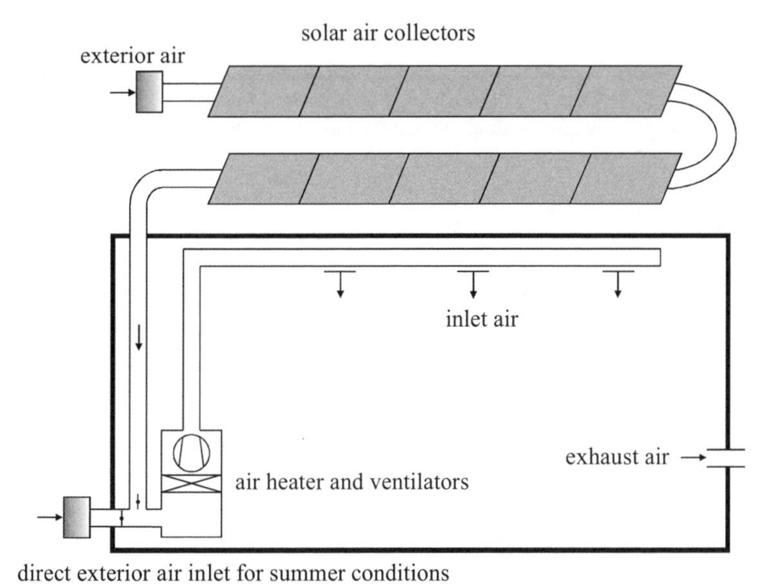

Figure 3.36: System sketch of the pre-heating of fresh air.

Direct air heating (fresh air/re-circulating air)

Apart from purely fresh air pre-heating, the warm air produced in air collectors can also be used for direct heating, if the outlet temperatures are at least 5 K above the room temperature. Systems for direct heating are usually operated with lower specific flow rates, to guarantee high rises in temperature even with low irradiances in winter (20–40 m³/m²h).

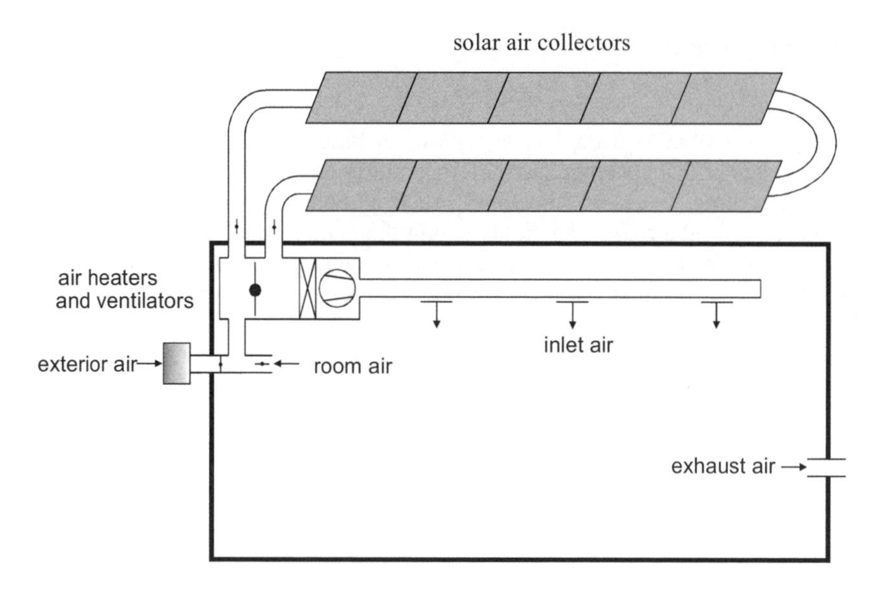

Figure 3.37: System sketch of air heating with air collectors and mixed-air operation.

Indirect air heating with hypocausts

When there is a high heating requirement in rooms, the amounts of air necessary to provide the heating can lead to uncomfortably high injection rates. To improve thermal comfort, the solar-heated warm air can be led into ceiling or wall cavities in a closed cycle (so-called hypocausts) and the room can be warmed by radiant heat.

Combination with heat recovery systems

The combination of an air collector system with heat recovery from the space exhaust air reduces the effectively possible heating energy savings by the air collector. During fresh air pre-heating the air collector and heat recovery system compete, with the potential saving being limited in total. With an air collector between the outside air inlet and the heat recovery unit, savings of 25–60 kWh per square metre of collector are possible. If the collector is placed behind the heat recovery unit to provide additional temperature rises of the room inlet air, the air collector system is energetically more favourable. In a low-energy building, between 60 and 110 kWh of heating energy per square metre of air collector surface can be saved in this way.

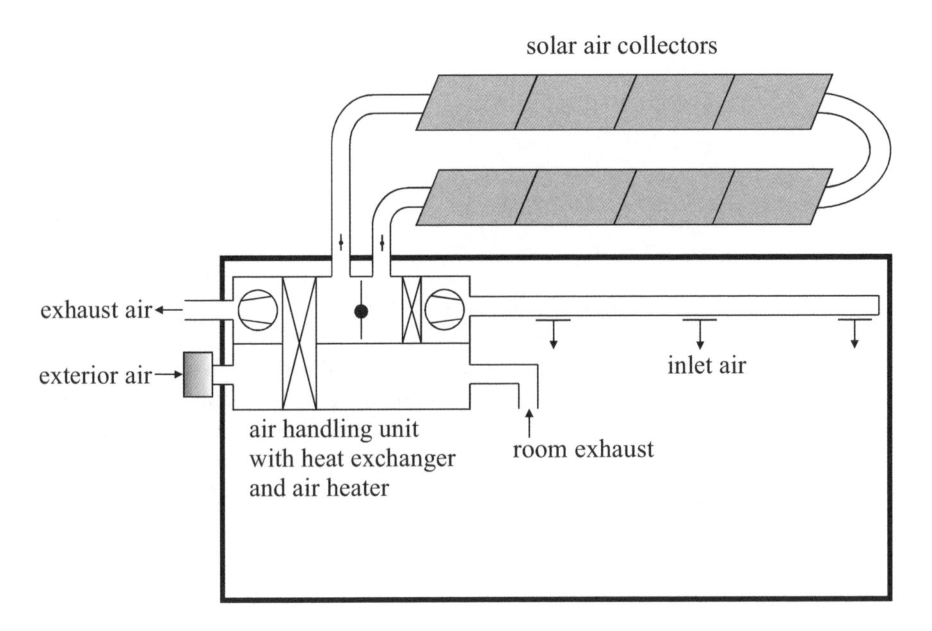

Figure 3.38: Combination of air collectors with heat recovery systems.

3.2.2 *Calculation of the available thermal power of solar air collectors*

3.2.2.1 Temperature-dependent material properties of air

In solar air heaters, rises in air temperature of 10–60 K usually occur in the direction of flow, and around 30 K (unribbed) to 10 K (ribbed) perpendicularly to the direction of flow between the absorber and rear wall, depending on the heat exchange surface.

The functions for calculating the kinematic viscosity, heat conductivity, specific thermal capacity and density of the fluid are polynomial fits to the numerical values of dry air at a constant pressure (p = 10^5 Pa) in the temperature range of 273 K to 373 K. The characteristic fluid properties are calculated as a function of the mean fluid temperature $T_{f,m}$ [K].

Kinematic viscosity: $$\nu = \left(0.09485\left(T_{f,m} - 273.15\right) + 13.278\right) \times 10^{-6} \quad [m^2/s] \tag{3.104}$$

Heat conductivity: $$\lambda = \left(0.02795\left(T_{f,m} - 273.15\right) + 24.558\right) \times 10^{-3} \; [W/mK] \tag{3.105}$$

Specific thermal capacity: $$c_p = 1006 + 0.05\left(T_{f,m} - 273.15\right) \quad [J/kgK] \tag{3.106}$$

The density ρ [kg/m^3] and the heat expansion coefficient β' [K^{-1}] are determined from the ideal gas equations.

$$\rho = \frac{p}{R\,T_{f,m}} = \frac{10^5\,N/m^2}{287.1\,J/kgK \times T_{f,m}} = \frac{348.3}{T_{f,m}} \tag{3.107}$$

$$\beta' = \frac{1}{T_{f,m}} \tag{3.108}$$

3.2.2.2 Energy balance and collector efficiency factor

Calculation of the thermal and optical characteristics of the transparent cover of an air collector with an under flowed absorber follows the method already described for water collectors, as the geometry of the glazing, the standing air layer and the absorber is identical. The heat transfer between the absorber and the fluid now takes place only perpendicularly to the absorber level. The geometry-dependent collector efficiency factor can thus be deduced from a stationary energy balance of three temperature nodes.

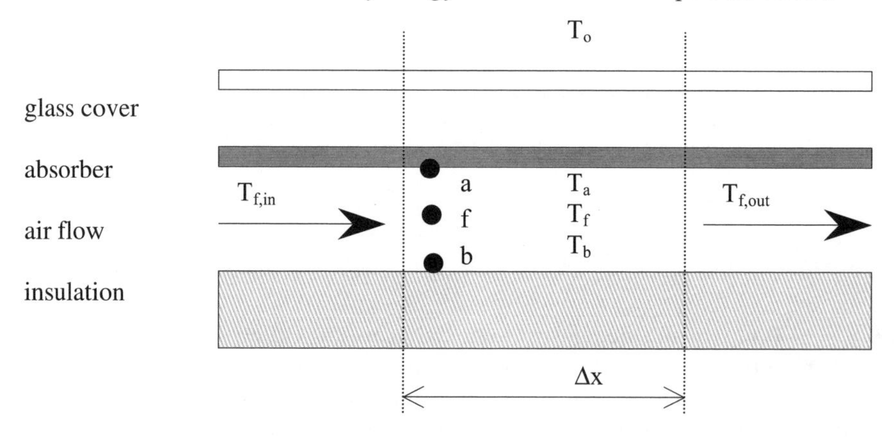

Figure 3.39: Designation of the nodes and temperatures in a solar air heater with an underflowed absorber (absorber width *B*).

At node *a* of the absorber, the irradiance *G* let through with the transmission coefficient τ is absorbed by the absorber sheet metal with the absorption coefficient α. The calorific losses through the transparent cover against the exterior temperature T_o are calculated by the heat transfer coefficient U_f. Convective heat with the heat transfer coefficient $h_{c,a\text{-}f}$ is transferred to the heat distribution medium fluid with temperature T_f, and heat is exchanged by the radiation heat transfer coefficient $h_{r,a-b}$ with the rear wall of the flow channel with temperature T_b.

$$G(\tau\alpha) - U_f\left(T_a - T_0\right) - h_{c,a-f}\left(T_a - T_f\right) - h_{r,a-b}\left(T_a - T_b\right) = 0 \tag{3.109}$$

At node *f* of the heat distribution medium fluid, the available power $\dot{Q}_u$ is produced over the collector width *B* and the distance Δx in the direction of flow. This power consists of the amount of heat transferred convectively from the absorber (with $h_{c,a\text{-}f}$) and from the rear (with $h_{c,b\text{-}f}$).

$$\dot{Q}_u - h_{c,a-f} B\Delta x \left(T_a - T_f\right) - h_{c,b-f} B\Delta x \left(T_b - T_f\right) = 0 \tag{3.110}$$

At node b of the rear wall of the flow channel, heat is radiated from the absorber sheet metal with a radiative heat transfer coefficient $h_{r,b\text{-}a}$, and at the same time heat is transferred convectively to the fluid. Calorific losses to the ambient air develop via the collector rear with the heat loss coefficient U_b.

$$h_{r,b-a}\left(T_a - T_b\right) - h_{c,b-f}\left(T_b - T_f\right) - U_b\left(T_b - T_0\right) = 0 \tag{3.111}$$

Equations (3.109) and (3.111) are solved for the absorber and rear wall temperatures T_a and T_b as a function of the fluid and ambient temperatures T_f and T_o, and these are inserted into the useful power Equation (3.110). Thus a conditional equation for the available power $\dot{Q}_u$ is obtained which depends only on the fluid and ambient temperatures,

$$\dot{Q}_u = AF' \left(G(\tau\alpha) - U_t\left(T_f - T_o\right)\right) \tag{3.112}$$

the collector efficiency factor F' being given by:

$$F' = \left(1 + \frac{U_t}{h_{c,a-f} + \left(\frac{1}{h_{c,b-f}} + \frac{1}{h_{r,a-b}}\right)^{-1}}\right)^{-1} \tag{3.113}$$

and U_t represents the total of the heat transfer coefficients over the front, back and sides. The functional dependency of all characteristics on temperatures and flow rates is discussed in the following section.

3.2.2.3 Convective heat transfer in air collectors

The collector efficiency factor is mainly dependent on the convective heat transfer coefficients. These vary over a wide range, approximately 5 to 50 W/m²K , depending on the flow rate (laminar or turbulent current) and the ribbing of the absorber. The order of magnitude of the convective heat transfer coefficient determines the possible thermal efficiency and is a crucial criterion in selecting a type of air collector, whether with flowed-through absorbers, back-ventilated absorbers (for example PV modules) with flat-parallel gap geometry or ribbed air collectors with very small gap dimensions of a few centimetres.

In active solar energy systems, the heat distribution medium air is moved by fans, so consideration of the forced convection has priority. In back-ventilated facade systems with large gap depths of between 0.1–1 m, the flow rates are often so small, however, that the lift term is not negligible and a proportion of free convection must be calculated.

In such systems with a large distance between the channel-limiting surfaces (cavity facade), heat transfer calculations for a separated individual plate geometry result in better results than for gap geometry. Besides, free convection between flat-parallel panels also dominates the thermal resistance of transparent collector covers, with which commercial air collectors are thermally insulated against outside air. The calculation of the convective heat transfer coefficients in the standing air layer between the absorber and glass cover is taken from section 3.1.10.4 in Chapter 3.

Asymmetrical heating of the air duct is common to active solar systems; the solar radiation-absorbing absorber side is clearly warmer than the thermally insulated rear, which closes the air duct either against outside air or, with warm facade constructions, from room air. The convective heat transfer coefficient h_c is directly proportional to the dimensionless Nußelt number Nu, which depends both on the material properties of the fluid and on forces of inertia and friction. For a range of geometries and temperature and flow conditions, experimentally determined correlations for Nußelt numbers Nu are available in the litreature, which enable calculation of the convective heat transfer coefficient using the heat conductivity of the fluid λ [W/mK] and a geometry-dependent characteristic length L [m].

$$h_c = \frac{Nu\,\lambda}{L} \tag{3.114}$$

For the usual case of air collector operation with a fan-driven forced current within the laminar or turbulent area, the characteristic length L is given by the hydraulic diameter of the flow channel d_h, defined as the relation of the quadruple duct cross-section A to the circumference U. For a rectangular channel of height (= panel distance) H and width W the result is:

$$d_{\mathrm{h}} = \frac{4A}{U} = \frac{4WH}{2(W+H)} = \frac{2WH}{(W+H)} \approx 2H \quad \text{for} \frac{H}{W} \ll 1 \tag{3.115}$$

The types of convective heat transfer in air collector systems, with in each case one application example, are clarified in Figure 3.40.

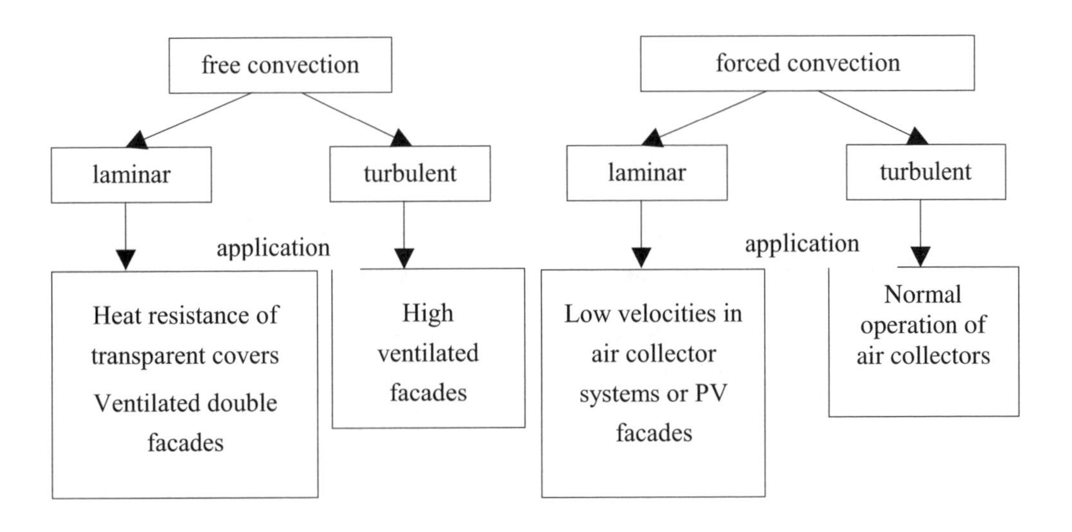

Figure 3.40: Convection mechanisms in air collectors.

In the following section, only classical air collector configurations with flat-parallel panels or ribbed absorbers will be analysed, whose heat transfer is dominated by forced convection (laminar or turbulent).

Laminar flow

For laminar forced flow in air ducts with one-sided heating, a set of Nußelt correlations can be found in the litreature. The Nußelt numbers are generally constant in a fully developed laminar current; with the boundary condition of constant heat flow, a value of $Nu = 5.4$ can be used. In the intake area of the flow channel, the Nußelt numbers are however clearly higher and are approximated as a function of the flow path x with functions of the type:

$$Nu = c_0 + c_1 \left(\frac{x}{d_h \,\mathrm{Re}\,\mathrm{Pr}} \right)^{-c_2} \tag{3.116}$$

The thermal intake length x_{th} is defined as the length within which the Nußelt number has fallen to a value 1.05 times higher than for the fully developed current. The intake lengths are generally very large with a laminar current.

$$x_{th} = x_{th}^{*} \,\mathrm{Re}\,\mathrm{Pr}\, d_h \tag{3.117}$$

According to Merker and Eiglmeier (1999) the factor x_{th}^{*} is between 0.0335 and 0.053 depending on the boundary condition. The Reynolds number $\mathrm{Re} = \mathrm{v}\, d_h / \nu$ is proportional to the flow velocity v, and the Prandtl number $\mathrm{Pr} = \nu c_p \rho / \lambda$ to the temperature-dependent material properties of the fluid.

Example 3.11

Calculation of the thermal intake length for an air collector channel with 0.0275 m hydraulic diameter and a flow velocity v of 1 m/s at a mean fluid temperature of 40°C . For the boundary condition of an intake developing both thermally and hydrodynamically, x_{th} = 0.053.

dynamic viscosity ν [m²/s]	1.707×10^{-5}
heat conductivity λ [W/mK]	0.02568
heat capacity c_p [J/kgK]	1008
density ρ [kg/m³]	1.1123
Re [–]	1610
Pr [–]	0.7455

This results in a thermal intake length x_{th} of 1.68 m.

The constants of the Nußelt correlation from Equation (3.116), c_0, c_1 and c_2 depend on the dimensionless flow path $x^* = x/(Re\ Pr\ d_h)$ (Shah and London, 1978). According to investigations into flat ribbed air heaters the relation of the flow channel width W to the channel height H is used as a parameter (Altfeld, 1985). As a boundary condition, a constant heat flow in the direction of flow and a constant temperature in the circumferential direction, i.e. along the rib, are assumed, and the critical Reynolds number for the transition from a laminar to a turbulent current is set at 3100.

Table 3.10: Coefficients for Nußelt correlations.

Parameter	c_0	c_1	c_2	*Reference*
$x^* \leq 0.00325$	0	0.5895	0.5	Shah and London
$0.00325 < x^* \leq 0.045$	0	2.614	0.24	Shah and London
$x^* > 0.045$	5.39	0	-	Shah and London
W/H = 0.5	4.11	0.0777	0.8212	Altfeld
W/H = 1.0	3.6	0.1028	0.7782	Altfeld

An extended equation by Merker and Eiglmeier is likewise taken up in the comparison of the Nußelt correlations represented in Figure 3.41. The average values of the Nußelt numbers, integrated from 0.1 to 2.5 m, are indicated in the legend and hardly differ.

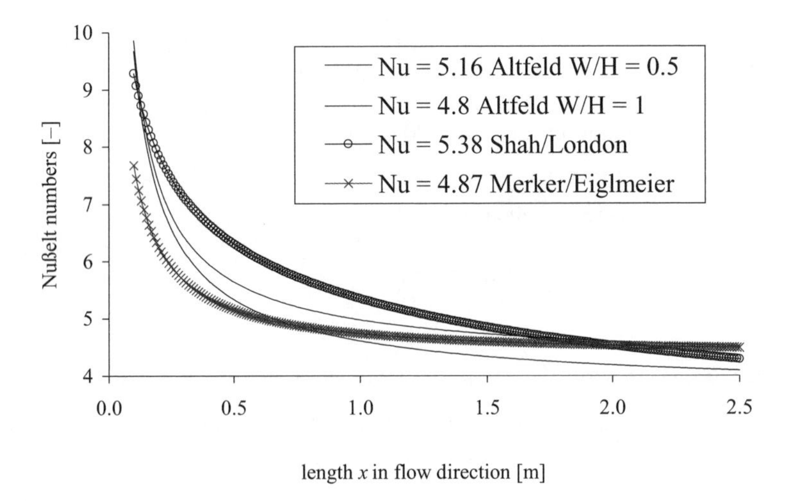

Figure 3.41: Nußelt numbers as a function of the distance x from the intake into an air collector for laminar current.

Since to calculate the convective heat transfer for the efficiency factor F', only the mean Nußelt number over the collector length is needed and the average values of the different correlations hardly differ, it is recommended that you calculate the integral for a correlation and determine directly the mean Nußelt number as a function of the collector length. If, for example, the Altfeld correlation is used with:

$$W/H = 0.5 \qquad Nu = 4.11 + 0.0777\left(\frac{x}{d_h \operatorname{Re}\Pr}\right)^{-0.8212} \tag{3.118}$$

the average value can be calculated directly from the integral:

$$Nu_m = \frac{\int_{L_0}^{L_k} Nu(x)\,dx}{L_k - L_0} = 4.11 + \frac{0.4345}{(d_h \operatorname{Re}\Pr)^{-0.8212}(L_k - L_0)}\left(L_k^{\,0.1788} - L_0^{\,0.1788}\right) \tag{3.119}$$

Since the Nußelt numbers for $x \to 0$ become infinitely large, for calculation of the mean Nußelt number, integration should only be carried out from, for example, $L_0 = 0.1$ m.

Example 3.12

Calculation of the mean Nußelt number for a collector of flow channel length 2.5 m for the collector in Example 3.11 at a flow rate of 1m/s.

At the Reynolds number of 1610, laminar flow conditions prevail. At integration limits of 0.1 m and 2.5 m, the mean Nußelt number results in $Nu_m = 5.758$. From this a mean heat transfer coefficient h_c of 5.38 W/m²K is calculated.

Forced turbulent flow

With turbulent flow, the influence of the intake area is smaller than with laminar current. The influence of the boundary conditions and the channel geometry is also small with turbulent flow. As a general equation for calculating the Nußelt numbers with turbulent flow, the modified Petukhov equation for Re > 3100 can be used:

$$Nu = \frac{(\mathrm{Re}-1000)\,\mathrm{Pr}\frac{f}{8}}{1+12.7\sqrt{\frac{f}{8}}\left(\mathrm{Pr}^{2/3}-1\right)}\left(1+\left(\frac{d_h}{L_c}\right)^{2/3}\right) \tag{3.120}$$

with $f = (0.79\ln\mathrm{Re}-1.64)^{-2}$ described as the pressure loss coefficient and the term $1+(d_h/L_c)^{2/3}$ is a correction term for short channels with a channel length L_c.

A simplified Nußelt correlation is indicated by Tan and Charters (1970) for asymmetrically heated flat-parallel panels and a fully developed turbulent current, which likewise contains a correction term for short channels.

$$Nu = 0.0158\,\mathrm{Re}^{0.8} + (0.00181\,\mathrm{Re} + 2.92)\exp(-0.03795\,L_c/d_h) \tag{3.121}$$

Both correlations produce sufficiently exact values for the convective heat transfer coefficients.

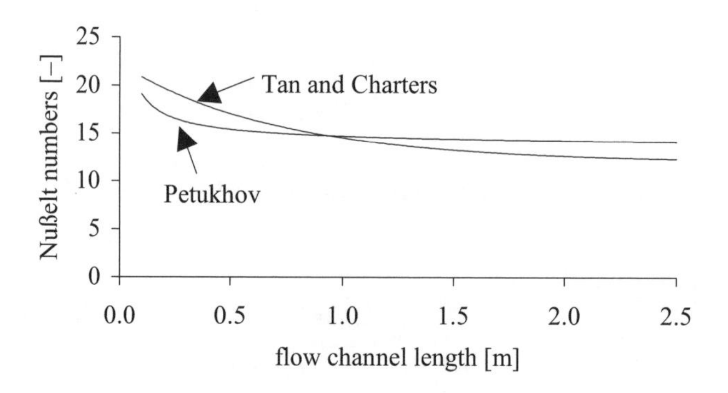

Figure 3.42: Nußelt correlations for forced convection.

With the above Nußelt correlations, the heat transfer coefficient of ribbed absorbers can be calculated.

Example 3.13

Calculation of the convective heat transfer coefficient h_c for the 2.5-m long collector in Example 3.11 at a flow rate of 2.5 m/s.

The Reynolds number for the flow channel of 0.0275 m hydraulic diameter is 4026. This results in a pressure loss coefficient $f = 0.041$ and a Nußelt number 14.53, based on the Petukhov equation. The convective heat transfer coefficient for a turbulent current is 13.57 W/m²K , i.e. almost three times as high as with the laminar current.

Rib efficiency

The heat transfer coefficient h_c calculated so far refers to the complete heat transfer surface of the air duct confinement surfaces A_g, which for each flow channel is composed of the absorber underside and channel rear wall of width W and the two lateral ribbed surfaces of height H. For a unit length in the direction of flow, the total area of the flow channel is:

$$A_g = 2(W + H) \times 1 \tag{3.122}$$

Since the surface reference for the available power calculation is the absorber surface, a surface normalisation for the convective heat transfer coefficient must be carried out. Furthermore, it is taken into account, by means of the so-called rib efficiency, that the temperatures of the ribs drop from the absorber temperature level.

The rib efficiency is deduced similarly to the temperature distribution calculation of the absorber sheet metal with water-throughflowed collectors, assuming ideal thermal contact of the rib to the absorber underside, a constant convective heat transfer coefficient and neglect of heat dissipation at the rib point.

$$\eta_{Ri} = \frac{\tanh(mH)}{mH} \tag{3.123}$$

with

$$m = \sqrt{\frac{h_c}{\lambda_{Ri}} \frac{U_{Ri}}{A_q}} \tag{3.124}$$

and h_c as the convective heat transfer coefficient related to the total area and λ_{Ri} as the heat conductivity of the sheet metal. The relation of the circumference of the rib U_{Ri} of rib thickness t to the cross-section of the rib A_q results from:

$$\frac{U_{Ri}}{A_q} = \frac{2(H + t)}{H\,t} \tag{3.125}$$

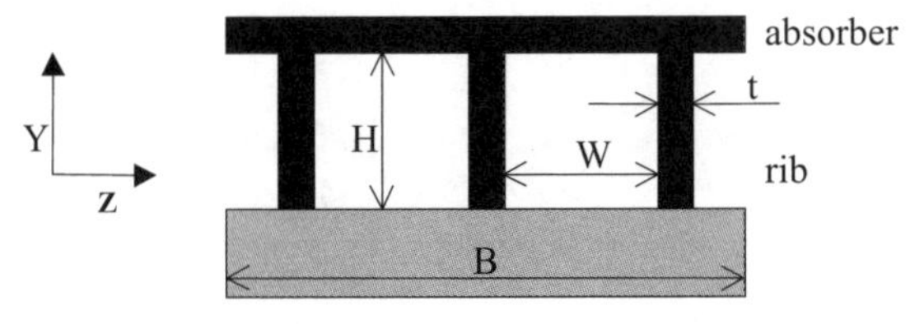

Figure 3.43: Geometry of a ribbed absorber.

The surface relation of the heat-transferring ribbed surface $A_{Ri}=2H\ x\ 1$ to the entire channel confinement surface A_g is defined with β, following Altfeld (1985).

$$\beta = \frac{A_{Ri}}{A_g} = \frac{2H}{2(W+H)} = \frac{H}{(W+H)} \tag{3.126}$$

and the relation of the surface between the ribs A_W to the absorber surface A is defined with γ.

$$\gamma = \frac{A_W}{A} = \frac{W}{W+t} \tag{3.127}$$

The heat flux from the absorber underside and rib to the fluid $\dot{Q}_a$, related to the absorber surface A, is described by the heat transfer coefficient $h_{c,a\text{-}f}$, which results from the rib efficiency and the surface factors from the already calculated h_c.

$$\frac{\dot{Q}_a}{A} = h_{c,a-f}(T_a - T_f) \tag{3.128}$$

with

$$h_{c,a-f} = \frac{\gamma}{1-\beta}\left(1+\beta\left(2\eta_{Ri}-1\right)\right)h_c \tag{3.129}$$

For the rear of the flow channel, the convective heat transfer coefficient is reduced by the factor γ, i.e. by the effective heat-transferring surface portion of the rear.

$$h_{c,b-f} = \gamma h_c \tag{3.130}$$

The two heat transfer coefficients $h_{c,a\text{-}f}$ and $h_{c,b\text{-}f}$ can be inserted into the equations of the collector efficiency factor.

Example 3.14

Calculation of the convective heat transfer coefficients $h_{c,a\text{-}f}$ and $h_{c,b\text{-}f}$ of the underflowed ribbed air collector already considered, with flow rates of 1 m/s or 2.5 m/s. To calculate the surface factors β and γ, the exact geometry of the flow channel is needed:

Rib height H of the channel:	0.028 m
Rib thickness t:	0.0014 m
Rib distance (= channel width) W:	0.027 m
Heat conductivity of the rib sheet metal λ_{Ri}:	238 W/mK

$\beta = 0.5091, \gamma = 0.9507$, $U_{Ri}/(\lambda_{Ri}\, A_q) = 6.3$

Flow velocity (m/s)	h_c [W/m²K]	*m*	*Rib efficiency* η_{Ri}	$h_{c,a-f}$ [W/m²K]	$h_{c,b-f}$ [W/m²K]
1	5.38	5.82	0.991	15.6	5.1
2.5	13.57	9.25	0.978	39.1	12.9

3.2.2.4 Thermal efficiency of air collectors

With the convective heat transfer coefficients determined above, the heat transfer coefficient for radiation h_r and the total heat transfer coefficient U_t between the absorber and environment, the efficiency factor F' can be calculated using Equation (3.113). The heat transfer coefficient for radiation is calculated on the simplified assumption that the absorber and gap rear wall can be described as infinitely expanded flat-parallel surfaces, and the ribs are not considered.

Due to the small emissivities of the duct confinement surfaces (ε_a, $\varepsilon_b = 0.04$–0.1) and the small temperature difference between the surfaces, typically 5K, these assumptions produce sufficiently exact results (T_a and T_b in Kelvin).

$$h_r = \frac{\sigma}{\frac{1}{\varepsilon_a} + \frac{1}{\varepsilon_b} - 1}\left(T_a^2 + T_b^2\right)\left(T_a + T_b\right) \tag{3.131}$$

The heat transfer coefficient U_t is calculated, as with the flat plate collector, in simplified fashion as the total of the front, side and rear wall losses, i.e. of the temperature node of the absorber to the environment; the side and rear wall losses U_s and U_b are temperature-independent and can be set as constant.

$$U_t = U_f + U_b + U_s \tag{3.132}$$

The heat transfer coefficient through the transparent cover takes into account wind influences ($h_{c,w}$) and radiation losses to the sky ($h_{g\text{-}sky}$) and is, as with the flat plate collectors, calculated iteratively as a function of the glas cover temperature T_g.

$$U_f = \frac{1}{\frac{1}{h_{c,a-g} + h_{r,a-g}} + \frac{1}{h_{c,w} + h_{r,g-sky}}} \tag{3.133}$$

Since all the heat transfer coefficients are temperature-dependent, first temperatures for all surfaces and the mean fluid temperature must be given. With the initial temperature field all coefficients are then calculated, the collector efficiency factor F' is determined and the available power is calculated as a function of the fluid input temperature:

$$\dot{Q}_n = AF_R\left(G(\tau\alpha) - U_t\left(T_{f,in} - T_o\right)\right) \tag{3.134}$$

with $$F_R = \frac{\dot{m}c_p}{AU_t}\left(1 - \exp\left(-\frac{U_t F' A}{\dot{m}c_p}\right)\right)$$

From the available power, the mean temperatures for the absorber T_a, flow channel rear wall T_b, fluid T_f and glass cover T_g are then calculated similarly to water-through flowed collectors.

$$\bar{T}_a = T_{f,in} + \frac{\dot{Q}_n}{AU_t F_R}(1 - F_R)$$

$$\bar{T}_f = T_{f,in} + \frac{\dot{Q}_n\left(1 - \frac{F_R}{F'}\right)}{A\,F_R U_t}$$

$$T_g = T_a - \frac{U_f\,(T_a - T_o)}{\left(h_{c,a-g} + h_{r,a-g}\right)}$$

The temperature of the flow channel rear wall is calculated by resolving the energy balance Equation (3.111) with the mean fluid temperature used for T_f.

$$T_b = \frac{h_{c,b-f}T_f + U_b T_o - h_{r,b-a}T_a}{h_{c,b-f} + U_b - h_{r,b-a}} \tag{3.135}$$

With these temperatures, in the next iteration new heat transfer coefficients are calculated. The iteration is continued until the change in the temperature field becomes negligibly small.

Example 3.15

Calculation of the available power, the outlet temperatures and the thermal efficiency for a facade-integrated air collector at 800 W/m² irradiance and an ambient temperature of 10°C. The ambient temperature equals the inlet temperature $T_{f,in}$ in the collector. The air duct geometry corresponds to the examples already calculated with a collector length of 2.5 m, and the heat transfer coefficient of the rear U_b is a constant 0.65 W/m²K (side losses ignored).

Table 3.11: Result table for air collector efficiencies.

No.	Boundary condition	U_t [W/ m²K]	F' [–]	F_R [–]	$\dot{Q}_u$ /A [W/m²]	$\bar{T}_a$ [K]	$\bar{T}_f$ [K]	$T_{f,out}$ [K]	η [-]
1	v =1m/s (v_{wind}=3m/s), ε_a=0.9	6.6	0.70	0.59	375	49.9	26.0	40.2	0.47
2	v =1m/s (v_{wind}=3m/s), ε_a=0.1	4.2	0.79	0.69	442	56.9	28.8	46.0	0.55
3	v =2.5m/s (v_{wind}=3m/s), ε_a=0.9	6.2	0.87	0.80	514	30.3	18.3	26.1	0.64
4	v =2.5m/s (v_{wind}=1m/s), ε_a=0.9	5.4	0.89	0.82	528	30.8	18.5	26.6	0.66
5	v =2.5m/s (v_{wind}=1m/s), ε_a=0.1	3.55	0.92	0.88	562	32.1	19.9	27.7	0.70

From the results, the efficiency rise with the change of the flow from laminar (No. 1, 2) into turbulent conditions (No. 3–5) is clearly evident. The first two simulations with laminar flow differ by the emission coefficient of the absorber ε_a, which in the case of the selective coating is 0.1 and with the black absorber 0.9. Due to the selective coating, the heat transfer coefficient falls from 6.6 to 4.2 W/m²K and the efficiency rises by 17%. At higher flow rates of 2.5 m/s, the influence of the outside wind velocity was examined. A reduction of 3 m/s to 1 m/s leads to a reduction of the U_t-value of 0.8 W/m²K, and an efficiency improvement of 3%. The selective coating brings a further 6% efficiency improvement.

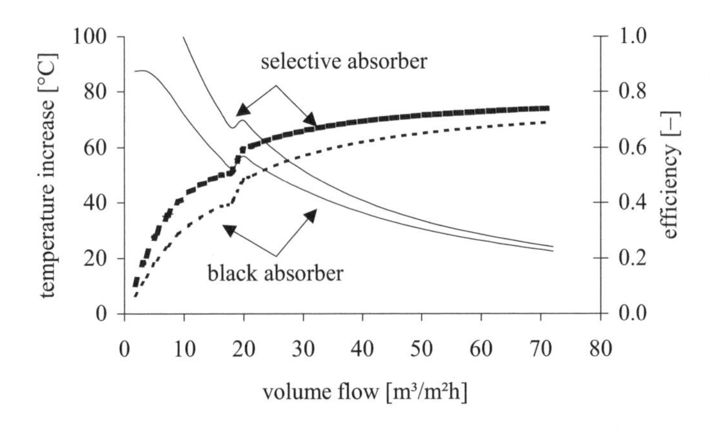

Figure 3.44: Rise in temperature (continuous lines) and efficiency (broken lines) of an air collector.

The boundary conditions for these temperature-rise or efficiency calculations as a function of the specific flow rate (in m³/h per m² of collector surface) are: irradiance of 800 W/m², 10°C ambient temperature and 3 m/s wind velocity.

3.2.3 Design of the air circuit

The volumetric air flow of the collector depends on the desired application. While during pure fresh air pre-heating, high flow rates of > 60 m³/m²h with good thermal efficiency are favourable, direct heating or hypocaust applications require high rises in temperature and thus low specific flow rates (20–40 m³/m²h).

With direct heating applications, the outlet temperature determined by the surface-specific flow rate must be limited for reasons of comfort, in houses to 45°C , in industrial applications to a maximum of 60°C. The temperature limitation can take place either by flow rate regulation or by the addition of cold air.

3.2.3.1 Collector pressure losses

For interconnecting the collectors it is best to select as long a collector series as possible, to reduce the connection channels and thus system costs. At the same time, by series connection at a given total volume flow $\dot{V}$ the flow velocity v in the air ducts and thus the convective heat transfer is increased, since the throughflowed duct cross-section surface A_q is small.

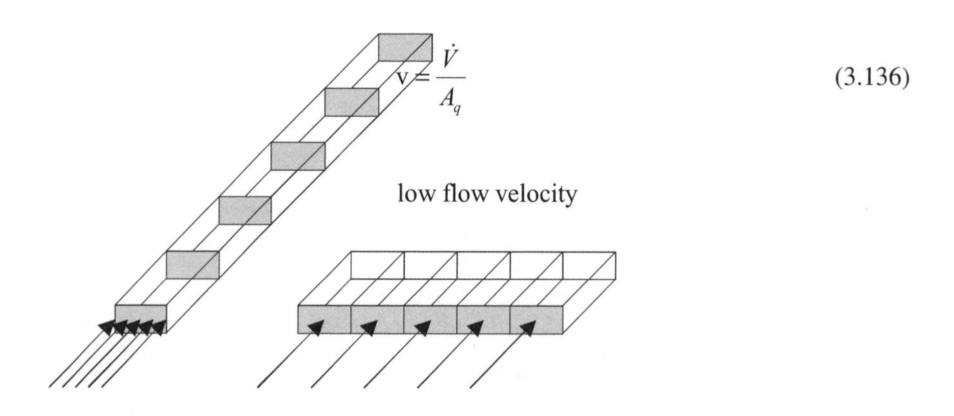

Figure 3.45: Series and parallel connection of collectors.

A limiting factor for the number of collectors switchable in series is the pressure loss by friction Δp_R, which at laminar flow rises linearly and with turbulent flow as a square of the flow velocity. The friction pressure loss is calculated from the coefficient of friction λ, the length l, the hydraulic diameter d_h and the dynamic pressure $\rho / 2\mathrm{v}^2$.

$$\Delta p_R = \frac{\lambda\, l}{d_h} \frac{\rho}{2} \mathrm{v}^2 \tag{3.137}$$

With laminar flow, the coefficient of friction λ is in inverse proportion to the Reynolds number Re, so the result is the linear connection between the pressure loss and the flow rate.

$$\lambda = \frac{64}{\mathrm{Re}} = \frac{64\nu}{\mathrm{v\,d_h}} \quad (3.138)$$

With turbulent flow, the coefficient of friction depends on the roughness of the wall. The ducts of a commercial air collector have, for example, an absolute roughness of ε = 0.15 mm, comparable to that of galvanised steel tubes. The pipe friction number is determined iteratively as a function of the roughness and the Reynolds number both for hydraulically smooth and rough pipes:

$$\frac{1}{\sqrt{\lambda}} = -2\log\left(\frac{2.51}{\mathrm{Re}\sqrt{\lambda}} + \frac{\varepsilon}{3.72\mathrm{d_h}}\right) \quad (3.139)$$

Example 3.16

Calculation of the pressure loss of the above air collector with d_h = 0.027 m at flow velocities of 1 m/s, 2.5 m/s and 5 m/s and a mean air temperature of 40°C .

velocity v [m/s]	*Reynolds number* Re [–]	*Friction coefficient* λ [–]	*pressure drop* Δp_R [Pa/m]
1	1583	0.04	0.82
2.5	3957	0.045	5.8
5	7914	0.038	19.6

3.2.3.2 Air duct systems

In air duct systems, besides pipe friction losses consideration must also be given to pressure losses through individual resistances Δp_Z resulting from changes of direction and cross-section and from branchings in the duct system. In a turbulent current, these flow resistances are caused by the formation of eddies and are proportional to the square of the mean flow velocity. The dimensionless coefficient of drag ζ is tabulated in heating and climate-technical manuals for all usual components, and varies over a wide value range of approximately 10^{-2} (steady narrowings or widenings) to 10 (non-return valve, branches etc.).

$$\Delta p_Z = \zeta\rho\frac{\mathrm{v}^2}{2} \quad (3.140)$$

The dimensions of the installations and of the air ducts result from given values for maximum flow velocities, which should not be exceeded due to sound-related problems and to high pressure losses, as shown in Table 3.12:

Table 3.12: Approximate values for flow velocities.

Duct type	*Velocitiy*
Small air systems in residential buildings up to 500 m³/h, tubes behind inlet/outlet valves	3–4 m/s
Medium air systems, connecting ducts and air distribution ducts	4–8 m/s
Large air systems and collecting channels	8 m/s

At a given flow rate and a given maximum velocity, the duct cross-section surface and the pipe diameter are calculated by $A_q = \pi d^2/4 = \dot{V}/v$. Typical distributions of pressure losses in large systems with short distribution pipes in the building are about 50% in the collector field itself, 15% in the collecting ducts and 35% in the building.

After determination of the total volume flow and the pressure loss calculation of the line with the highest losses (main line), the electrical power required by the fan can be determined. This is proportional to the flow rate and pressure loss, and depends on the fan and motor efficiency.

$$P_{el} = \frac{\dot{V}\Delta p}{\eta} \tag{3.141}$$

Table 3.13: Approximate values for ventilator efficiencies (Grammer, 2002).

Volume flow rate [m³/h]	*Ventilator efficiency* η_V [–]	*Motor efficiency* η_M [–]	*Total efficiency* η [–]
up to 300	0.4 – 0.5	0.8	0.32 – 0.4
400 – 1000	0.6 – 0.7	0.8	0.48 – 0.56
2000 – 5000	0.7 – 0.8	0.8	0.56 – 0.64
6000 – 10000	up to 0.85	0.82	up to 0.7

The total pressure difference produced by the fan is used up in the connected duct system by duct friction and individual resistances. With parallel air strands, duct cross-sections and installations must be dimensioned in such a way that the pressure loss is as high as in the main line. If this basic requirement of the calculation is not adhered to, the flow rates adjust in such a way on operation of the system that the requirement of equal pressure losses is fulfilled; duct systems are therefore self-regulating. This automatic modification of the flow rates has the consequence that in inaccurately calculated duct systems the design flow rates do not flow in the individual parallel strands, and the modified total volume stream can lead to power modifications in central humidifiers, air heaters and the like.

4 Solar cooling

To cover the air-conditioning and cooling requirement in buildings, both electrically and thermally driven cooling machines are available. Conventional refrigeration technology is dominated by compression coolers; about 100 million stationary compression systems are built annually world-wide (Reichelt, 2000), plus about 35 million mobile systems. Absorption cooling devices of medium and large power are manufactured in comparatively small numbers, around 10 000 systems per year world-wide, of which 85% are produced in Asia; in Germany today about 100 systems are installed annually. Annual sales of cooling equipment in Europe during 1998 reached about 3 billion Euro, an increase of about 30% since 1996. In Greece for example, cooling machine sales during 2000 reached 117 Million Euro (210 000 compression type units).

Solar technologies in buildings can supply photovoltaically (PV) produced electricity for compression coolers or solar-thermally produced heat for absorption or adsorption coolers. Coupling a photovoltaic generator to a compression cooler does not create special planning demands, since additional energy is always available from the electricity mains. Via the average summer electricity requirement for air-conditioning of an administrative buildings, for example 50 kWh/m^2 under German climatic conditions, the necessary surface area of the photovoltaic generator can be easily estimated; with an annual PV generation of around 120 kWh/m^{2}a in that climate, the cooling power requirement of 2.4 m^2 of office surface can be covered by one square metre of photovoltaics. For 1000 m^2 of effective area, 420 m^2 of PV surface is necessary for air-conditioning alone; in addition there are electricity requirements for lighting and equipment. Due to the very high capital outlays for photovoltaic systems, approximately 750 €/m^2, this version of solar cooling is not common at present. Technologies for thermal cooling with low-temperature heat sources, which can use solar energy and also waste heat, are economically more viable.

The market is dominated by absorption refrigeration technologies with the chemical pairs water–lithium bromide (LiBr) or ammonia - water, which produce cold by a closed cyclic process. In Germany around 1000 absorption refrigeration systems with a total cooling capacity of 1000 MW are installed, of which about half are in industrial companies using cheap waste heat.

The evaporator temperature can be lowered in ammonia coolers to –60°C, so industrial cooling processes are possible. When using water as a refrigerant, the evaporator temperature is limited to temperatures above the freezing point of at least 4–5°C. In absorption refrigeration technology the refrigerant (water or ammonia) is absorbed in a liquid solvent (LiBr or water), desorbed by direct or indirect heating in a generator at high temperatures, and brought to the required condenser pressure. During absorption, solution heat is released, which must be removed via a cooling circuit. The drive temperatures for desorption are between 90 and 140°C, depending on the technology.

If ammonia is used as a refrigerant, a high vapour pressure of 4.85×10^5 Pa develops at evaporator temperatures of +5°C. To liquify the refrigerant at condenser temperatures of 40°C, the ammonia pressure in the generator must be brought to about 15×10^5 Pa, i.e. ammonia absorption coolers must be constructed for high system pressures.

The refrigerant water, on the other hand, evaporates at +5°C with an extremely low vapour pressure of 872 Pa and condenses at +40°C at 7375 Pa, so a LiBr water absorption cooler must be vacuum-operated. While the construction requirements are clearly lower, with the small water vapour pressure extremely high flow rates must circulate, to produce the cooling output, and large cross-sections must be used to reduce pressure losses. In a 100 kW cooler, 145 kg of water must be evaporated per hour, which at the low system pressure of under 1000 Pa corresponds to a flow rate of 21 300 m^3/h. In contrast, 286 kg of ammonia would have to be evaporated for the same cooling output, which at the high evaporator pressure of about 5×10^5 Pa corresponds to a flow rate of only 80 m^3/h.

In adsorption technology the refrigerant water is physically adsorbed to a solid such as silica gel, with release of adsorption and condensation heat. With increasing accumulation of water molecules, the heat of adsorption tends towards zero, so only condensation heat has to be removed. The desorption of the accumulated water and the compression for condensation take place at low drive temperatures of 60–70°C, so this technology is particularly suited to the application of solar energy. Likewise, chilled water of at least 5–6°C is produced by the cyclic process in closed adsorption coolers.

Open adsorption plants use the supply air directly as a cooling carrier. The physical adsorption of water by silica gel or LiCl serves to dry the air in this process. Thereafter, cooling takes place by direct evaporative humidification of the dried air, which has been pre-cooled with room exhaust air via a heat exchanger. The thermal driving energy is necessary for the regeneration of the adsorbent, i.e. for the desorption of the adsorbed water. With open adsorption, the process limits air temperatures to about 16°C minimum, so the area of application is air-conditioning. In this process too, the drive temperatures can be very low (60–70°C).

Coefficients of performance

Single-stage absorption coolers produce about 0.6–0.7 kW of cold per kW of assigned amount of heat (coefficient of performance COP = 0.6–0.7). In water/lithium bromide absorption systems, two-stage generators with a high-temperature section are available on the market for direct natural gas heating and use of condensation heat for a low-temperature generator. In a two-stage process, the coefficient of performance rises to 1.1–1.3. With ammonia–water absorption coolers a two-stage process is not possible due to extremely high system pressures, so the performance figures remain limited to approximately 0.6. The coefficient of performance of closed adsorption coolers depends on the available coolant temperature and can also achieve values between 0.6–0.7. With open sorption-supported air-conditioning, the state of the outside air influences the possible coefficient of performance. With dry outside air the air-conditioning system can be operated purely by evaporative cooling, so no thermal energy is necessary and the COP tends towards infinity. With very damp outside air, the drying potential of the sorption material is insufficient, and conventional cooling must be added. Typical coefficients of performance are between 0.5 and 1.0.

Table 4.1: Outline of solar-thermally powered cooling and air-conditioning processes.

Technology	Absorption cooling with water–lithiumbromid	Absorption cooling with ammonia–water	Closed cycle adsorption water–silica gel	Open cycle desiccant cooling
Refrigerant	H_2O	NH_3	H_2O	-
Sorbent	LiBr	H_2O	silica gel	silica gel
Chilling carrier	water	Water–glykol	water	air
Chilling temperature	6–20°C	–60° up to +20°C	6–20°C	16–20°C
Heating temperature	80–110°C	100–140°C	55–100°C	55–100°C
Cooling water temperature	30–50°C	30–50°C	25–35°C	not applicable
Cooling power range	35–7000 kW	10–10 000 kW	50–430 kW	20–350 kW
COP [–]	0.6–0.75	0.6–0.7	0.3–0.7	0.5–1.0
Approximate investment costs per kW cooling power	550 €/kW [200 kW] 1000 €/kW [50 kW]	800 €/kW [200 kW] 500 €/kW [1000 kW]	500–1000 €/kW	≥1000 €/kW

4.1 Open cycle desiccant cooling

4.1.1 Introduction to the technology

Desiccant cooling systems (DCS) are a mature technology for air-conditioning buildings and are particularly suitable for the application of thermal solar energy, due to the low temperature demands of around 60–80°C.

The technology is based on the principle of outside air dehumidication by an adsorbent such as silica gel or lithium chloride. After pre-cooling, the dried fresh air with maximally humidified room exhaust air, subsequent evaporative cooling produces the desired supply air temperatures of 16–18°C. The desiccant cooling process can be continuously operated with slowly rotating sorption wheels, where the outside air humidity taken up in the adsorbent is transferred to the exit air heated by supply of solar or waste heat.

For process reasons, with open cycle sorption technology no chilling water circuit with the usual supply temperatures of 6–10°C can be obtained. The chilling carrier in open sorption is the moistened air, which is injected directly into the space. Due to the limited dehumidifying performance of the adsorbents used (about 6 g of water per kilogram of dry air) in very damp climate zones desiccant cooling systems must be coupled with compression or absorption coolers, to avoid direct supply air humidification.

Sorption systems are also used purely to dehumidify outside air, with the very energy-intensive fall below the dew point in a compression cooler not occurring. The removal of sensible heat in the space can then take place via surface cooling (usually cooled ceilings).

As an air-led system with cooling load removal only by cooled outside air, the application is particularly suitable if there is a high fresh-air requirement in the building. In winter the sorption system with a sorption wheel and a heat exchanger can be used as a highly efficient heat recovery system, and the thermal solar plant can be used for heating support.

Thermal solar energy or waste heat is used to heat regeneration air. In closed exhaust air discharge, the exhaust air is warmed up after the heat recovery device by the thermal collector, led through the sorption wheel after being heated, and expelled as waste air.

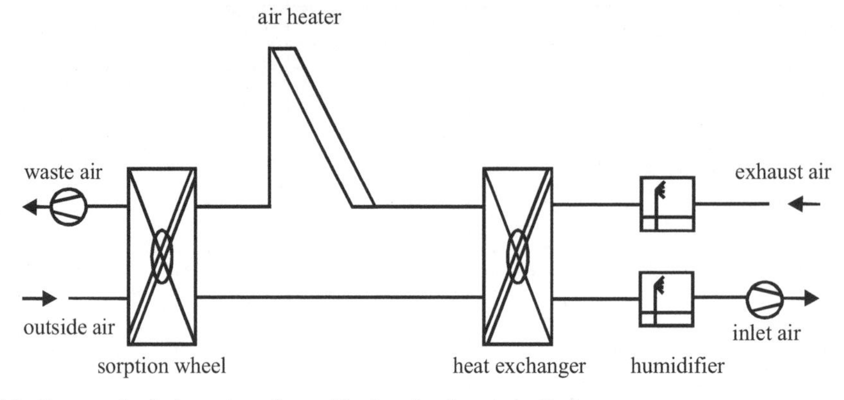

Figure 4.1: Open cycle desiccant cooling with closed exhaust air discharge.

During open exit air discharge the exhaust air is expelled after the heat recovery device as waste air. In the collector, outside air is sucked in, warmed up to regeneration temperature and expelled after the sorption wheel. This version is usually selected for practical reasons, when either the space exhaust air is too strongly contaminated or the air circulation possibilities are spatially limited, since here an air pipe is saved from the cooling machine to the collector. However, the outside temperature, which is the input temperature into the collector, is lower than the temperature after the heat recovery device so that more heating energy is required.

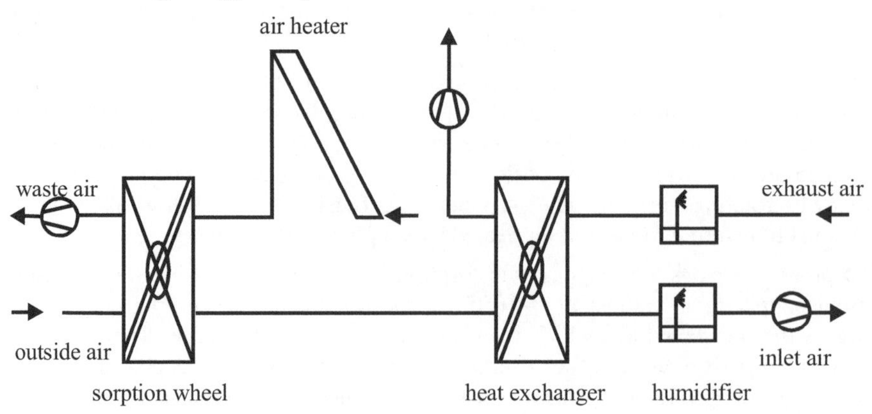

Figure 4.2: Open cycle desiccant cooling with open exhaust air discharge.

The process steps in the sorption wheel, heat recovery device, inlet air and exhaust air humidifier, and also the regeneration air heater can be understood through the enthalpy-humidity or Mollier diagram.

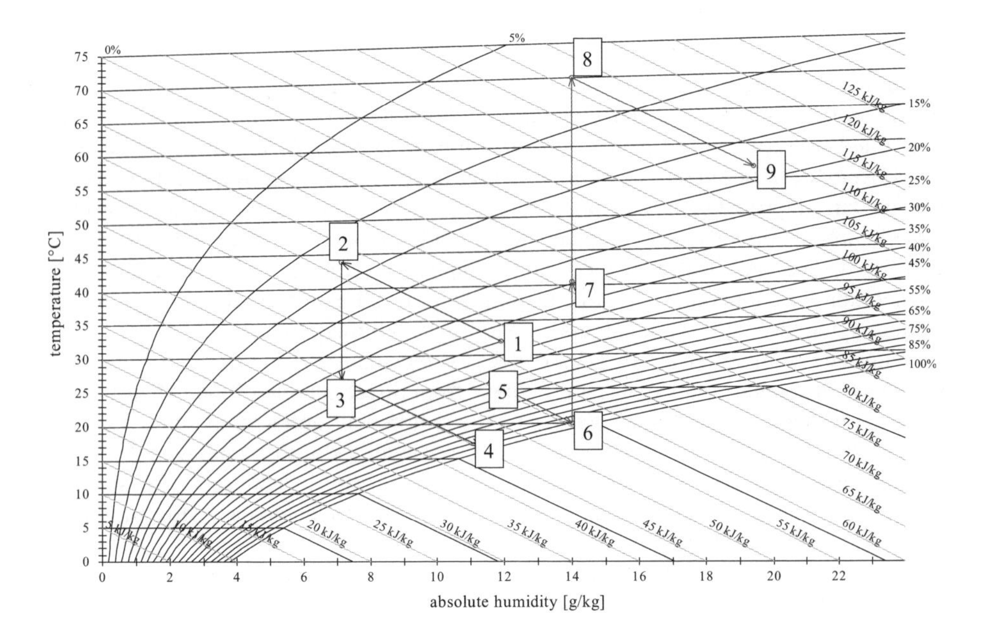

Figure 4.3: Process steps of desiccant cooling systems in the enthalpy–absolute humidity (h–x) diagram.

Outside air (1) is dried in the sorption wheel (2), pre-cooled in the heat recovery device with the additionally humidified cool space exhaust air (3) and afterwards brought to the desired supply air status by evaporative cooling (4). The space exhaust air (5) is maximally humidified by evaporative cooling (6) and warmed in the heat recovery device by the dry supply air (7). In the regeneration air heater the exhaust air is brought to the necessary regeneration temperature (8), takes up in the sorption wheel the water adsorbed on the supply-air side, and is expelled as warm, humid exhaust air (9). Typical temperature and humidity conditions at design condition of 32°C outside temperature and 40% relative humidity are shown in the system scheme:

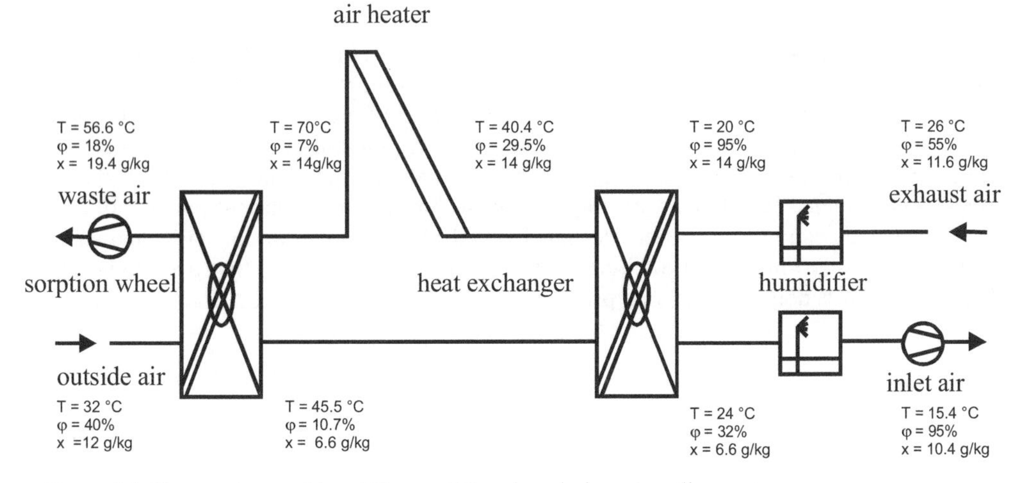

Figure 4.4: Temperature and humidity conditions in a desiccant cooling system.

4.1.2 Coupling with solar thermal collectors

Due to the low regeneration air temperatures for dehumidifying the sorption wheel, solar thermal collectors are particularly suitable for this air-conditioning technique. Both with air-led and water-based flat plate collectors, temperatures of 70°C can easily be achieved. Air collectors can be integrated simply and on large surfaces into the building shell, since leakages are unproblematic and no frost protection is necessary. Solar air collectors are flowed through directly with outside air or with pre-heated exhaust air after the heat recovery device, and heated to the required regeneration temperature. Fluctuations in the regeneration air temperature due to changing solar irradiation are balanced by an auxiliary heating system or by the storage masses of the air-conditioned space.

With the use of water-led thermal collectors, the solar heat is transferred via an air-to-water heat exchanger to the exhaust air. Apart from direct heating of the exhaust air, buffer stores are used here to store the solar heat. At a required air outlet temperature of 70°C (i.e. around 40°C temperature difference from the surroundings in the summer), efficiencies of approximately 60% are achieved with air collectors at full irradiance (1000 W/m^2); with selectively coated absorbers even 70%. Here the air collectors are flowed through at surface-related flow rates of 35–40 m^3/m^2h.

Water-led collectors, which today almost exclusively use selectively coated absorbers, need around 5–10K higher fluid outlet temperatures to achieve the same regeneration temperature of 70°C by means of the air–water heat exchanger. The efficiency of a flat plate collector is comparable at these temperatures to that of an air collector.

From these efficiencies and typical coefficients of performance of a desiccant cooling system of 0.9, the minimum collector area requirement can be estimated. At full irradiance around 600 W/m^2 of usable collector energy and thus 540 W/m^2 of cooling output are produced. A minimum of 1.85 m^2 of collector surface is necessary per kW of cooling output. With a kilowatt of cooling output, some 20 m^2 of office surface can be air-conditioned in a typical administrative building, with 50 W/m^2 cooling load occurring.

To produce the necessary regeneration heat, combinations of back-ventilated photovoltaic facades and solar air collectors can also be used. Such energy facades supply about 80 kWh/m^2 of electricity annually, apart from thermal energy.

4.1.3 Costs

In general, investment in the cooling machine of an air-conditioning system is only about 30% of the total cost. In cost comparisons between compression coolers and sorption-supported air-conditioning systems, therefore, the pure device costs may not be compared, since in the air-led sorption system dehumidifying, humidifying and heat recovery functions as well as supply and exhaust air fans are already integrated. The DCS device costs can be divided into thirds, for components, machine construction costs and control.

In a cost allocation for a system designed with 100 m^2 of solar air collectors and an air capacity of 18 000 m^3/h with total costs of 185 000€, it can be seen that even today the DCS device costs dominate the total price.

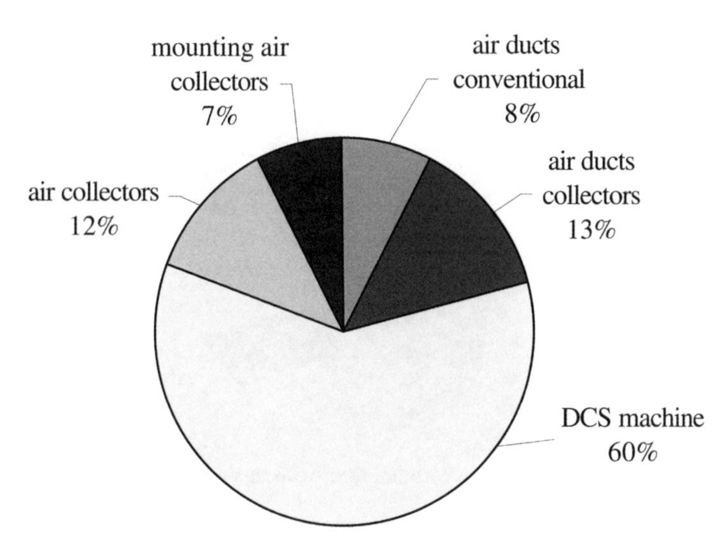

Figure 4.5: Cost allocation of a solar powered desiccant cooling air-conditioning system (18 000 m^3/h) for a production hall in Althengstett, Germany (Eicker *et al.*, 2002).

4.1.4 Physical and technological bases of sorption-supported air-conditioning

4.1.4.1 Technology of sorption wheels

Continuously operating adsorption air dehumidifiers are slowly rotating, hygroscopic storage masses which are flowed through on one side by outside air and on the other by heated regeneration air. As solid sorption materials, silica gel and hygroscopic salts such as lithium chloride are commonly used. They are applied in continuously operating systems to a rotating substrate or used as fixed bed systems for intermittent operation. As a substrate for silica gel rotors, glass or ceramic fibres are used, for LiCl a cellulose matrix. Typical dehumidifying performances at regeneration temperatures of 70°C are around 4–6 g/kg of dry air (Heinrich, 1997; Eicker *et al.*, 2002).

To obtain optimal dehumidifying performance, the number of revolutions of the sorption wheel must be adapted to the regeneration air temperature and to the respective humidity conditions. Too high regeneration temperatures warm up the rotary sorption wheel after desorption so strongly that the sorption material at first can hardly take up moisture on the supply–air side, but has to be cooled down first (so-called heat inhibition).

The separation between regeneration air and outside air is ensured by sealing strips, which leave a minimal gap between the seal and the sorption rotor. Maximum leakage rates between process and regeneration air should be under 3%. Such values have been measured even for small rotors of 1 m diameter. Pressure drops should not be above 150 Pa. By diverting a small section of dried outside air into the regeneration air flow, the typically 3–5% of exhaust air remaining in the storage mass is reduced by flushing to 0.5%, and in addition heat inhibition is reduced by cooling of the storage mass.

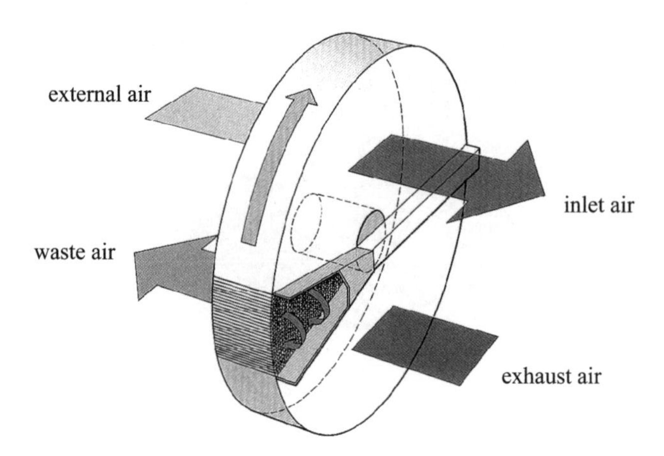

Figure 4.6: Schematic representation of a sorption rotor with flushing zone.

For winter operation the sorption wheel is adjusted from dehumidifying operation with 10–15 revolutions per hour to enthalpy recuperation with revolutions of 10 min^{-1}.

4.1.4.2 Air-status calculations

When calculating the sorption process, the temperature and humidity conditions have to be determined for each process step. For this, the functional relations between absolute (x) and relative (φ) humidity, temperature T and total pressure p have to be determined: $x(\varphi,T,p)$ or $\varphi(x,T,p)$. For the energy balances, the enthalpies $h(x,T)$, i.e. the amounts of heat at constant pressure p, are balanced. The water vapour content of the air describes the mass ratio of the water vapour m_w to the dry air m_a.

$$x = \frac{m_w}{m_a} \quad \left[\frac{kg}{kg}\right] \tag{4.1}$$

To calculate the adsorption process, a new temperature and relative humidity must be calculated for each value of the absolute humidity and enthalpy, the relative humidity φ being defined as the ratio of the water vapour partial pressure p_w [Pa] to the temperature-dependent saturation vapour pressure $p_s(T)$ [Pa].

$$\varphi = \frac{p_w}{p_s(T)} \tag{4.2}$$

If water vapour with pressure p_w and dry air with pressure p_a are regarded as ideal gases, following Dalton's law the partial pressures add up to the total pressure p.

$$p = p_a + p_w \tag{4.3}$$

By the ideal gas equations, the water vapour content of the air can then be calculated as a function of the total pressure as well as of the water vapour partial pressure.

$$p_w V = \frac{m_w}{M_w} RT$$
$$p_a V = (p - p_w) V = \frac{m_a}{M_a} RT \tag{4.4}$$

From this, results can be found for the absolute water content of the air with the mol mass of water vapour $M_w = 18 \times 10^{-3}$ kg/mol and the mean mol mass of air $M_a = 28.97 \times 10^{-3}$ kg/mol:

$$x = \frac{m_w}{m_a} = \frac{M_w}{M_a}\frac{p_w}{p - p_w} = 0.622\frac{p_w}{p - p_w} = 0.622\frac{\varphi p_s}{p - \varphi p_s}\left[\frac{kg}{kg}\right] \tag{4.5}$$

Conversely, the relative humidity φ can be calculated as a function of the saturation water vapour pressure, the total pressure and the absolute water vapour content (in kg of water vapour/kg dry air):

$$\varphi = \frac{x}{x + 0.622}\frac{p}{p_s(T)} \tag{4.6}$$

The temperature-dependent saturation vapour pressure of water $p_s(T)$ is calculated by the Clausius–Clapeyron equation, which describes the equilibrium between liquid and vapour.

$$\frac{dp_s}{p_s} = \frac{h_e(T)}{RT^2} dT \tag{4.7}$$

The solution depends on the selected approximation for the temperature dependence of the evaporation enthalpy $h_e(T)$. This describes the enthalpy difference of vapour h_v and liquid h_l.

$$h_e(T) = h_v(T) - h_l(T) \tag{4.8}$$

The decrease in evaporation enthalpy $h_e(T)$ [kJ/kg] with temperature T [°C] results from the following approximation equations (Glück, 1991):

$$h_v(T) = 2501.482 + 1.789736\,T + 8.957546 \times 10^{-4} T^2 - 1.300254 \times 10^{-5} T^3 \tag{4.9}$$

$$h_l(T) = -2.25 \times 10^{-2} + 4.2063437\,T - 6.014696 \times 10^{-4} T^2 + 4.381537 \times 10^{-6} T^3 \tag{4.10}$$

For the temperature range 10° < T < 200°C, the error in this approximation is less than 0.04%. The saturation water vapour pressure is then approximated for the range 0°C < T < 100°C, with an error of under 0.02% (*T* in [C], *p* in [Pa]):

$$p_s(T) = 611\exp\begin{pmatrix} -1.91275\times10^{-4} + 7.258\times10^{-2}T - 2.939\times10^{-4}T^2 \\ +9.841\times10^{-7}T^3 - 1.92\times10^{-9}T^4 \end{pmatrix} \quad (4.11)$$

The density of the humid air ρ depends on the total pressure p [Pa], the water vapour pressure p_w [Pa] and the temperature T [K]:

$$\rho = \frac{m_a + m_w}{V} = 10^{-4}\left(34.8\frac{p}{T} - 13.2\frac{p_w}{T}\right) \left[\frac{kg}{m^3}\right] \quad (4.12)$$

The enthalpy of the air–vapour mixture h [kJ/kg] is composed of the sensible heat of the dry air c_aT and the water vapour xc_wT, and of the evaporation enthalpy of the water $xh_e(T)$. The temperature-dependent evaporation enthalpy based on Equation (4.9) usually clearly dominates the enthalpy total.

$$h = h_a + xh_w = (c_a + xc_w)T + xh_e(T) \quad (4.13)$$

with

heat capacity of water vapour c_w	1.875 kJ/kg K
heat capacity of dry air c_a	1.004 kJ/kg K

4.1.4.3 Dehumidifying potential of sorption materials

The water vapour uptake and release by the sorption material are described by the corresponding sorption isothermal. At low temperatures, hygroscopic sorption materials take up high water-vapour quantities even at low water-vapour partial pressure of air flowing through. The highest dehumidifying rates at low outside humidity are achieved with zeolites, which require regeneration temperatures of 150–200°C and are thus not suitable for solar operation. More crucial for sorptive air-conditioning is the water vapour uptake at medium to high humidities of the outside air, at which silica gel can take up 0.2–0.3 kg per kilogram of water vapour (maximum loading is around 0.5 kg/kg). The water vapour uptake of cellulose material soaked in lithium chloride depends strongly on the degree of soaking and can achieve higher values than silica gel.

At high temperatures, water vapour receptiveness decreases, and regeneration of the sorption material takes place.

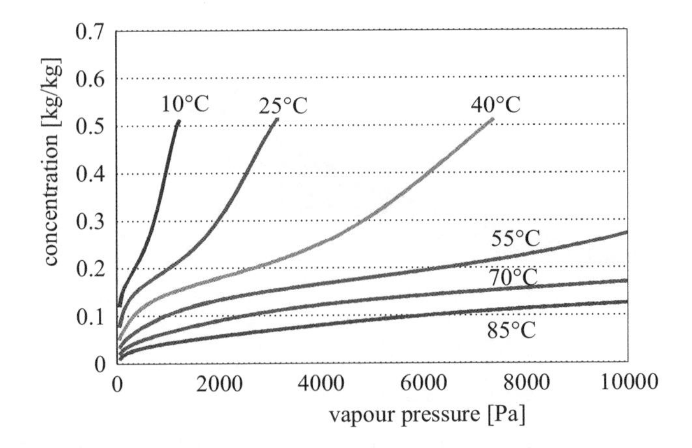

Figure 4.7: (a) Measured concentration of water adsorbed on silica gel (sorption isothermal) as a function of water vapour pressure within air (Henning, 2001).

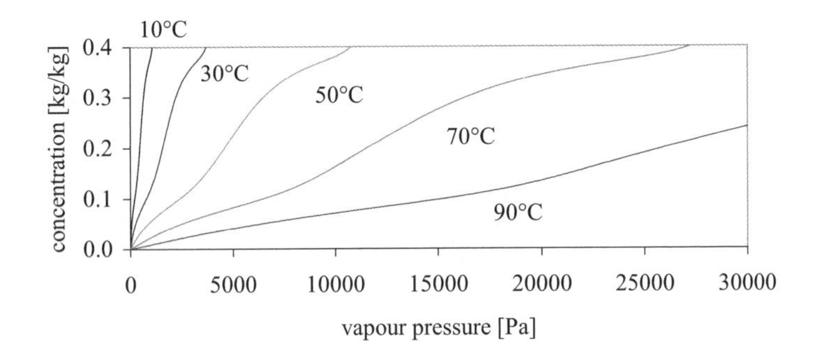

Figure 4.7: (b) Calculated sorption isothermals of silica gel as a function of water vapour pressure.

If the load is represented as a function of relative humidity, i.e. a coordinate transformation of the *x*-axis is carried out, the sorption isotherms almost coincide. High water vapour partial pressure at high temperatures corresponds to a low relative humidity and the load is small. Low partial pressure at a low temperature corresponds to a high relative humidity and the associated high load value shifts to the right.

Even at a very low relative humidity of approximately 5%, the minimum load is about 0.1 kg of water per kilogram of silica gel. This minimum load value is determined by the relative humidity of the regeneration air, and the outside air to be dried cannot fall below it.

Assuming that in a simple dehumidifying model the sorption isotherms coincide, represented as a function of the relative humidity, the outside air humidity at a given loading value can be lowered to a minimum of the regeneration air humidity. An exact knowledge of the functional process of the sorption isotherms is then not necessary.

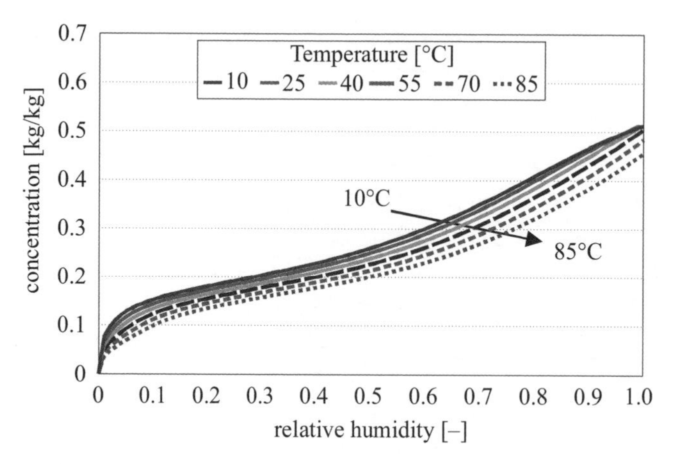

Figure 4.8: Measured sorption isotherms as a function of the relative humidity (Henning, 2001).

For the final state of the dried air in the simple model the relative humidity, i.e. the regeneration humidity, is known–as well as the enthalpy (enthalpy of the outside air with isenthalpic dehumidification). Although both depend on temperature and absolute humidity, the equations cannot be solved analytically and the final state must be iterated. For this the absolute water vapour content of the outside air x_o is reduced until the final state corresponds to the given humidity condition of the regeneration air. As a check, the new temperature and relative humidity values are determined after each dehumidifying step, until supply air humidity and regeneration humidity correspond.

Since in the rotating sorption wheel no stationary status occurs, and the sorption isotherms do not coincide exactly, this minimum value of the supply air humidity with a vapour content x_{dry}^{ideal} is not achieved in practice. Non-ideal dehumidifying at an effective supply air humidity value x_{dry}^{eff} is covered by the dehumidifying efficiency η_{dh}.

$$\eta_{dh} = \frac{x_o - x_{dry}^{eff}}{x_o - x_{dry}^{ideal}}$$

$$x_{dry}^{eff} = x_o - \eta_{dh}\left(x_o - x_{dry}^{ideal}\right) \quad \left[\frac{g}{kg}\right] \tag{4.14}$$

Apart from dehumidifying efficiency, deviations of the sorption wheel parameters from the nominal conditions can also be covered by correction factors. Detailed investigations of correction factors of LiCl sorption rotors have been carried out by Heinrich (1999). The most important influence parameters for dehumidifying behaviour are rotor speed, air speed and the flow rate ratio between the regeneration air and outside air. For rotor speeds under the rated speed of $n = 22\ \text{h}^{-1}$, the dehumidifying performance falls constantly to approximately 80% at $n = 7\ \text{h}^{-1}$ (correction factor c_1). A reduction of the air flow velocity (at the same flow rate ratio) from a nominal rate of v = 2.5 m/s to 1.5 m/s leads to an increase in dehumidifying performance of up to 15% (correction c_2). If only the regeneration air speed is lowered at processing-air nominal speed, the dehumidifying performance likewise rises by up to 20% (with flow rate reduction to 60%) (correction c_3).

Measurements on different sorption technologies (LiCl and silica gel) showed that the optimum rotation speed of the sorption wheel is dependent on the technology. A titanium silicate rotor has its optimal dehumification at higher rotation speeds of 80–100 rotations per hour, whereas the LiCl rotor is best at low rotation rates of about 20 h^{-1}.

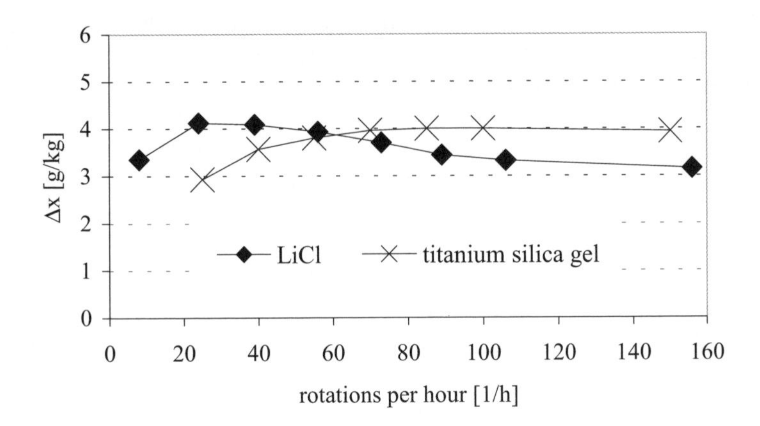

Figure 4.9: Comparison of dehumidification rates as a function of sorption wheel rotation speed (Eicker, 2002).

Taking these corrections into account, the result is the effective supply–air water content x_{dry}^{eff}:

$$x_{dry}^{eff} = x_o - \eta_{dh} c_1 c_2 c_3 \left(x_o - x_{dry}^{ideal} \right) \quad \left[\frac{g}{kg} \right] \tag{4.15}$$

It is generally recommended, therefore, to use sorption rotors with as large cross-sectional areas as possible, so that at a given flow rate the flow speeds remain small and dehumidifying performance becomes high.

4.1.4.4 Calculation of the sorption isotherms and isosteres of silica gel

For a more exact view of the sorption behaviour, first calculate the load concentration C of the sorption material as a function of the relative humidity for a constant temperature: $C|_{T=const} = f(\varphi)$.

In the Brunauer–Emmett–Teller theory the water vapour receptiveness of silica gel is described by accumulation of water vapour in multi-molecular layers. For the 40°C isotherm, the following approximation for the load concentration C_0 is given by Kast (1988):

$$\frac{C_0}{C_m} = \frac{\varphi}{1-\varphi} + \frac{2(b-1)\varphi + 2(b-1)^2\varphi^2 + \left(Nb^2 + Nh - N^2b^2\right)\varphi^N \ldots}{2(1 + 2(b-1)\varphi + (b-1)^2\varphi^2 + \left(b^2 + h - 2b - Nb^2\right)\varphi^N \ldots.}$$
$$\frac{\ldots + \left(2b + N^2b^2 + 2Nb - 2b^2 - Nb^2 - 2h - 2Nh\right)\varphi^{N+1} + \left(Nh + 2h\right)\varphi^{N+2}}{\ldots + \left(Nb^2 + 2b - 2b^2 - 2h\right)\varphi^{N+1} + h\varphi^{N+2}))} \tag{4.16}$$

with the coefficients $C_m = 0.11$, $b = 11$, $N = 7.2$, $h = 19000$ for the combination silica gel–water. The equation is valid up to a maximum load of $C_{max} = 0.4$ kg/kg.

The conversion of the sorption isotherms to other temperatures is made by analysing the vapour pressure over the sorbens at a given load concentration C. If the adsorption process is regarded as a phase transition, the vapour pressure above the sorption material and from this the relative humidity can be calculated from the Clausius–Clapeyron equation, which normally describes the equilibrium between vapour pressure and fluid in a one-phase system as a function of the temperature.

The adsorption heat h_{ads} as a total of binding enthalpy (h_b) and evaporation enthalpy (h_e) corresponds here to the enthalpy difference between vapour and fluid in the one-phase system. The binding enthalpy decreases with rising load concentration C; the, evaporation heat of the water falls with rising temperature.

$$h_{ads}(C,T) = h_b(C) + h_e(T) \tag{4.17}$$

The adsorption heat and the binding heat as a difference of evaporation heat and adsorption heat can be approximated, following Otten (1989), by quadratic functions of the load concentration. Up to a load concentration of $C = 0.1934$ kg/kg, these equations are valid:

$$\begin{aligned} h_{ads} &= h_b + h_e = h_o\left(1 + aC + bC^2\right) \\ h_b &= h_{ads} - h_e = h_o\left(1 + aC + bC^2\right) - h_e \end{aligned} \tag{4.18}$$

with the constants $a = -2.34$, $b = 6.05$ and $h_o = 3172$ kJ/kg for silica gel. The evaporation enthalpy h_e of water is selected for the calculation of the binding enthalpy as a constant with 2453 kJ/kg (20°C). Above a load concentration of $C = 0.1934$ kg/kg, the binding enthalpy is set at $h_b = 0$ and the adsorption enthalpy is equal to the evaporation enthalpy h_e. At $C = 0$, i.e. perfectly dry material, the binding enthalpy is 719 kJ/kg.

Other authors (Henning, 1994; Gassel, 1998) have produced approximation formulae with clearly higher binding enthalpies, which at zero loading lie between 1000 and 2000 kJ/kg. According to Gassel the binding enthalpy can be calculated with the simple linear equation $h_b = 1000\frac{kJ}{kg} - C \times 4000\frac{kJ}{kg}$.

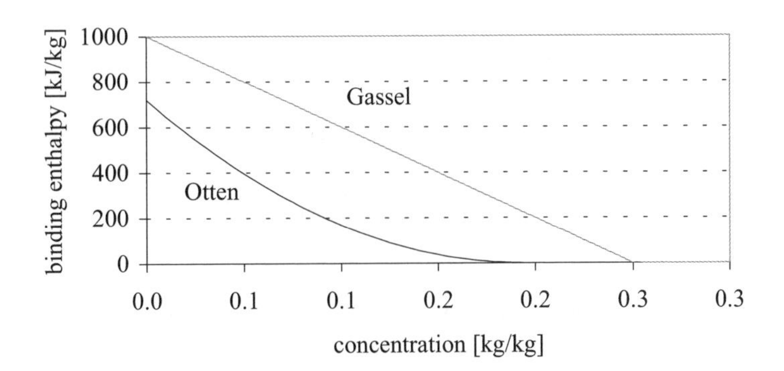

Figure 4.10: Binding enthalpy of water vapour to silica gel as a linear or quadratic function of the load concentration C.

Only at high load concentrations and binding enthalpy of zero is the change in status of air isenthalpic. At a low concentration, heating of the outside air is greater, due to the dehumidifying process through the uptake of the binding heat, than in a purely isenthalpic change in status.

For water vapour partial pressure p_w over the sorbens, the following equation is valid, based on Clausius–Clapeyron:

$$\frac{dp_w}{dT} = \frac{h_{ads}(C,T)}{R_w T^2} p_w \Leftrightarrow \frac{d \ln p_w}{d(1/T)} = \frac{-h_{ads}(C,T)}{R_w} \tag{4.19}$$

with $R_w = 461$ J/kgK as the specific gas constant of water vapour.

If the vapour pressure p_w is plotted logarithmically against the inverse temperature $1/T$, then the heat of adsorption for a given load concentration results from the gradient. The load-dependent vapour pressure curves are called isosteres (C = const). The smaller the load concentration of the silica gel, the lower is the vapour pressure over the sorbens at a given temperature. For the completely saturated sorbens the binding enthalpy is zero and the well-known Clausius equation for the saturation vapour pressure of water is obtained as a function of the temperature.

$$\frac{dp_s}{dT} = \frac{h_e}{R_w T^2} p_s \tag{4.20}$$

The integration of the Clausius equation over a temperature range T_1 to T_2 enables the calculation of the isosteres, if the pressure p_1 at temperature T_1 is known. To simplify the integral, the adsorption enthalpy is assumed to be temperature-independent.

$$\int_{p_{w1}}^{p_{w2}} \frac{dp_w}{p_w} = \int_{T_1}^{T_2} \frac{h_{ads}(C)}{R_w T^2} dT$$
$$\ln\left(\frac{p_{w2}}{p_{w1}}\right) = \frac{h_{ads}(C)}{R_w}\left(\frac{1}{T_1} - \frac{1}{T_2}\right) \tag{4.21}$$
$$p_{w2} = p_{w1} \exp\left(\frac{h_{ads}(C)}{R_w}\left(\frac{1}{T_1} - \frac{1}{T_2}\right)\right)$$

From the well-known 40°C isotherms, the vapour pressure p_{w1} can be determined as the still unknown parameter of the isosteres equation for each load concentration. For a given load concentration C_0, the relative humidity φ_0 of the 40°C isotherms (i.e. T_1 = 313 K) is determined iteratively from equation (4.16). From the relative humidity, the vapour pressure p_{w1} can then be calculated to $p_{w1} = \varphi_0 \times p_s(40°\text{C})$. For the load concentration $C = C_0$, the constants of the vapour pressure equation p_{w1} and T_1 are then known.

Example 4.1

Calculation of the isostere parameters of water vapour over silica gel at a load concentration of C = 0.1 kg/kg.

For C = 0.1 kg/kg the adsorption heat is h_{ads} = 2622 kJ/kg. The relative humidity resulting in C_0 = 0.1 kg/kg at 40°C is 18.75%.

From it the vapour pressure $p_{w1} = \varphi_0 \times p_s(40°\text{C}) = 0.1875 \times 7384 Pa = 1384.5 Pa$ results, so the complete isostere can be calculated with the following equation:

$$p_w = 1384.5 \exp\left(\frac{2622}{0.461}\left(\frac{1}{313} - \frac{1}{T}\right)\right).$$

The two Clausius equations for pure water and water–silica gel are divided by each other at the same temperature T, so that from the relation of the vapour pressure p_w to the saturation vapour pressure p_s, the relative humidity and finally the functional connection of relative humidity and load concentration can be determined.

$$\frac{dp_s}{p_s} = \frac{h_e}{h_{ads}} \frac{dp_w}{p_w} \tag{4.22}$$

Equation (4.22) is integrated

$$\int_{p_{s,0}}^{p_s} \frac{dp_s}{p_s} = \int_{p_{w,0}}^{p_w} \frac{h_e}{h_{ads}} \frac{dp_w}{p_w}$$
$$\frac{p_s}{p_{s,0}} = \left(\frac{p_w}{p_{w,0}}\right)^{\frac{h_e}{h_{ads}}} = \left(\frac{\varphi p_s}{p_{w,0}}\right)^{\frac{h_e}{h_{ads}}} \tag{4.23}$$

and solved for the relative humidity φ:

$$\varphi = \frac{p_{w,0}}{p_{s,0}^{\frac{h_{ads}}{h_e}}} \frac{p_s^{\frac{h_b+h_e}{h_e}}}{p_s} = const\ p_s^{\frac{h_b(C)}{h_e(T)}} \quad (4.24)$$

This equation provides the sought connection of the sorption isotherms between relative humidity, temperature (via the saturation vapour pressure p_s) and load concentration C (via the binding enthalpy h_b).

With the same loading, i.e. the same binding enthalpy h_b, conversion is thus possible from a single known sorption isotherm (with temperature T_o) to other temperatures. For simplification the evaporation heat is set temperature-independently.

$$\frac{\varphi_o}{\varphi} = \frac{p_s(T_o)^{\frac{h_b(C_o)}{h_e(T_o)}}}{p_s(T)^{\frac{h_b(C_o)}{h_e(T)}}} \approx \left(\frac{p_s(T_o)}{p_s(T)}\right)^{\frac{h_b(C_o)}{h_e}} \quad (4.25)$$

For each relative humidity φ_0 of the known 40°C sorption isotherm $C_0(\varphi_0)$ from Equation (4.16), the associated relative humidity φ at temperature T can be determined. This corresponds to the load concentration $C = C_0$, and the value pair (C, φ) is obtained.

$$\varphi = \varphi_0 \times \left(\frac{p_s(T)}{p_s(T_0)}\right)^{\frac{h_b(C_o(\varphi_o))}{h_e}} \quad (4.26)$$

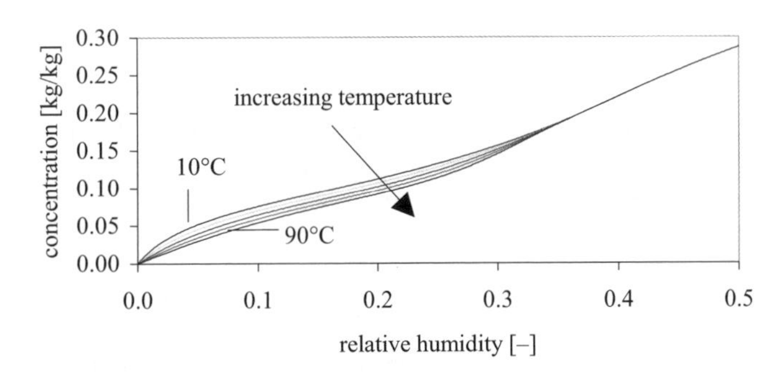

Figure 4.11: Sorption isotherms as a function of the relative humidity for temperatures between 10°C and 90°C, calculated from the known 40°C isotherm.

As soon as the binding enthalpy becomes zero, i.e. for load concentrations greater than 0.1934 kg/kg, all isotherms coincide exactly.

In practice the problem is mostly that for a given humidity φ, the associated load concentration C must be determined. Thus for example the regeneration humidity φ_{reg} gives the minimum load concentration to which the outside air can finally be dehumidified. To determine this concentration, Equation (4.26) must now be solved for the relative humidity φ_0, for which the load concentration C_0 is known from the 40°C isotherm.

$$\varphi_o = \varphi \times \left(\frac{p_s(T_o)}{p_s(T)} \right)^{\frac{h_b(C_o(\varphi_o))}{h_e}} \tag{4.27}$$

Since the load concentration C_0 depends on φ_0, an implicit equation is obtained, which must be solved iteratively. After the end of the iteration the load concentration $C = C_0$ is known, which results from the given relative humidity φ and temperature T.

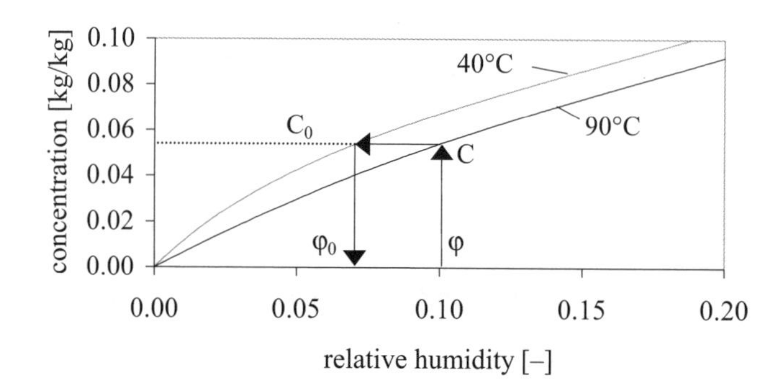

Figure 4.12: Determination of the load concentration C for a given humidity φ over the known sorption isotherm at 40°C, by determining the relative humidity φ_0 at the same load $C_0 = C$.

Thus with the extended model the relative humidity φ_{reg} of the regeneration air is given for the determination of the dehumidifying performance of a sorption rotor. Via the iterative method from Equation (4.27) the load concentration C_{reg} of the regenerated sorption material is determined. If outside air is now gradually dehumidified, the associated load concentration is determined for each new temperature and relative humidity φ_{dry} of the dried outside air after the dehumidifying step, until C_{dry} is equal to C_{reg}.

4.1.4.5 Calculation of the dehumidifying performance of a sorption rotor

The climatic boundary conditions chosen for an example system calculation correspond to German standards and are 32°C outside air temperature, 40% relative humidity or 12 g/kg of water vapour, at a total pressure of 1.013×10^5 Pa. In the following the air statuses of the dried outside air are to be calculated, when regeneration air is at 70°C and contains 14 g/kg of water vapour. First the relative humidity φ of the regeneration air is calculated at 70°C using Equation (4.6) with the saturation vapour pressure p_s from Equation (4.11), which determines the minimum relative humidity of the supply air to be dehumidified.

$$\varphi = \frac{14\times10^{-3}\,\frac{\text{kg}}{\text{kg}}}{14\times10^{-3}\,\frac{\text{kg}}{\text{kg}}+0.622\frac{\text{kg}}{\text{kg}}}\times\frac{101300Pa}{31158Pa}=0.07$$

The enthalpy of the regeneration air based on Equation (4.13) with the evaporation enthalpy based on equation (4.8) is given by:

$$h_{\text{reg}} = \left(1.004\frac{\text{kJ}}{\text{kgK}}+14\times10^{-3}\,\frac{\text{kg}}{\text{kg}}\times1.875\frac{\text{kJ}}{\text{kgK}}\right)70°\text{C}+14\times10^{-3}\,\frac{\text{kg}}{\text{kg}}\times2334\frac{\text{kJ}}{\text{kg}}=104.8\,\frac{\text{kJ}}{\text{kg}}$$

The value is not required for the calculation of the supply air status.

The enthalpy of the outside air determines the temperature of the air after dehumidifying, when the process is isenthalpic.

$$h_o = \left(1.004\,\frac{\text{kJ}}{\text{kgK}}+12\times10^{-3}\,\frac{\text{kg}}{\text{kg}}\times1.875\frac{\text{kJ}}{\text{kgK}}\right)32°\text{C}+12\times10^{-3}\,\frac{\text{kg}}{\text{kg}}\times2425\frac{\text{kJ}}{\text{kg}}=62\frac{\text{kJ}}{\text{kg}}$$

The iterative calculation of the outside air drying process is solved with fast convergence with the Regula falsi algorithm. In this, the initial water content $x_{o,0}$ of the outside air (relative humidity $\varphi_{o,0}$) in the first iteration step is reduced by say 50%, and the temperature T_1 and new relative humidity φ_1 are calculated via the given constant enthalpy at the new water content x_1. At the above outside air status of 32°C, 12 g/kg of vapour content, dehumidification is 6 g/kg in the first step. The temperature of the dried outside air is calculated using equation (4.13) at 46.7°C. Via the saturation vapour pressure of p_s = 10445 Pa, based on equation (4.11), the relative humidity φ_1 at 6 g/kg vapour content based on Equation (4.6) is obtained as 9.3% . The evaporation enthalpy is kept constant during iteration. The relative humidity φ_1 is above the regeneration humidity of 7%, so further dehumidification is necessary.

To determine the next water content x_2, the ratio of the humidity difference $\varphi_{o,0}-\varphi_1$ to the water-content difference $x_{o,0}-x_1$ of the first dehumidifying step (i.e. the gradient of the humidity function) is equated to the ratio of the humidity difference between the outside air and regeneration air $\varphi_{o,0}-\varphi_{reg}$ to the water-content difference $x_{o,0}-x_2$. The new water content x_2 is thus equated approximately to the absolute humidity of the final state.

$$\frac{\varphi_{o,0}-\varphi_1}{x_{o,0}-x_1}=\frac{\varphi_{o,0}-\varphi_{reg}}{x_{o,0}-x_2} \tag{4.28}$$

This results in the new water content of the next iteration x_2.

$$x_2 = x_{o,0} - \left(\varphi_{o,0} - \varphi_{reg}\right)\frac{x_{o,0} - x_1}{\varphi_{o,0} - \varphi_1}$$

$$x_2 = 12\times10^{-3}\frac{\text{kg}}{\text{kg}} - \left(0.4 - 0.07\right)\frac{12\times10^{-3}\frac{\text{kg}}{\text{kg}} - 6\times10^{-3}\frac{\text{kg}}{\text{kg}}}{0.4 - 0.093} = 5.55\times10^{-3}\frac{\text{kg}}{\text{kg}} \tag{4.29}$$

The following is generally valid:

$$x_n = x_{n-1} - \left(\varphi_{n-1} - \varphi_{reg}\right)\frac{x_{n-2} - x_{n-1}}{\varphi_{n-2} - \varphi_{n-1}} \tag{4.30}$$

For the new water content $x_2 = 5.55$ g/kg, the new relative humidity $\varphi_2 = 0.081$ is obtained via the new temperature $T_2 = 47.8°C$; it is still over the limiting regeneration humidity φ_{reg}. As the next water content, x_3 is obtained:

$$x_3 = 5.55\times10^{-3}\frac{\text{kg}}{\text{kg}} - \left(0.081 - 0.07\right)\frac{6\times10^{-3}\frac{\text{kg}}{\text{kg}} - 5.55\times10^{-3}\frac{\text{kg}}{\text{kg}}}{0.093 - 0.081} = 5.14\times10^{-3}\frac{\text{kg}}{\text{kg}}$$

The water content x_3 of 5.14 g/kg corresponds to a relative humidity of 7.1% at an outlet temperature of 48.8°C, i.e. in the second iteration the humidity of the regeneration air state is almost achieved.

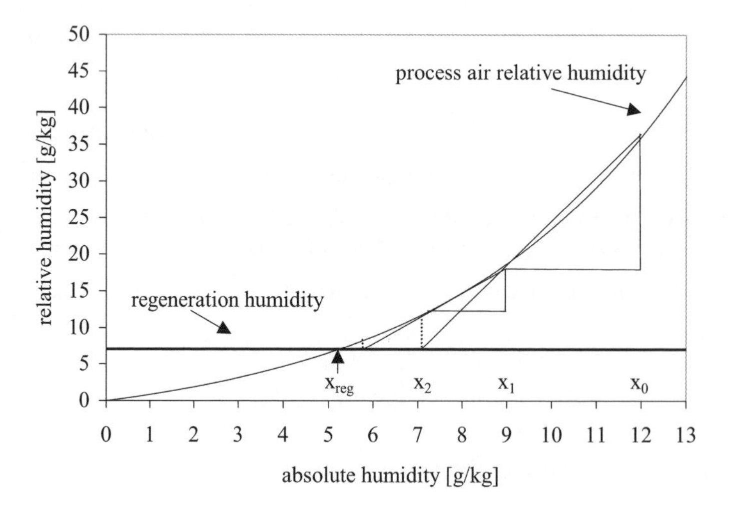

Figure 4.13: Relative humidity as a function of absolute humidity with constant enthalpy of the outside air of 62 kJ/kg for the calculation of dehumidifying.

The value of the regeneration relative humidity is fixed at 7% and gives the final state of the outside air to be dried. For better illustration of the Regula Falsi algorithm, in the first dehumidifying step only 3 g/kg has been dehumidified. Nevertheless in the second iteration the final humidity is almost achieved. The final value of the iteration is x_∞ = 5.1g/kg, i.e. of the initial value x_o = 12 g/kg, the outside air has been dehumidified by 6.9 g/kg. The non-ideal dehumidifying is now considered by means of the dehumidification efficiency, which is set at 80%. Using Equation (4.14) the residual moisture is then 6.5 g/kg and the associated temperature of the dried outside air is 45.5°C. The associated relative humidity is 10.6%.

The status of the regeneration air after water vapour take-up in the sorption wheel can be calculated similarly. The total humidity taken up by the silica gel, ideally 6.9 g/kg (non-ideally 5.5 g/kg), is now given off with isenthalpic humidification of the regeneration air. The absolute humidity rises from 14 g/kg to 20.9 g/kg (or 19.5 g/kg).

Temperature of the damp exhaust air:	53.7°C	(56.9°C)
Relative humidity:	22%	(17.9%)

4.1.5 The technology of heat recovery

In sorption-supported air-conditioning systems, predominantly regenerative rotary heat recovery devices are used. These achieve, depending on the incident-flow velocity, good heat recovery efficiencies between 70 and 90%. As a heat accumulator, corrugated aluminium plate is often used, wound into heat exchanger wheels of up to 5m diameter. Volume flow rates are between 1000 and 150 000 m^3/h, with pressure losses of 50 to 200 Pa. In partial-load operation the heat recovery efficiency rises due to the reduced incident-flow velocities, i.e. slightly oversized heat exchangers lead to better system efficiencies. The rotation rate of the heat recovery wheels is between 5 and 15 revolutions per minute, though an increase in the storage mass stream leads to an improvement of the heat recovery efficiency.

The hygroscopic sorption wheels are likewise used in winter for heat recovery and achieve, with a corresponding increase in the figure of revolutions, similar heat recovery efficiencies to the non-hygroscopically coated storage masses. At typical air incident-flow velocities of 3m/s and a supply-air to exhaust-air ratio of 1, heat recovery efficiencies between 75% and 85% are achieved, depending on the depth of the sorption rotor. Since the model for the rotary regenerative heat-transfer agent is based on a cross-flow heat exchanger, the calculation is executed first for recuperators (partitioned wall heat exchangers).

4.1.5.1 Recuperators

In a recuperator, heat is transferred from the warmer fluid (with mass flow $\dot{m}_1$ and thermal capacity c_1) convectively, with a heat transmission coefficient of h_{c1}, to the partition and after thermal conduction by the partition material convectively with h_{c2} to the colder fluid (with mass flow $\dot{m}_2$ and thermal capacity c_2).

For a high transmission rate of a heat exchanger, as high a heat transition coefficient U as possible is necessary. The heat transition coefficient is dominated by the convective transition resistances. The thermal resistance $R = s / \lambda$ of the plate material (with plate thickness s [m] and heat conductivity of the plate λ [W/mK]) is normally negligible. The heat transfer coefficient of a recuperator with an even partition surface is given by

$$U = \left(\frac{1}{h_{c1}} + R + \frac{1}{h_{c2}} \right)^{-1} \tag{4.31}$$

The proportion of the radiation to the heat transfer is negligible due to the virtually identical temperatures of the individual partition surfaces.

The heat transfer coefficient by convection h_c is determined in practice by model tests. These test results can then be transferred to other geometrically and hydrodynamically similar heat transfer conditions.

$$h_c = \frac{\mathrm{Nu}(\mathrm{Re}, \mathrm{Pr})\lambda}{L} \tag{4.32}$$

Nu: Nußelt number [–]
λ: heat conductivity of the fluid [W/mK]

The characteristic length L depends on the respective geometry of the heat exchanger. The most important Nußelt correlations as a function of the Reynolds and Prandtl numbers (Re, Pr), of the overall length of the heat-transferring gap/tube/channel l, and of the characteristic length L, are summarised in Table 4.2. The calculation of the material properties of air as well as of the Reynolds and Prandtl numbers can be found in Chapter 3.2 (solar air collectors).

Table 4.2: Relevant Nußelt correlations for heat exchanger calculations.

Geometry	*Flow condition*	*Nußelt correlation*	*Characteristic length*	*Reference*
gap/ channel	laminar (Re < 2300)	$Nu = \left[7{,}541^3 + 1{,}841^3 \sqrt{\mathrm{Re}\,\mathrm{Pr}\frac{L}{l}} + \left(\frac{2}{1+22 \cdot \mathrm{Pr}} \right)^{1/6} \left(\mathrm{Re}\,\mathrm{Pr}\frac{L}{l} \right)^{0,5} \right]^{1/3}$	$L = 2h$	Al Amouri, 1994
	turbulent (Re > 8000)	$Nu = 0{,}116 \cdot (\mathrm{Re}^{2/3} - 125) \cdot \mathrm{Pr}^{1/3} \left[1 + \left(\frac{L}{l} \right)^{2/3} \right]$	$L = \frac{4 \cdot A_{free}}{C}$	Gregorig, 1959
tube	laminar (Re < 2320)	$Nu = \left(49{,}0 + 4{,}17 \cdot \mathrm{Re} \cdot \mathrm{Pr} \cdot \frac{L}{l} \right)^{1/3}$	$L = d_i$	Hering *et al.*, 1997
	turbulent (Re > 2320)	$Nu = 0{,}116 \cdot \left(\mathrm{Re}^{2/3} - 125 \right) \cdot \mathrm{Pr}^{1/3} \cdot \left[1 + \left(\frac{L}{l} \right)^{2/3} \right]$	$L = d_i$	Gregorig, 1959

h: distance between heat-transferring surfaces (gap/channel)
A_{free}: free area for flow
C: circumference
d_i: inside diameter of the pipe

The amount of heat $\dot{Q}$ given off by the warmer fluid is taken up completely by the colder fluid flow, ignoring heat losses to the external environment, and is calculated from the heat transfer rate of the heat exchanger. This results from the product of the two-dimensional elements dA, the heat transfer coefficient U, and the locally varying temperature difference $T_1 - T_2$ between the two fluid flows.

$$\dot{Q} = \int U \times \left(T_1(x,y) - T_2(x,y)\right) dA = \dot{m}_1 c_1 \left(T_{1,in} - T_{1,out}\right) = \dot{m}_2 c_2 \left(T_{2,out} - T_{2,in}\right) \quad (4.33)$$

Normally only the inlet temperatures of the warm and cold fluids into the heat exchanger are known, for example in the sorption system the temperature of the warm dried supply air ($T_{1,in}$) and the temperature of the colder space exhaust air ($T_{2,in}$). To calculate the transferred heat as a function of the inlet temperatures, the heat recovery efficiency η_{hx} is introduced, which is also known as an operational characteristic and is defined as the ratio of actually transferred power to maximum transferred power.

$$\begin{aligned} \dot{Q} &= \dot{m}_1 c_1 \left(T_{1,in} - T_{2,in}\right)\Phi = \dot{m}_1 c_1 \left(T_{1,in} - T_{1,out}\right) = \dot{m}_2 c_2 \left(T_{2,out} - T_{2,in}\right) \\ \Phi &= \frac{\dot{m}_1 c_1 \left(T_{1,in} - T_{1,out}\right)}{\dot{m}_1 c_1 \left(T_{1,in} - T_{2,in}\right)} = \frac{\dot{m}_2 c_2 \left(T_{2,out} - T_{2,in}\right)}{\dot{m}_1 c_1 \left(T_{1,in} - T_{2,in}\right)} \end{aligned} \quad (4.34)$$

Only at identical thermal capacity streams of the two sides $\dot{m}_1 c_1 = \dot{m}_2 c_2$ are the two temperature difference ratios (defined by the heat recovery efficiency) the same.

$$\Phi = \frac{\left(T_{1,in} - T_{1,out}\right)}{\left(T_{1,in} - T_{2,in}\right)} = \frac{\left(T_{2,out} - T_{2,in}\right)}{\left(T_{1,in} - T_{2,in}\right)} \quad (4.35)$$

The heat recovery efficiencies of the most important recuperators (same-, counter- and crossflow heat exchangers) depend functionally on the ratio of heat transfer performance UA and thermal capacity stream $\dot{C} = \dot{m}c$, which are called NTU (number of transfer units).

$$NTU = \frac{UA}{\dot{C}} \quad (4.36)$$

The heat recovery efficiency for a counter-current heat exchanger with the thermal capacity stream $\dot{C}_1 < \dot{C}_2$ is given by Bosnjakovic (1951):

$$\Phi = \frac{1 - e^{-\left(1 - \frac{\dot{C}_1}{\dot{C}_2}\right)\frac{UA}{\dot{C}_1}}}{1 - \frac{\dot{C}_1}{\dot{C}_2} e^{-\left(1 - \frac{\dot{C}_1}{\dot{C}_2}\right)\frac{UA}{\dot{C}_1}}} \quad (4.37)$$

At identical mass flows on the warm and cold sides, the heat recovery efficiency is at most 0.8, with a high ratio of transmission rate UA to thermal capacity stream (> 4). The heat recovery efficiency for $\dot{C}_1 = \dot{C}_2$ is

$$\Phi = \frac{\dfrac{UA}{\dot{C}_1}}{1 + \dfrac{UA}{\dot{C}_1}} \tag{4.38}$$

The operational characteristic improves with unequal thermal capacity streams.

Direct current heat exchanger:

$$\Phi = \frac{1 - e^{-\left(1+\frac{\dot{C}_1}{\dot{C}_2}\right)\frac{UA}{\dot{C}_1}}}{1 + \dfrac{\dot{C}_1}{\dot{C}_2}} \tag{4.39}$$

In crossflow heat exchangers the directions of flow of the two fluids run perpendicular to each other. The heat recovery efficiency is obtained by an infinite series which depends on $UA/\dot{C}$.

Pure cross-current plate heat exchanger ($\dot{C}_1 < \dot{C}_2$):

$$\Phi = \frac{1}{\dfrac{UA}{\dot{C}_2}} \sum_{n=0}^{\infty} \left(1 - \exp\left(-\frac{UA}{\dot{C}_1} \right) \sum_{p=0}^{n} \frac{\dfrac{UA}{\dot{C}_1}}{p!} \right) \times \left(1 - \exp\left(-\frac{UA}{\dot{C}_2} \right) \sum_{p=0}^{n} \frac{\dfrac{UA}{\dot{C}_2}}{p!} \right) \tag{4.40}$$

In the infinite series it is sufficient to calculate the terms n = 0 to n = 5.

A pure crossflow heat exchanger is defined by the fact that no lateral mixing of the individual fluid lines is possible, and occurs in practice with heat exchangers whose heat-transferring surface consists of flat or corrugated plates (plate-type heat exchangers). Typical gap widths for a plate-type heat exchanger are between 5–10 mm.

If in a tube heat exchanger the fluid in the pipes is flowed around perpendicularly by another fluid over the whole cross-section, a mixing of the fluid lines of the outside fluid can occur transverse to the direction of flow, and a so-called one-side agitated crossflow heat exchanger is the result. The larger the number of the tubing rows, the stronger is the approximation to the pure cross current.

One-side agitated cross current: shell and tube heat exchanger:
Current $\dot{C}_1$ remains unmixed, current $\dot{C}_2$ is agitated (with $\dot{C}_1 < \dot{C}_2$)

$$\Phi = \frac{1-\exp\left(-\frac{\dot{C}_1}{\dot{C}_2}\left(1-e^{-\frac{UA}{\dot{C}_1}}\right)\right)}{\frac{\dot{C}_1}{\dot{C}_2}} \tag{4.41}$$

Current $\dot{C}_2$ remains unmixed, current $\dot{C}_1$ is agitated:

$$\Phi = 1-\exp\left(-\frac{\dot{C}_2}{\dot{C}_1}\left(1-e^{-\frac{UA}{\dot{C}_2}}\right)\right) \tag{4.42}$$

Example 4.2

Calculation of the heat recovery efficiency of a counter-current plate-type heat exchanger for a sorption system with 20 000 m^3/h of both supply air and exhaust air flow rate.

	Geometry		
H =	1.5	[m]	Height of recuperator
B =	1.5	[m]	Width of recuperator
l =	1.5	[m]	Length of channel
n =	250	[–]	Number of plates
s_{Pl} =	0.0002	[m]	Thickness of individual plates
λ_{Pl} =	229	[W/mK]	Heat conductivity of plate material
d_g =	0.0058	[m]	Gap width (distance between individual plates)
A_c =	1.09	[m²]	Free cross section of recuperator (one direction)
A_{Pl} =	2.25	[m²]	Area of individual plates (length x width)
A_{free} =	0.009	[m²]	Free flow cross section (one channel)
d_h =	0.012	[m]	Hydraulic diameter
L =	0.012	[m]	Characteristic length
A_{hx} =	562.50	[m²]	Heat transferring surface area

	Warm air:	**Cold air:**		
V/t =	5.56	5.56	[m³/s]	Volume flow
T =	45.50	20.00	[°C]	Temperature
λ_{air} =	0.0258	0.0251	[W/m K]	Heat conductivity air
ρ_{air} =	1.0933	1.1884	[kg/m³]	Density air
$c_{p.air}$ =	1008.3	1007.0	[J/kg K]	Heat capacity air
υ_{air} =	1.76E-05	1.52E-05	[m²/s]	kinematic viscosity air
v_g =	5.11	5.11	[m/s]	Mean gap velocity

Pr =	0.751	0.723	[–]	Prandtl number
Re =	3345	3878	[–]	Reynold number
Nu =	10.805	13.176	[–]	Nußelt number
hc =	24.25	28.75	[W/m²K]	Convective heat transfer coefficient
U =	13.16	[W/m² K]		Heat transfer coefficient
C_1 =	6128.87	[W/K]		Smaller heat capacity flow
C_2 =	6653.58	[W/K]		Larger heat capacity flow
ϕ =	**0.56**	[–]		Heat recovery efficiency

$T_{1.in}$ =	45.50°C	$T_{1.out}$ =	31.25°C
$T_{2,out}$ =	33.13°C	$T_{2,in}$ =	20.00°C

4.1.5.2 Regenerative heat exchangers

The degree of heat transfer of a regenerative heat exchanger depends on the incident-flow velocity v_a of the air and on the number of revolutions n of the wheel. As with the recuperators, the heat recovery efficiency increases with the rising ratio $UA/\dot{C}$, i.e. at a given transfer rate with a falling incident-flow velocity.

The concept for a rotary heat exchanger is based on the heat transfer of crossflow heat exchangers. In this, the storage mass is simulated as a plate flowed around on both sides by air and moving perpendicularly to the air flow direction, thus cross-current. After heat uptake in the warm phase, the storage mass of the wheel moves into the cool air section and gives off the heat taken up. With this model, the steady-state temperature gradient is obtained iteratively. The heat recovery efficiency improves, the greater the storage mass stream, i.e. the number of revolutions of the wheel.

For calculating the local temperature gradients in the regenerator and the outlet temperatures of the air, an imaginary partition is introduced between the air flow and the storage mass stream flowing perpendicularly to it; heat transfer takes place through this. The partition is therefore situated in the level of the thin flow channels of the regenerator and is limited by the building depth of the regenerator.

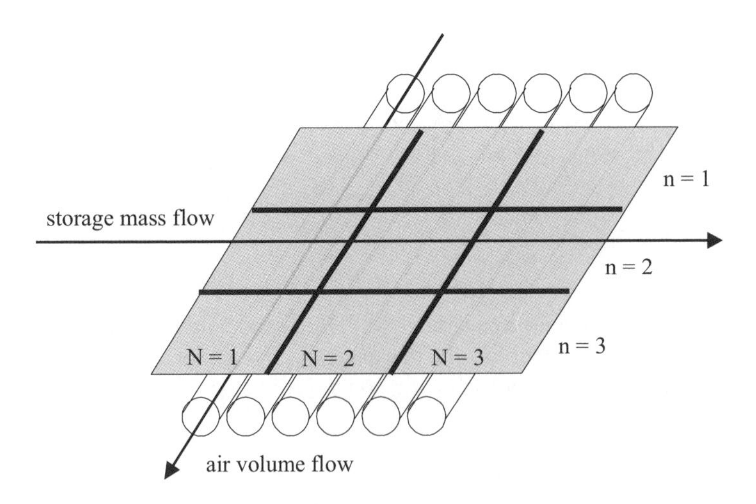

Figure 4.14: Imaginary partition surface between air flow and the storage mass stream, which is given by the mass of the channel walls.

To calculate the temperature distribution on the entry side of the air flow, the partition surface is divided into N sections and subdivided on the entry side of the storage mass stream into n sections. The surface of a partition element results from the total area $A_g/(Nn)$. The heat flow $\dot{Q}$ transferred by a partition element results then from the thermal capacity stream of the air $\dot{C}_a = \rho \dot{V} c_{p,a}$ per segment N and the temperature difference between element entry $T_{a,1}$ and exit $T_{a,2}$, or from the thermal capacity stream of the storage mass $\dot{C}_S$ per segment n and the associated temperature difference $T_{S,2} - T_{S,1}$.

$$\dot{Q} = \frac{\dot{C}_a}{N}(T_{a,1} - T_{a,2}) = \frac{\dot{C}_S}{n}(T_{S,2} - T_{S,1}) \tag{4.43}$$

The heat removal from the air, or the heat uptake of the storage mass in the warm phase, is equal to the convective heat flow between the mean air temperature and the mean storage mass temperature of the respective element. For a partition element of surface $A_g/(Nn)$ and a heat transfer coefficient between air and storage mass of h_c, the result is:

$$\dot{Q} = h_c \frac{A_g}{Nn}\left(\frac{T_{a,1} + T_{a,2}}{2} - \frac{T_{S,1} + T_{S,2}}{2}\right) \tag{4.44}$$

From Equations (4.43) and (4.44), there follows for the outlet temperatures:

$$\begin{aligned} T_{a,2} &= T_{a,1} - E_w\left(T_{a,1} - T_{S,1}\right) \\ T_{S,2} &= T_{S,1} + F_w\left(T_{a,1} - T_{S,1}\right) \end{aligned} \tag{4.45}$$

E_w and F_w are abbreviations for the following expressions:

$$E_w = \frac{\frac{h_c A_g}{\dot{C}_a} \frac{1}{n}}{1 + \frac{1}{2}\left(1 + \frac{\dot{C}_a}{\dot{C}_S} \frac{n}{N}\right) \frac{h_c A_g}{\dot{C}_a} \frac{1}{n}} \tag{4.46}$$

$$F_w = E_w \frac{\dot{C}_a}{\dot{C}_S} \frac{n}{N}$$

The calculation of the cold phase takes place in a similar way. Since the volumetric air flow rates of the inlet and exit air need be not identical, the thermal capacity stream of the air and of the convective heat transmission coefficient can change on the cold side.

The calculation process:
The calculation begins with the warm phase.

As the first partition element (1,1) the cold storage mass element ($N = 1$) is selected, which first comes in contact with the entering warm air flow ($n = 1$). The second partition element ($N = 1$, $n = 2$) is then situated in the direction of flow of the warm air being cooled. Depending on the number of subdivisions n, first all temperatures $T_{1\,n}$ of the storage mass elements $N = 1$ are calculated. Next the second air flow channel $N = 2$ with all subdivisions n is calculated. The result is, for example, for N and n from 1 to 3 the following calculation order:

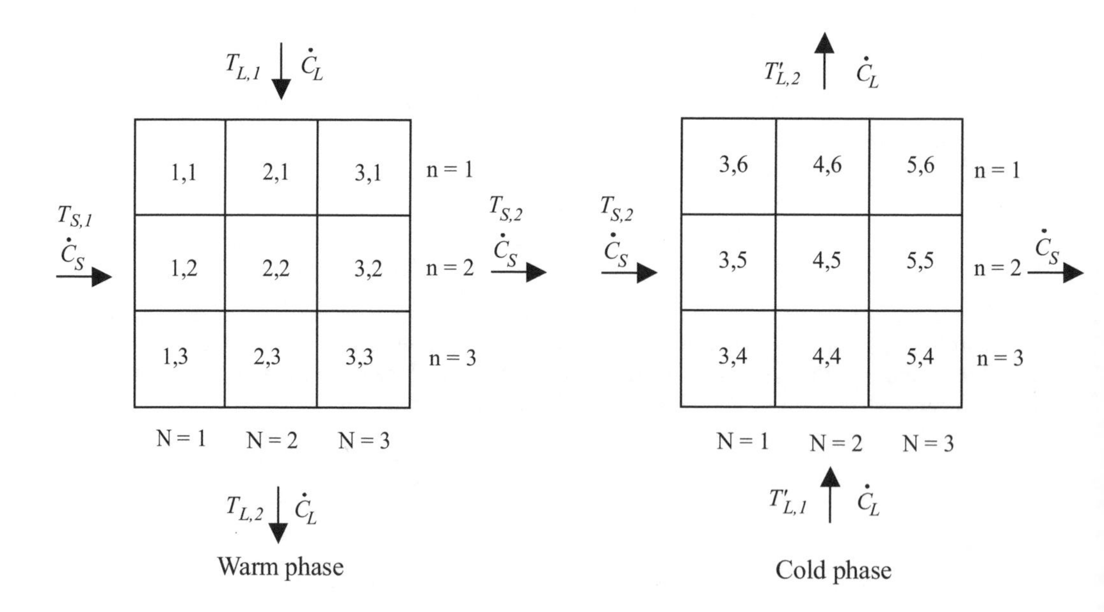

Figure 4.15: Calculation order for regenerative heat exchanger

The thus determined temperature profile of the storage mass stream on exiting the warm period equals the entry profile into the cold period. With the temperature profile of the storage mass stream on exiting the cold phase, calculation of the warm phase again begins from the start, until the outlet temperature profile of the storage mass stream no longer changes and the transferred amounts of heat are equal:

$$\dot{Q} = \dot{Q}'$$
$$\dot{C}_a \left(T_{a,1} - T_{a,2}\right) = \dot{C}'_a \left(T'_{a,2} - T'_{a,1}\right) \tag{4.47}$$

The operational characteristic can be calculated with the calculated average values of the outlet temperatures of the two gas flows. If the warmer gas flow represents the smaller thermal capacity stream, then this applies:

$$\Phi = \frac{\bar{T}_1 - \bar{T}_2}{\bar{T}_1 - \bar{T}'_1} \tag{4.48}$$

If the colder gas flow represents the smaller thermal capacity stream, then this applies:

$$\Phi = \frac{\bar{T}'_2 - \bar{T}'_1}{\bar{T}_1 - \bar{T}'_1} \tag{4.49}$$

Example 4.3

Calculation of the heat recovery efficiency of a regenerator with a wheel diameter of 90.5 cm, and a flow rate of 3000 m^3/h.

Regenerator

U =	10	$[min^{-1}]$	Rotations per min
l =	0.3	[m]	Depth of wheel
D =	0.905	[m]	Diameter of wheel
m =	40	[kg]	Rotating storage mass
A_{warm}/A_{cold} =	50%	[%]	Area ratio of warm air sector
Matrix			
d_i =	0.0019	[m]	Inner diameter of capillary tubes (corresponds to hydraulic diameter)
A_{free}/A_{total} =	91%	[%]	Area ratio of free (open) cross-section to total cross-section
A_{hx} =	366.3	[m²]	Heat exchanging total surface of regenerator (internal surface)
c_s =	870	[J/kg K]	Specific heat capacity of storage mass

A_{free} =	0.2911	0.2911	[m²]	Free cross-section
	Warm air:	**Cold air:**		
λ_a =	0.0279	0.0260	[W/m K]	Heat conductivity air

ρ_a =	1.0938	1.1890	[kg/m³]	Density of air flowing through regenerator
$c_{p.a}$ =	1008.3	1007.0	[J/kg K]	Heat capacity air
υ_a =	1.76E-05	1.52E-05	[m²/s]	Kinematic viscosity
V/t =	0.83	0.83	[m³/s]	Volume flow
T =	45.50	20.00	[°C]	Temperature
V =	2.86	2.86	[m/s]	Mean flow velocity
Pr =	0.696	0.698	[–]	
Re =	309	358	[–]	
Nu =	3.796	3.817	[–]	
h_c =	55.69	52.27	[W/m²K]	Heat transfer coefficient
A_g =	183.15	183.15	[m²]	Heat exchange area
C_a =	918.66	997.35	[W/K]	Heat capacity flow of air
C_S =	5800.00		[W/K]	Heat capacity flow of storage mass
$T_{air, mean}$ =	23.39	40.36	[°C]	

Φ =	**0.87**	[–]	Heat recovery efficiency

T_{a1} =	45.50°C	T_{a2} =	23.39°C
T_{a2}' =	40.36°C	T_{a1}' =	20.00°C

4.1.6 Humidifier technology

To achieve the evaporative cooling effect central to sorption-supported air-conditioning, only humidification systems which bring liquid water but not vapour into the air can be used. The selection of the humidification system depends both on the water quality available at the location and on capital outlays, humidifier overall lengths and pressure losses. While spray humidifier systems produce very good humidification efficiencies with small pressure losses (95–100% at pressure losses of around 50 Pa), the capital outlays and the overall lengths are clearly higher than in simple contact humidifiers. These get along with a length of barely 60 cm, and achieve humidification efficiencies of over 90%, though the pressure losses vary according to manufacturers' data between 50 and 150 Pa. Water quality is an important issue, both for hygienic and cost reasons. Contact evaporators in contrast to spray evaporators do not transmit aerosols into the process air stream, but still require low mineral contents of the water.

The humidification efficiency η_h is defined as the ratio of the achieved absolute humidity increase (from x_{in} to x_{out}) to the maximum possible humidification to 100% relative humidity (corresponding to a maximum absolute humidity x_{max}).

$$\eta_h = \frac{x_{in} - x_{out}}{x_{in} - x_{\max}} \tag{4.50}$$

To calculate the maximum absolute humidity x_{max}, it is easiest to assume isenthalpic humidification, gradually increase the absolute humidity, calculate the new temperature and relative humidity, until 100% relative humidity is reached. More precise is the consideration of the sensible heat of the water added to the air.

4.1.7 Design limits and climatic boundary conditions

4.1.7.1 Demands on room temperatures and humidities

With the models described for sorption wheels, heat recovery devices and humidifiers, the attainable supply air statuses for different outside air statuses can be calculated. Thus both the limitations of pure evaporative humidification and of sorption technology can be determined. At very high outside air humidities, procedure combinations with closed cycle refrigerant plants for supply air cooling must be examined.

According to the relevant norms for thermal comfort, a rise in the perceived temperature up to 27°C is allowed at high outside temperatures in the summer and with only briefly occurring high thermal loads. The upper boundary of the humidity content is set at 11.5 g/kg or a maximum of 65% relative humidity. At a design room temperature of 26°C (which corresponds to the space exhaust air status), 11.5 g/kg of absolute humidity results in a relative humidity of 55%.

For supply air conditioning a further specification applies, that the temperature difference between supply and room air must exceed ΔT = 10K to prevent draughts and thermal discomfort. If for energy-saving reasons the room temperature is set at 26°C, the upper boundary of the comfort field, a minimum supply air temperature of 16°C results.

The maximum supply air humidity depends on the humidity loads of the space which are to be removed. If dehumidification is not necessary, the supply air can be moistened to the maximum admissible room value of 11.5 g/kg. From the outside air status of 32°C, 12 g/kg humidity, only 0.5 g/kg of humidity must then be removed, an energetically favourable change in status (see section 4.1.8, case 1). In air-conditioning of administrative buildings, 8.45 g/kg of supply air humidity is assumed to allow a removal of humidity loads, an accordingly energy-intensive process (case 2).

4.1.7.2 Regeneration temperature and humidity

The regeneration temperature determines the maximum dehumidification of the outside air. At high outside temperatures of over 30°C, 70°C regeneration air temperature can be achieved easily with flat-plate air or water collectors. The relative humidity of the regeneration air depends on the process: if the collector is flowed through by space exhaust air, the humidity is given by the exhaust air humidity plus the evaporative humidification of the exhaust air. For cost reasons, space gain and contamination of exhaust air, however, sucking in outside air for the collector is often selected (only one air duct from the collector is necessary). The absolute humidity of the regeneration air is then equal to the outside air humidity.

4.1.7.3 Calculation of supply air status with different climatic boundary conditions

To evaluate the application possibilities of open adsorption air-conditioning systems, attainable supply air temperatures must be compared with the demands on the room air status. In the following, the temperatures are to be calculated for a design value of 32°C, 40% humidity (a temperate climate) and a damp climate with 35°C, 50% humidity for a sorption system with closed exit air discharge.

The methodology for the calculation of the air statuses is described as follows:

1. Specification of the outside air status: temperature and humidity.
2. Specification of the space exhaust air status: admissible temperature and humidity.
3. Specification of the regeneration air temperature.
4. Calculation of the exit air humidity after the evaporative humidifier as a function of the humidification efficiency (from manufacturers' data).
5. Determination of the regeneration humidity: equal to the exit air humidity in closed systems or equal to the outside air humidity with open suction of regeneration air.
6. Iterative calculation of the supply air status (temperature and humidity) after sorptive dehumidifying, as a function of the relative humidity of the regeneration air, taking into account a dehumidifying efficiency.
7. Calculation of the supply air temperature after the heat recovery device, as a function of the heat recovery efficiency.
8. Calculation of the supply air temperature and humidity after the evaporation humidifier, as a function of the humidification efficiency.
9. Checking the room air humidity and possibly reducing the supply air humidification.

For the two outside air statuses mentioned, the process is calculated in the following examples. The parameters to be given by the planner, such as efficiences of humidification, dehumidifying, heat recovery etc. are printed in italics.

1. Specification of outside air status

Status A:	*32°C, 40% relative humidity (12 g/kg)*
Status B:	*35°C, 50% relative humidity (17.8g/kg)*
2. Specification of space exhaust air status:	*26°C, 11.5 g/kg (55%)*
3. Specification of regeneration temperature:	*70°C*
4. Calculation of the exhaust air humidity	
Humidification efficiency:	*95%*

The maximum absolute humidity of the exhaust air at 100% relative humidity and isenthalpic humidification is 14.1 g/kg, i.e. in total 2.6 g/kg could be added to the exhaust air. At a humidification efficiency of 95%, 2.5 g/kg can be effectively added. The humidification takes place adiabatically and the temperature after humidification can be calculated from the enthalpy of the exhaust air (54.9 kJ/kg). For the vapour content of 11.5 g/kg + 2.5 g/kg = 14 g/kg after humidification, the result is a new temperature of 20°C.

5. Determination of the regeneration air humidity

The relative humidity of the regeneration air at 70°C and 14 g/kg is 7%. The relative humidity of the regeneration air is independent of the outside air status for closed exhaust air-circulation. If outside air has been sucked in, the result for status A with 12 g/kg of vapour content is a relative humidity of 6%, and for status B with 18 g/kg, 9.1%.

6. Iterative calculation of the supply air status after the drying process

Dehumidification efficiency: *80%*
Status A: The relative humidity of the regeneration air determines the maximum dehumidification: to bring the supply air to ideally 7% relative humidity, 6.9 g/kg must be adsorbed from the outside air. The effective dehumidifying performance is about 5.5 g/kg and the temperature is 45.5°C, so a relative humidity of the dried supply air of 10.6% or absolute humidity of 6.5 g/kg results.

Status B:
The effective dehumidifying performance is about 7.6 g/kg (residual moisture 10.2 g/kg) at a temperature of 54°C.

7. Calculation of the supply air temperature after the heat recovery wheel

Heat recovery efficiency: *80%*
Status A: The supply air temperature after the heat recovery device is 25.1°C.
Status B: Supply air temperature 26.8°C.

8. Calculation of the supply air temperature and humidity after the humidifier

Humidification efficiency: 95%
Status A: With a humidification of the dried, precooled supply air to 95% of the maximum, a supply air status of 15.4°C and 10.4 g/kg humidity can be achieved.
Status B: The supply air status with humidification of 3.1 g/kg is 19°C and 13.3 g/kg humidity.

9. Comparison of maximum admissible room air humidity with given supply air statuses
Status A:
The calculated 10.4 g/kg of absolute humidity is related to the desired room air temperature of 26°C, resulting in a relative humidity of 50%, thus under the limit value of 55%.
Status B:
The relative humidity of the supply air related to the room temperature is 63%, thus higher than the desired maximum value of 55%. An increase in the regeneration air temperature to 80°C would improve the supply air status to 18.3°C and 12.5 g/kg humidity, which is however still above the humidity limit value for the room air status. Only if the direct humidification of the supply air is reduced can the humidity condition be met. However, the supply air temperatures are then so high that the cooling is insufficient: at a regeneration air temperature of 80°C the result is, for the space status with 55% humidity, a supply air temperature of 20.6°C. For such humid climates it is often necessary to further cool the supply air without adding extra humidity.

4.1.7.4 Limits and application possibilities of open sorption

Desiccant cooling air-conditioning is thus suited to temperate and warm climates with not too high air humidities (under about 15 g per kg of dry air). Only in extremely dry climates can the desired air conditioning be achieved without a sorption wheel, via pure evaporative cooling.

Under design criteria of 32°C and 40% relative humidity, supply air statuses under 15°C and 9.5 g/kg of humidity content can be achieved with a sorptive cooling system with a dehumidifying performance of 6 g/kg, but without a sorptive drying process barely 20°C and 13 g/kg. This humidity content is already clearly over the maximum admissible supply air value of 11.6 g/kg. Real sorption systems with humidifier efficiencies under 95% and heat recovery efficiencies of between 70% and 75%, achieve under the above design criteria supply air statuses of around 17–19°C.

In a Mediterranean climate with mean monthly maximum temperatures of 36°C and 13 g/kg humidity content, a supply air status of only 20°C and 14 g/kg humidity content can be achieved without drying the air. The use of a sorption wheel, however, enables supply air statuses of 16°C and about 11 g/kg.

Concerning control of the system, the different outside air statuses must always be considered in order to achieve maximum energy efficiency. When cooling begins, first only the heat recovery device and exhaust air humidifier are switched on, and only at higher cooling loads are the sorption wheel, regeneration air heater and supply air humidifier used.

4.1.8 Energy balance of sorption-supported air-conditioning

4.1.8.1 Usable cooling power of open sorption

Sorption-supported air-conditioning systems are driven with pure fresh air. The cooling capacity $\dot{Q}_c$ is therefore calculated from the enthalpy difference between the outside air status h_o and the supply air status h_{in}. The removable cooling load from the room $\dot{Q}_l$, however, is given by the enthalpy difference between supply air h_{in} and space exhaust air h_r, with the space exhaust air temperature usually several Kelvin under the outside temperature. What proportion of the cooling capacity produced is usable depends in particular on the required dehumidifying performance as well as on the necessary fresh air flow rate, which must in every case be cooled from the outside air status, even in conventional air-conditioning systems.

$$\begin{aligned}\dot{Q}_c &= \rho\dot{V}\left(h_o - h_{in}\right)\\ &= \rho\dot{V}\left(\left(c_a + x_o\, c_v\right)T_o + x_o\, h_e - \left(\left(c_a + x_{in}\, c_v\right)T_{in} + x_{in}\, h_e\right)\right)\end{aligned} \tag{4.51}$$

$$\begin{aligned}\dot{Q}_l &= \rho\dot{V}\left(h_r - h_{in}\right)\\ &= \rho\dot{V}\left(\left(c_a + x_r\, c_v\right)T_r + x_r\, h_e - \left(\left(c_a + x_{in}\, c_v\right)T_{in} + x_{in}\, h_e\right)\right)\end{aligned} \tag{4.52}$$

For the three most important applications of sorption-supported air-conditioning the cooling power can be determined by Equation (4.51) and the removed cooling load from the room by Equation (4.52).

CASES:

1. Pure cooling of the outside air with minimum dehumidifying to 11.5 g/kg.
2. Cooling of the outside air to 16°C supply air temperature with dehumidifying to 8.5 g/kg.
3. Pure dehumidifying of the room air to 8.5 g/kg without additional cooling, i.e. the supply air temperature equals 26°C.

The enthalpy of the outside air remains constant at 62 kJ/kg here, at design criteria of 32°C and 40% relative humidity (12 g/kg). The enthalpy of the space exhaust air is 26°C, with 55% relative humidity (11.5 g/kg) at 54.9 kJ/kg.

Case 1–pure cooling with minimum dehumidifying:
With the inlet air humidity specification of 11.5 g/kg, at 95% humidification and good heat recovery efficiencies of 80% a minimum supply air temperature of 17°C is possible. The enthalpy difference between the outside air and supply air is

$$h_o - h_{in} = 62 - 45.7 = 16.3 \quad [kJ/kg].$$

From this a cooling capacity for 1000 m^3/h flow rate results:

$$\rho \dot{V}\left(h_o - h_{in}\right) = 1.18\ kg/m^3 \times 1000\ m^3/3600s \times 16.3 \times 10^3\ J/kg = 5343\ W$$

With a flow rate of 1000 m^3/h, however, only a sensible cooling load of 3 kW can be removed.

$$\rho \dot{V}\left(h_r - h_{in}\right) = 1.19\ kg/m^3 \times 1000\ m^3/3600s \times \left(54.9 - 45.7\right) kJ/kg = 3041\ W$$

Thus if only a sensible cooling load is to be removed, without a fresh air requirement existing, the sorption system must produce 1.8 times more cold than is needed as cooling output; an energetically unfavourable application.

Case 2–Cooling with dehumidification:
If humidity loads of the space must be removed (here for example 3 g/kg from 11.5 g/kg, to 8.5 g/kg), the energy expenditure for air-conditioning clearly becomes higher. At such high air-drying performance of the sorption wheel, the supply air temperature must now be limited to a minimum value, since the usual 95% humidification would produce supply air temperatures far below 16°C.

The enthalpy difference rises to

$$h_o - h_{in} = 62 - 37.3 = 24.7 \quad [kJ/kg]$$

and thus the cooling capacity to 8.1 kW per 1000 m^3/h flow rate.

The cooling load of the space now consists of sensible heat and latent heat of dehumidifying, and the enthalpy difference is

$$h_r - h_{in} = 54.9 - 37.3 = 17.6 \quad [kJ/kg]$$

The total cooling power is still 1.4 times higher than the cooling load removal from the room of 5.8 kW per 1000 m^3/h, so here too a high fresh air requirement offers a favourable initial position for open sorption cooling.

Case 3–pure dehumidifying:
To dehumidify the room air to 8.5 g/kg, i.e. by 3 g/kg, without cooling, the outside air at 32°C and 12 g/kg must be dehumidified by 3.5 g/kg.

$$h_o - h_{in} = 62 - 47.3 = 14.7 \quad [kJ/kg]$$

The necessary cooling performance per 1000 m^3/h volumetric air flow is 4.9 kW.

If dehumidifying were carried out with recirculating air, i.e. not fresh air but space exhaust air, the enthalpy difference would be reduced to

$$h_r - h_{in} = 54.9 - 47.3 = 7.6 \quad [kJ/kg]$$

The removable cooling load is around 2.5 kW.

The different energy expenditures on the basis of the constant outside air status are summarised in Table 4.3. The outside air status is given with 32°C, 40% relative humidity and an enthalpy of 62 kJ/kg.

Table 4.3: Total cooling power and cooling load removal of open sorption-supported air-conditioning with different applications.

Case	*Inlet air status: temperature and absolute humidity*	*Enthalpy inlet air* (kJ/kg)	*Enthalpy difference outside air to inlet air* (kJ/kg)	*Cooling power per 1000 m³/h* (kW)	*Enthalpy difference room exhaust air to inlet air* (kJ/kg)	*Removable cooling load per 1000 m³/h* (kW)
1	17°C 11.5 g/kg	45.7	16.3	5.3	9.2	3.0
2	16°C 8.5 g/kg	37.3	24.7	8.1	17.6	5.8
3	26°C 8.5 g/kg	47.3	14.7	4.9	7.6	2.5

An optimal field of deployment for sorption-supported air-conditioning is found in applications with a high fresh air requirement. At high space-cooling loads with dehumidifying needs but little fresh air requirement, a combination of sorption systems with closed cycle coolers is suitable for separating dehumidifying from load removal. From the ratio of cooling power to regeneration heat, the energy performance figures of open sorption can be determined in what follows.

4.1.8.2 Coefficients of performance and primary energy consumption

To make available a kilowatt-hour of cold, compression refrigerators with a coefficient of performance (COP) of 3 require a total of 0.33 kWh of electricity. The coefficient of performance is generally defined as the ratio of produced cooling power to the supplied power, either electrical power or heat.

$$COP = \frac{\dot{Q}_{cooling}}{\dot{Q}_{supply}} \tag{4.53}$$

At an average primary energy conversion efficiency η_{con} for electricity production of 35%, 0.95 kWh of primary energy must is used for 0.33 kWh of electricity. The primary energy efficiency η_{pe} as a ratio of the produced cooling energy to the supplied primary energy results from the product of COP and the conversion efficiency of the respective energy carrier.

$$\eta_{pe} = \frac{\dot{Q}_{cooling}}{\dot{Q}_{primary\,energy}} = COP \times \eta_{con} \tag{4.54}$$

In electrical compression coolers, η_{pe} is around $3.0 \times 0.35 = 1.05$. If compression coolers are operated in a full air-conditioning system, reheating is often necessary after dehumidifying due to the low dew-point temperature, and the mean primary energy efficiency falls to 0.6, i.e. for a kWh of cold, 1.7 kWh of primary energy are used.

With sorption-supported air-conditioning systems, both thermal energy for regeneration and electricity for fans and auxiliary aggregates such as humidifier pumps and wheel drives must be supplied. First the purely thermal COP should be considered, i.e. the ratio of cooling power produced to the necessary regeneration heat.

$$COP = \frac{\dot{Q}_{cooling}}{\dot{Q}_{regeneration}} = \frac{h_{outside\,/\,exhaust} - h_{supply}}{h_{reg} - h_{after\ HX\,/\,outside}} \tag{4.55}$$

If total cooling power from outside air is considered, the enthalpy difference between outside and supply air ($h_{outside} - h_{supply}$) must be selected. If only room loads are removed, the difference between room exhaust air and supply air ($h_{exhaust} - h_{supply}$) is to be used.

The regeneration power in the denominator is calculated as the enthalpy difference between entry into the (solar) regeneration air heater and exit from the heater with enthalpy h_{reg}. For closed exhaust air systems, the entry air into the heater is the exit air after humidification and the heat exchanger (h_{HX}); for open exhaust systems, outside air is used with enthalpy $h_{outside}$. The enthalpy after heating depends on the regeneration temperature necessary for the respective application.

As an example, the respective performance figures for the three applications of open sorption can be calculated, for closed exhaust air systems. As a boundary condition, an outside air status of 32°C and 40% relative humidity is selected. The specification is the desired supply air status (cooling with or without drying); the regeneration temperature is calculated as a function of the air to be supplied.

Table 4.4: Performance figures of open sorption-supported air-conditioning with a closed exhaust air system.

Case	*Temperature and humidity supply air* [°C]	*Temperature regeneration air* [°C]	*Enthalpy regeneration* [kJ/kg]	*Enthalpy exhaust after HX* [kJ/kg]	*Enthalpy increase regeneration* [kJ/kg]	*COP from outside air* [–]	*COP room load* [–]
1	17°C 11.5 g/kg	53.3	88	70.9	17.1	0.93	0.53
2	16°C 8.5 g/kg	95.1	129.7	79.2	50.5	0.48	0.35
3	26°C 8.5 g/kg	48.3	83.1	69.3	13.8	1.05	0.55

With a falling regeneration temperature, less dehumidifying occurs and the necessary amount of heat falls. At very low outside air humidities, air-conditioning can take place by energy-neutral evaporative cooling alone, so the thermal performance figure becomes infinite in extreme cases without dehumidifying.

However, there is a low limit to the regeneration air temperature dependent on external humidity. For example, at an air status of 32°C and 40%, the performance figure ($COP_{thermal}$) does rise with a falling regeneration temperature. If the regeneration temperature falls below 52°C, there is so little dehumidification that in the supply air humidifier humidifying to 95% can no longer take place (due to the maximum admissible room air humidity). This results in an unadmissible rise in the supply air temperatures.

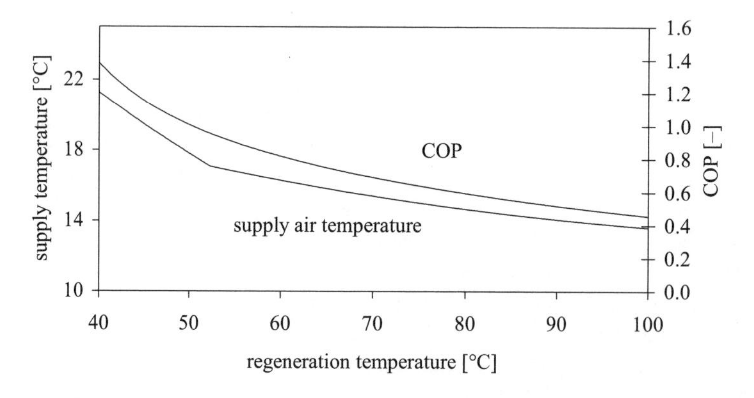

Figure 4.16: Supply air temperatures and cooling performance figures (COP) as a function of the regeneration air temperature at constant outside air statuses of 32°C and 40% relative humidity.

Although energetically very interesting, pure evaporative cooling is limited to dry outside air statuses and is only possible for a limited number of hours of operation. By using thermal solar energy, however, the regeneration heat can likewise be produced primary- energy neutrally at full sorption operation.

For a total energy balance, the additional pressure losses through the sorption wheel, heat recovery device and humidifier, and the associated electrical power increase, must be considered. At a typical flow velocity of 3 m/s in the sorption system, pressure losses of about 150–200 Pa result in the sorption wheel and heat recovery device respectively, and in the humidifier between 100 and 250 Pa, depending on the design. The total of supply-side and exhaust side pressure losses is between 800 and 1300 Pa. For a 100 m^2 air collector field as a regeneration air heater, pressure losses of about 250 Pa can be expected.

From the total pressure losses Δp the electrical power P_{el} of the fans is calculated as a function of the fan efficiency η. At an efficiency of a large fan of 70%, the result is thus an electrical power demand of 417–615W per 1000 m^3/h of volumetric air flow, with total pressure losses between 1050 and 1550 Pa. In addition there are about 100 W per 1000 m^3/h for electric drives of the components (circulation pumps, wheel drive etc.).

$$P_{el} = \frac{\dot{V}\Delta p}{\eta} = \frac{1000m^3/3600s \times 1050Pa}{0.7} = 417W \tag{4.56}$$

Altogether, therefore, the result is connected electrical loads of some 500–700 W per 1000 m^3/h of flow rate, i.e. about 1.4–2 kW primary energy requirement. Thus a cooling capacity of between 4.9–8.1 kW can be produced, depending on the application, i.e. the electrical primary energy efficiency is between 2.4–5.8. This value contains the pressure losses both for the heat recovery and the humidification function. These must also be considered during conventional cooling by compression refrigerant plants as part of a full air-conditioning system. If the heat is supplied either primary energy-neutrally by solar energy or waste heat is used, the desiccant cooling process is primary-energetically clearly superior to electrical compression refrigerant plants.

4.2 Closed cycle adsorption cooling

4.2.1 Technology and areas of application

Closed adsorption coolers operate similarly to open cycle adsorption systems with silica gel and water, the refrigerant water being led in the closed cycle. At low pressure, heat is extracted from the environment by evaporation of the water (i.e. usable cold is produced). The compression of the water vapour to the pressure in the condenser necessary for liquefaction takes place via a thermal compressor: the water vapour is first adsorbed on silica gel (suction function) and afterwards desorbed by the heat supply and brought to the necessary pressure. Cold water at 5–15°C inlet temperature is produced by the closed refrigerant circulation, and can then be distributed in the building with small pipe cross-sections, a substantial advantage over purely air-led systems with large air pipes. Cooled ceilings with high cooling inlet temperatures of around 15°C can be operated with performance figures of 0.7, due to the low temperature rise.

The adsorption refrigerant plants manufactured so far in small numbers are available only in large power ranges between 50 and 500 kW, and due to the small numbers of items require high capital outlays at present (around 500 €/kW of cooling capacity in large systems). In comparison to closed absorption systems on a water/lithium bromide basis, the substantial advantage of adsorption technology lies in the possible use of low heating temperatures below 90°C. With solar energy or waste heat at heating temperatures around 60°C, usable cold can be produced by closed adsorption independently of the climatic boundary conditions. The performance of all closed cycle technologies critically depends on a low cooling water temperature.

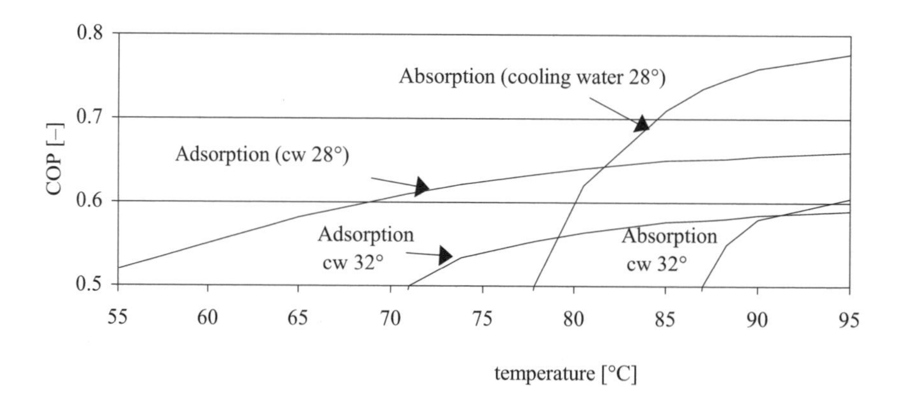

Figure 4.17: Coefficients of performance of closed adsorption and absorption coolers (Gassel, 2000).

For the design, the temperature boundary conditions for the hot water circuit (advance and return temperature), the cooling circuit (advance and return) and the cold water circuit must be known. Generally it is favourable to operate the cooler at high cold-water temperatures, since per Kelvin rise in temperature the performance rises by around 8%.

For high performance figures, low cooling water temperatures for adsorber and condensor heat removal produced by a cooling tower are of special importance. To avoid contamination of the cooling water circuit, only closed cooling towers or open cooling towers with a heat exchanger can be used. In the cooling tower system, in contrast to LiBr water absorption systems, no lower temperature limit is necessary since no crystallisation of the sorbens can occur. At too low temperatures in the evaporator (under 4°C) the lower evaporator area is warmed up by the heating water circuit to avoid freezing of the evaporator heat exchanger.

Between adsorption of the refrigerant water and desorption of the sorption material silica gel, switching takes place cyclically between two chambers, so that quasi-stable operating conditions are achieved. The adsorption plant is controlled via the cycle duration as a function of the cold water inlet temperature at the evaporator: at too high inlet temperatures the adsorption process is terminated and switched to the chamber with dry sorption material. Due to the higher refrigerant flow rates into this chamber, the cooling capacity rises. If the cold water return temperature falls below a desired value, the cycle is extended, the adsorption rate sinks with increasing saturation of the sorption material and the cooling capacity falls.

Before initial operation, the four process chambers (evaporator, condenser and two sorption chambers) are evacuated with a small vacuum pump to an operating pressure of approximately 1000 Pa. The pump is operated briefly every 60 operational hours to remove desorbed gases from the materials or leakage air from the armatures.

When planning, the time-variable cold inlet temperature must be considered, as it fluctuates during a cycle by about ±3 K around the desired value. A cold store of approximately 1/40th of the hourly cooling flow rate effectively buffers the fluctuations.

4.2.2 Costs

The capital outlay for a thermally operated adsorption cooler is clearly higher than the cost of a conventional air-cooled cold water set. For a 350 kW system, a capital outlay of approximately 160 000 € must be expected, compared with 60 000 € for a compression refrigerant plant. With a high annual operating time (6000 full-use hours yearly) and waste-heat use or already available solar thermal plants with low heat costs (0.01 €/kWh) however, economical operation can be achieved, according to manufacturers' information, due to the low operating cost (GBU mbH, 1998).

4.2.3 Operational principle

An adsorption cooler consists of two chambers filled with silica gel which are used alternately for water vapour adsorption and desorption, and enable a quasi-continuous process. The heat of adsorption, or the heat necessary for desorption, is removed or supplied by heat exchangers, whose ribs are tightly packed with silica gel for good thermal contact.

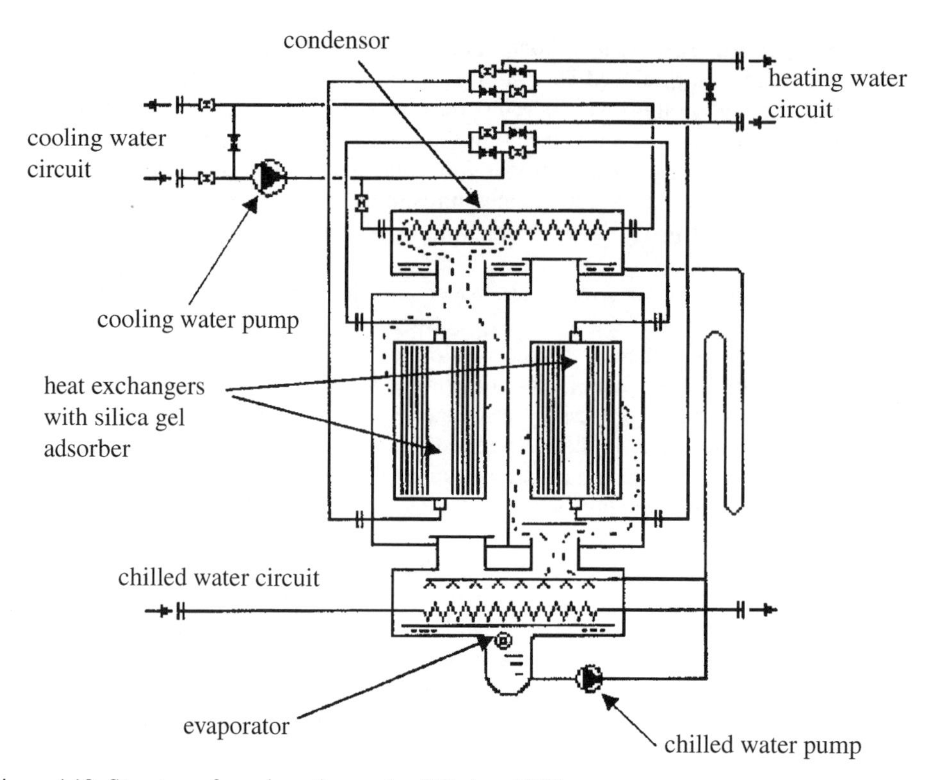

Figure 4.18: Structure of an adsorption cooler (Albring, 2001).

The two adsorbent chambers contain heat exchangers surrounded by the adsorption material silica gel, so heat of adsorption can be removed and heat supplied for desorption. The refrigerant pump only pumps evaporated water back to the spray nozzles of the evaporator. All further pumps for the cold water cycle as well as the cooling and heating circuit of the cooler are fitted externally. Each silica gel chamber is connected by two controllable valve flaps either to the evaporator or condenser. The process consists of two strokes and a short switching phase between the two strokes.

In stroke 1 the lower valve flap of one of the two silica gel chambers is opened to the evaporator (on the right in Figure 4.18), and the water vapour produced in the evaporator is adsorbed on the dry, pre-cooled silica gel. The upper valve flap to the condenser of this chamber is closed. Loading takes place at low evaporator pressure (e.g. 1000 Pa at 5°C), and the freed adsorption enthalpy is removed by the cooling water. The possible water vapour load of the silica gel rises with falling coolant temperature $T_{c,in}$, which thus determines the end of adsorption. In the second chamber (left in Figure 4.18), the valve flap is closed to the evaporator and the flap to the condenser opened during the first stroke. By heat supply, the water vapour collected in the previous stroke is expelled and liquified at condenser pressure.

In stroke 2 the valve flaps operate in exactly opposite directions. The adsorbed water vapour of the first chamber is now expelled by heat supply into the condenser (valve flap to the condenser open, to the evaporator closed). In the second chamber the dried silica gel is used to adsorb water vapour from the evaporator.

Between the strokes, there is a switching phase of approximately 20 seconds for heat recovery, both chambers being flowed through alternately with cooling or heating water. The heating water is used to pre-heat the adsorption chamber of the preceding stroke, and the cooling water pre-cools the hot desorption chamber. A typical cycle lasts 400 seconds, so including the switching phase, a stroke lasts 7 minutes.

4.2.4 *Energy balances and pressure conditions*

The process steps of a closed adsorption cooler can be clarified in an isostere (constant concentration) diagram, in which the vapour pressure of the refrigerant water p_r is plotted logarithmically as a function of the inverse temperature $1/T$ [K^{-1}]. The diagrams result from the Clausius equation of the adsorption of water vapour by silica gel already discussed for open sorption systems.

$$p_r = p_{r1}(C)\exp\left(\frac{h_{ads}(C)}{R_D}\left(\frac{1}{T_1}-\frac{1}{T}\right)\right) \tag{4.57}$$

The parameters of the Clausius equation p_{r1} (vapour pressure at temperature T_1) and the adsorption enthalpy h_{ads} depend on the load concentration C and are calculated iteratively from the sorption isotherm at 40°C (i.e. at T_1 = 313K; see section 4.1).

Table 4.5: Adsorption enthalpy and vapour pressure p_{r1} as a function of the load concentration of water on silica gel (kg_{H2O}/kg_{sor}).

Load concentration C [kg_{H2O}/kg_{sor}]	*Adsorption enthalpy h_{ads}* [kJ/kg]	*Vapour pressure p_{r1} at 40°C* [Pa]
0.05	2849	465
0.10	2622	1392
0.15	2490	2185
0.20	2453	2738
0.25	2453	3259
0.30	2453	3879
0.35	2453	4896
0.40	2453	6445

The adsorption enthalpy results from the total concentration-dependent binding enthalpy and assumed temperature-independent evaporation enthalpy at 20°C. The vapour pressure of the refrigerant water (C = 1) is calculated most precisely with the saturation vapour pressure formula for water from section 4.1.

The vapour pressure over the sorption material rises with temperature and the load concentration. In the isostere diagram, therefore, the straight line of highest pressure is the saturation vapour pressure of water at maximum load concentration C = 1. On this saturation vapour pressure curve are the status points 1 and 2 of the closed adsorption process. The pure refrigerant water is injected in the evaporator, evaporated there at low pressure (1), compressed by the thermal adsorption/desorption process and finally liquefied in the condenser at a higher pressure (2). The higher the evaporator temperature, the higher the vapour pressure of status point 1, which determines the low pressure level of the adsorption cooling process.

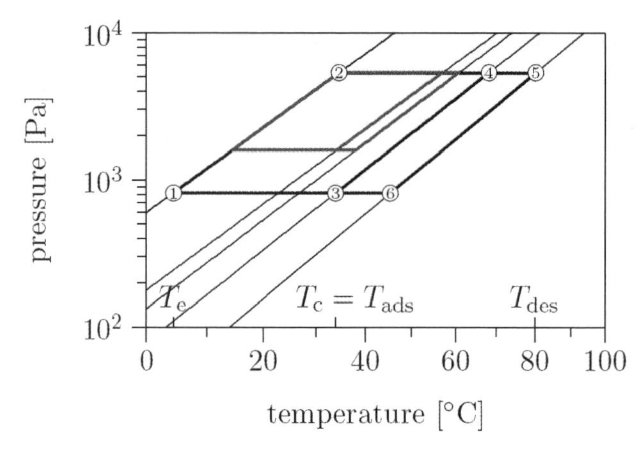

Figure 4.19: Water vapour pressure over silica gel as a function of the load concentration C (isosteres).

Illustrated are two processes with a very low evaporator temperature (4°C) for cold water production and a higher evaporator temperature (14°C) for cooled ceiling applications. The high pressure level (status points 2, 4 and 5) is determined by the temperature of the condenser T_c, which due to a common cooling circuit also corresponds to the temperature of the adsorbent T_{ads}. The maximum process temperature is the desorption temperature T_{des}, at which the refrigerant from the silica gel is expelled (status point 5). From the process comparison it is evident that at low evaporator temperatures (numbered process) high desorption temperatures are necessary (here 80°C), while at high evaporator temperatures desorption temperatures under 60°C suffice.

The pressure conditions and energy balances are represented for the evaporator and condenser first, and afterwards the cyclic process of adsorption and desorption of water vapour by silica gel is analysed.

4.2.4.1 Evaporator

In the evaporator the saturation vapour pressure p_e prevails; it depends on the design temperature of the chilled water circuit (status point 1 on the $\ln p - 1/T$ curve for pure water). Since heat is removed from the chilling circuit by the evaporator through a heat exchanger, the temperatures in the evaporator must be about 2–5K under the cold water inlet temperature.

The circulating chilling water quantity $\dot{m}_{ch}$ results from the desired cooling capacity of the machine. The evaporation enthalpy of the refrigerant water at evaporator temperature T_e is used to cool the chilling circuit from return temperature $T_{r,ch}$ at the evaporator entrance to advance temperature $T_{a,ch}$ at the evaporator exit (useful energy), and to cool the liquid, still warm refrigerant from the condenser (not useful). The outlet temperature from the condenser T_c is around 5 K over the cooling water return temperature, depending on the heat exchanger sizing of the cooling water circuit. The mass flow of the refrigerant water $\dot{m}_r$ is calculated from the sum of the useful cooling power and the required condensor pre-cooling power. The performance balance of the evaporator therefore reads:

$$\dot{m}_r\left(h_v - h_l\right) = \dot{m}_{ch}c_p\left(T_{r,ch} - T_{a,ch}\right) + \dot{m}_r c_p\left(T_c - T_e\right) \tag{4.58}$$

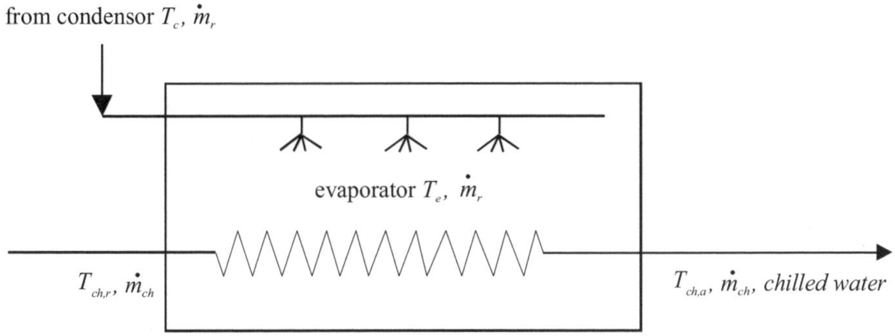

from condensor T_c, $\dot{m}_r$

evaporator T_e, $\dot{m}_r$

$T_{ch,r}$, $\dot{m}_{ch}$

$T_{ch,a}$, $\dot{m}_{ch}$, *chilled water*

Figure 4.20: Temperatures and mass flows at the evaporator.

Example 4.4

An adsorption cooler is to be operated with 200 kW of cooling capacity at a cooling water advance temperature of 29°C (return 33°C), and the cold water and evaporation mass flows necessary for this, plus the heat of evaporation, are to be calculated. The cold water set is to produce 6°C advance temperature and 12°C return temperature. In the evaporator a temperature of 4°C must be delivered at a temperature difference at the heat exchanger of 2 K.

For simplicity, calculation is based in all cycles on the material values of pure water.

The chilling water mass flow results from:

$$\dot{m}_{ch} = \frac{\dot{Q}_{ch}}{c_p\left(T_{r,ch} - T_{a,ch}\right)} = \frac{200\text{kW}}{4.190\frac{\text{kJ}}{\text{kgK}}6\text{ K}} = 7.955\frac{\text{kg}}{s} \triangleq 28.6\frac{\text{m}^3}{h}$$

The outlet temperature from the condenser is about 38°C, at a cooling water return temperature of 33°C and a temperature difference over the heat exchanger of 5 K.

The evaporation mass flow is then calculated by dissolution of Equation (4.58). The evaporation enthalpy of water at 4°C is 2492 kJ/kg.

$$\dot{m}_r = \frac{\dot{m}_{ch}c_p\left(T_{r,ch} - T_{a,ch}\right)}{\left(h_v - h_l\right) - c_p\left(T_c - T_e\right)} = \frac{7.955\frac{\text{kg}}{s}4.190\frac{\text{kJ}}{\text{kgK}}6\text{ K}}{2492\frac{\text{kJ}}{\text{kg}} - 4.19\frac{\text{kJ}}{\text{kgK}}(38-4)\text{K}} = 0.086\frac{kg}{s}$$

To cool the still warm condenser water, therefore,

$\dot{m}_r c_p (T_c - T_e) = 0.086 \frac{\text{kg}}{s} 4.19 \frac{\text{kJ}}{\text{kgK}} (38-4)\text{K} = 11.6$ kW of cooling capacity must be used, which has to be produced by the evaporator in addition to the 200 kW of cooling capacity for cold water production.

The above energy balance of the evaporator assumes that the refrigerant mass flow $\dot{m}_r$ is constant. However, in a real adsorption machine with a limited amount of adsorbens, the removal of refrigerant from the evaporator by the suction of the adsorption chamber slows down with time, as the adsorption material starts to saturate. Only at short cycle times can the effect be ignored, otherwise the dynamics of the adsorption process need to be considered.

4.2.4.2 Condenser

In the condenser (status point 2) the desorbed refrigerant water $\dot{m}_r$ is condensed at high pressure, and the condensation heat has to be be removed by a cooling water circuit with mass flow $\dot{m}_{cool,1}$. The lowest attainable condenser temperature T_c is at a cooling water inlet temperature of $T_{cool,a}$, plus the temperature difference ΔT_{hx}, typically 5 K, necessary at the heat exchanger.

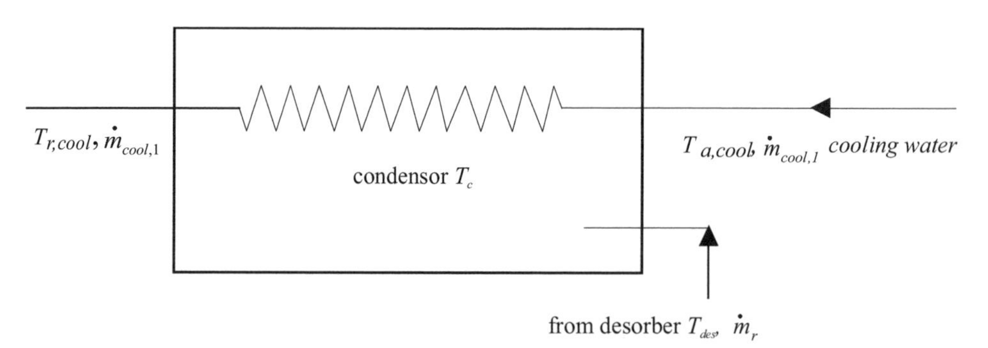

Figure 4.21: Temperatures and mass flows at the condenser.

The pressure level in the condenser determines the switching point for the desorption process. As soon as the pressure necessary for condensation is achieved in the heating phase, the valve between the desorption chamber and condenser is opened and heated further at constant pressure to the final desorption temperature T_{des}, in order to desorb the refrigerant. The lower the cooling water inlet temperature $T_{a,cool}$, the lower is the pressure necessary in the condenser.

The evaporation enthalpy and the sensible heat of the water vapour heated to the desorption temperature must be removed by the cooling water, with the cooling water with mass flow $\dot{m}_{cool,1}$ being heated to the return temperature $T_{r,cool}$. Typical temperature spreads of the cooling water are about 4 K.

$$\dot{m}_{cool,1} c_p \left(T_{r,cool} - T_{a,cool}\right) = \dot{m}_r \left(h_v - h_l\right) + \dot{m}_r c_p \left(T_{des} - T_{a,cool} + \Delta T_{hx}\right) \quad (4.59)$$

Example 4.5

Calculate the necessary condensor cooling-water mass flow for the cooling machine of Example 4.4 for a final desorption temperature T_{des} of 90°C.

The circulating refrigerant mass flow $\dot{m}_v$ must first be cooled from the desorber to the condensor temperature, resulting in a cooling power requirement of

$$0.086 \frac{\text{kg}}{s} 4.19 \frac{\text{kJ}}{\text{kgK}} (90 - 29 + 5)\text{K} = 23.8\text{kW}.$$

With a condensation enthalpy of 2420 kJ/kg at 34°C condenser temperature the total power to be removed is 208 kW + 23.8 kW = 231.8 kW and the resulting necessary cooling water mass flow at a temperature spread of 4 K is

$$\dot{m}_{cool,1} = \frac{231.8\text{kW}}{4.19 \frac{\text{kJ}}{\text{kgK}} 4\text{ K}} = 13.8 \frac{\text{kg}}{s} = 50 \frac{\text{m}^3}{h}$$

The pressure level in the condenser at a condenser temperature of 34°C is 5324 Pa. This pressure must be produced by heating the sorption material during the desorption process.

4.2.4.3 The adsorption process

In the adsorber, in the ideal process, the same pressure appears as in the evaporator (status point 3). The load level of the silica gel (the isostere) is given by the temperature of the adsorbent material T_{ads}, which is at a minimum of the cooling water inlet temperature (plus heat exchanger temperature difference ΔT_{hx}). The status point thus results from the intersection of the evaporator pressure line and the coolant temperature. The lower the coolant water temperature, the more water vapour can be adsorbed.

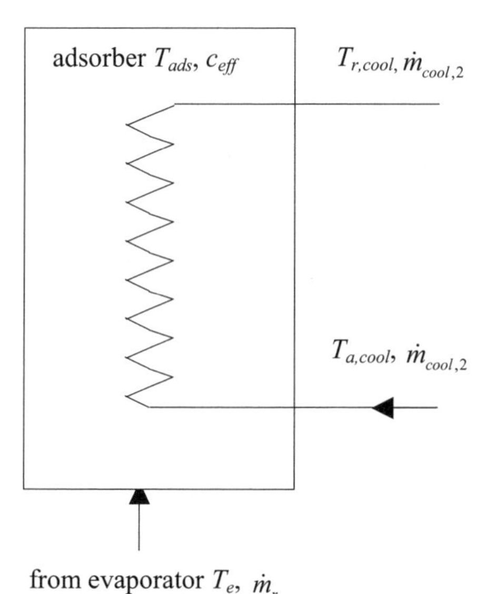

Figure 4.22: Temperatures and mass flows in the adsorber.

The pressure level in the adsorbent is determined by the vapour pressure in the evaporator, which is very low for the refrigerant water. At an evaporator temperature of 4°C, for example, the refrigerant vapour pressure p_r is only 813 Pa. At this pressure the sorption material has to adsorb, the load concentration being determined by the sorption isotherm for silica gel. For this, the relative humidity in the adsorbent is calculated, which results from the ratio of evaporator pressure and saturation vapour pressure at adsorbent temperatures.

$$\varphi_{ads} = \frac{p_e}{p_s\left(T_{ads}\right)} \tag{4.60}$$

From the relative humidity, the load concentration is then determined using the procedure from section 4.1, i.e. by conversion to the known 40°C sorption isotherm.

Example 4.6

Calculate the load concentration in the adsorbent at a coolant temperature of 29°C, a temperature difference at the heat exchanger of 5K and thus an adsorbent temperature of 34°C at an evaporator temperature of 4°C.

The saturation vapour pressure in the evaporator at 4°C is 813 Pa. Related to the saturation vapour pressure in the adsorbent at 34°C of 5324 Pa, a relative humidity of 15.3% results. The load concentration is then 0.09 kg/kg, which is very low.

If due to low evaporator pressure or high coolant temperatures, only low load concentrations are achieved, a large mass of sorption material is necessary to take up the circulating refrigerant flow rate. The adsorption machines are then correspondingly large and heavy. The necessary sorption mass can, however, only be calculated when the load concentration after desorption is known. The difference in load concentrations is called the degassing width and indicates the water vapour quantity per kg of sorption material that can be effectively adsorbed. For a high degassing width, high evaporator temperatures are favourable (and thus high vapour pressures) as well as low adsorbent temperatures (and thus high relative humidities in the adsorbent). Figure 4.23 shows the connection between load concentration and evaporator temperature at different adsorbent temperatures.

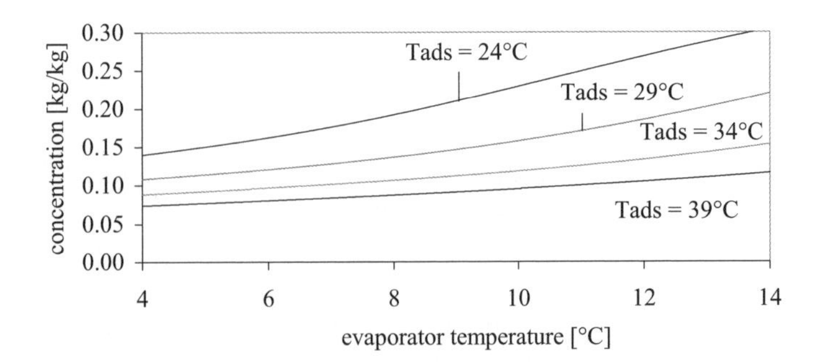

Figure 4.23: Load concentration as a function of the evaporator temperature with the adsorbent temperature as a parameter.

In the real process, a differential pressure between the evaporator and adsorbent must exist, to compensate for the pressure losses between the two chambers at the necessary flow rates. The load concentration at the end of the adsorption process is therefore lower than the value corresponding to that of the vapour pressure in the evaporator.

The heat of adsorption is removed by the cooling water $\dot{m}_{cool,2}c_p\left(T_{r.cool}-T_{a,cool}\right)$ and by the heating of the cold refrigerant vapour from the evaporator to the silica gel temperature $\dot{m}_r c_p\left(T_{ads}-T_e\right)$. Additionally, the heat from the preceding desorption step remaining after switching and pre-cooling must be removed. The silica gel temperature equals the temperature of the cooling water at the end of the adsorption process (plus the temperature difference of the heat exchanger ΔT_{hx}) . The effective thermal capacity c_p^{eff} covers both the thermal capacity of the sorbens material c_p^{sor} (of about 1.0 kJ/kgK) and also that of the heat exchanger c_p^{hx} related to the sorbens mass m_{sor} (e.g. 0.385 kJ/kgK for copper).

$$\dot{m}_{cool,2}c_p\left(T_{r,cool}-T_{a,cool}\right)+\dot{m}_r c_p\left(T_{ads}\left(t\right)-T_e\right)=\dot{m}_r h_{ads}+c_p^{eff}m_{sor}\frac{T_{ads}\left(t\right)-T_{a,cool}+\Delta T_{hx}}{\Delta t} \tag{4.61}$$

with

$$c_p^{eff}=c_p^{sor}+\frac{m_{hx}}{m_{sor}}c_p^{hx}$$

At a typical mass ratio of heat exchanger to sorbens of approximately 2.0, an effective thermal capacity of $c_p^{eff}=1\frac{\text{kJ}}{\text{kg}}+2\times0.385\frac{\text{kJ}}{\text{kg}}=1.77\frac{\text{kJ}}{\text{kg}}$ results.

Example 4.7

Calculate the necessary cooling water mass flow of the adsorbent for the above cooler under steady state conditions, i.e. after achieving the final adsorption temperature T_{ads} (= cooling water inlet temperature +5 K, here 29°C + 5°C = 34°C). Under steady-state conditions the last term of Equation (4.61), the temperature rise of the adsorbent material with time, is zero, so the effective thermal capacity need not be known.

The heat of adsorption is 2650 kJ/kg at the low load concentration of around 0.1 kg/kg. The outlet temperature from the evaporator is equated to the cold water return temperature of 12°C. The cooling water mass flow is

$$\dot{m}_{cool,2} = \frac{0.086\frac{\text{kg}}{s}\left(2650\frac{\text{kJ}}{\text{kg}} - 4.19\frac{\text{kJ}}{\text{kgK}}(34-12)\text{K}\right)}{4.19\frac{\text{kJ}}{\text{kgK}}(33-29)\text{K}} = 13.2\frac{\text{kg}}{s} = 47.4\frac{\text{m}^3}{\text{h}}$$

4.2.4.4 Heating phase

After adsorption has ended, both valve flaps of the adsorption chamber are closed and heated from the final temperature of adsorption T_{ads} until the steam pressure of the condenser is achieved, at an initially constant load concentration (change in status 3 to 4). The load concentration is given by the adsorption process and thus by the evaporator and coolant temperature.

$$\dot{m}_{heat} c_p \left(T_{a,heat} - T_{r,heat}\right) = \left(m_{sor} c_p^{eff} + m_{H_2O} c_p\right)\left(T_H(t) - T_{ads}\right) / \Delta t \tag{4.62}$$

The final temperature T_H of the heating phase at constant loading results from the condenser temperature. The higher this is, the higher is the pressure necessary to liquefy refrigerant, and the higher the temperatures which must be produced during the heating phase and the subsequent desorption phase.

Example 4.8

Calculate the final temperature of the heating phase T_H at a condenser temperature of 34°C (cooling water 29°C + ΔT_{hx}) and an evaporator temperature of 4°C.

The condenser temperature determines the pressure level. To be able to liquefy water at 34°C, the vapour pressure must be 5324 Pa. The evaporator temperature determines the load concentration of the adsorbent at a given adsorbent temperature (here 34°C), which taking Example 4.6 is 0.09 kg/kg. For this isostere the temperature can now be determined with the Clausius parameters from

Table 4.5, for which at the calculated load concentration the vapour pressure over the sorbens corresponds to the condenser pressure.

$$p_r = 1392 \text{ Pa} \times \exp\left(\frac{2622\text{kJ/kg}}{0.461\text{kJ/kgK}}\left(\frac{1}{313K} - \frac{1}{T}\right)\right) = p_c = 5324\ Pa$$

$$\Rightarrow T = T_H = 338 \text{ K}$$

At 65°C the valve between the silica gel chamber and the condenser is opened, and liquefaction of the refrigerant begins. In order that, at a decreasing load concentration, the necessary vapour pressure is still produced, the temperature must be increased to the final desorption temperature.

4.2.4.5 The desorption process

At a constant condenser pressure, the valve flap of the desorber is now opened to the condenser, and water vapour is expelled by a rise in temperature from the heating temperature T_H to the final desorption temperature T_{des} (4→5). The isostere of the minimum load concentration results from the intersection of the condenser pressure and the given desorption final temperature.

The aim of the desorption process is to reduce as far as possible the load concentration of the sorption material, to receive large degassing widths. To achieve low load concentrations, the relative humidity in the desorber must be as low as possible. This results from the ratio of the vapour pressure in the desorber, which is equal to the condenser pressure, and the saturation vapour pressure at the desorption temperature T_{des}:

$$\varphi_{des} = \frac{p_c}{p_s(T_{des})} \tag{4.63}$$

At low cooling water temperatures the condenser temperatures and the vapour pressure p_c are low, and a low relative humidity in the desorber results, enabling a deep unloading of the sorption material. More effective, however, is a high desorption temperature, since the saturation vapour pressure rises exponentially with the temperature, and a rise in temperature in the desorber leads to the reduction of the relative humidity than a falling coolant temperature by the same temperature difference.

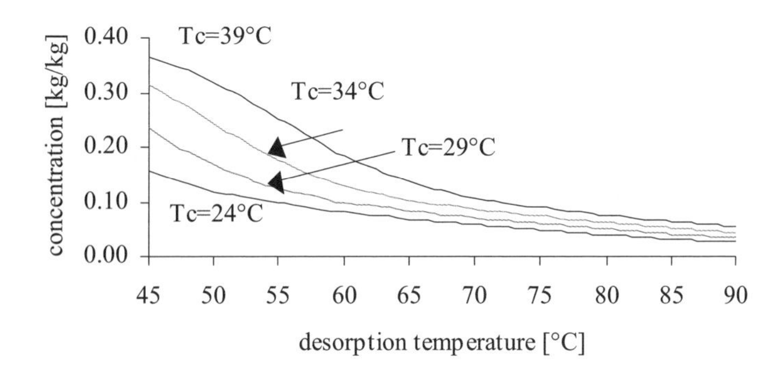

Figure 4.24: Load concentration at the end of desorption as a function of the desorption temperature at different condenser temperatures.

If a desorption temperature T_{des} is now established, the degassing width of the adsorption/desorption process can be calculated, and the mass and effective thermal capacity of the sorption material can be determined. From Figure 4.24 it is clear that even at high final desorption temperatures and low condenser temperatures, a load concentration much below 0.1 kg/kg can be attained only with difficulty. For a condenser temperature of 34°C, a value of 0.06 kg/kg at T_{des} = 80°C is the minimum load concentration. In the real process, the vapour pressure in the desorber must be somewhat higher than in the condenser, to compensate for the pressure losses at the flaps between the chambers, i.e. the desorber temperatures are somewhat higher than in the ideal process. If the adsorption machine is to be operated at very low evaporator temperatures (for example, 4°C), the degassing width ΔC as the concentration difference between adsorption C_{ads} and desorption C_{des} is correspondingly small.

Example 4.9

Calculation of the degassing width at $T_{des} = 80°C$, $T_{ads} = 34°C$, $T_c = 34°C$ and $T_e = 4°C$.

The load concentration at the end of adsorption and a vapour pressure of 813 Pa at 4°C is 0.09 kg/kg (see Example 4.6). At the end of desorption (T_{des} = 80°C) the load concentration is 0.06 kg/kg. The degassing width ΔC = 0.09 kg/kg – 0.06 kg/kg = 0.03 kg/kg is thus very low.

The necessary sorption mass m_{sor} can be calculated from the circulating refrigerant mass flow $\dot{m}_r$, the degassing width ΔC and the cycle time t_c.

$$m_{sor} = \frac{\dot{m}_e t_c}{C_{ads} - C_{des}} \tag{4.64}$$

Example 4.10

Calculation of the sorbens mass for the above cooler at a stroke duration t_c of 200 s. During 200 s, a total of 17.2 kg of water is adsorbed in the 200 kW cooler with an evaporation mass flow of 0.086 kg/s. The sorbens mass at the degassing width of 0.03 kg/kg is 573 kg:

$$m_{sor} = \frac{0.086\text{kg}_{H_2O}/s \times 200s}{0.09\text{kg}_{H_2O}/\text{kg}_{sor} - 0.06\text{kg}_{H_2O}/\text{kg}_{sor}} = 573\text{kg}_{sor}$$

With the known masses and thermal capacities of the sorbens material and adsorbed water, the necessary amount of heat for the desorption process can be calculated. The heating output is used to provide the desorption heat $\dot{m}_r h_{ads}$, and to heat the masses from the heating temperature T_H to the final desorption temperature T_{des}.

$$\dot{m}_{heat} c_p \left(T_{a,heat} - T_{r,heat}\right) = \dot{m}_r h_{ads} + \left(m_{sor} c_p^{eff} + m_{H_2O} c_p\right)\left(T_{des}(t) - T_H\right)/\Delta t \tag{4.65}$$

Example 4.11

Calculation of the heating output for the desorption process at an effective thermal capacity of 1.77 kJ/kgK and a heating duration of 200 seconds.

$$\dot{Q}_{heat} = \underbrace{0.086\frac{\text{kg}}{s}2622\frac{\text{kJ}}{\text{kg}}}_{225.5\text{ kW}} + \underbrace{\left(573\text{ kg} \times 1.77\frac{\text{kJ}}{\text{kgK}} + 17.2\text{ kg} \times 4.19\frac{\text{kJ}}{\text{kgK}}\right)(80-65)\text{K} \times \frac{1}{200\text{ s}}}_{81.5\text{ kW}} = 307\text{ kW}$$

At a temperature spread of 10 K, the result is a mass flow of

$$\dot{m}_{heat} = 7.3\frac{\text{kg}}{\text{s}} = 26.4\frac{\text{m}^3}{\text{h}}$$

4.2.4.6 Cooling phase

After termination of desorption all valve flaps are closed, and by switching the heating water circuit to a cooling water operation the silica gel chamber is cooled at constant low loading until at temperature T_C the evaporator pressure is achieved (change in status 5 to 6). The energy balance corresponds to the heating phase.

$$\dot{m}_{cool}c_p\left(T_{r.cool}-T_{a,cool}\right)=\left(m_{sor}c_p^{eff}+m_{H_2O}c_p\right)\left(T_{des}-T_C(t)\right)/\Delta t \tag{4.66}$$

4.2.5 *Coefficients of performance*

The performance figure of the closed adsorption cooler can be calculated from the energy balances of the process steps discussed. The coefficient of performance is defined as the ratio of chilling capacity $\dot{Q}_{ch}$ produced (or energy Q_{ch} produced in the cycle) in the evaporator, to the necessary heating power $\dot{Q}_{heat}$ for the heating process and desorption (changes in status $3 \rightarrow 4$ and $4 \rightarrow 5$). Since during the changeover process from desorption to adsorption ($5 \rightarrow 6$), heat can be recovered with heat recovery efficiency η, the heating output can be reduced by this amount.

$$COP=\frac{\dot{Q}_{ch}}{\dot{Q}_{heat}}=\frac{Q_{ch}}{Q_{heat}}=\frac{Q_{ch}}{\underbrace{Q_{3\rightarrow 4}}_{\text{heating}}+\underbrace{Q_{4\rightarrow 5}}_{\text{desorption}}-\eta\underbrace{Q_{5\rightarrow 6}}_{\text{cooling}}} \tag{4.67}$$

For the cooler considered so far the result, for one kilogram of evaporated refrigerant for example, is calculated as follows:

The usable cold results from the evaporation enthalpy of the refrigerant water:
$Q_{ch}=h_e\left(T_e=4°\text{C}\right)=2492\text{ kJ}$

During the heating phase, the adsorbent temperature T_{ads} is raised to the heating temperature T_H, which corresponds to the condenser pressure. The mass of the sorbens material per kilogram of evaporated water results from the reciprocal value of the degassing width, and is about 33.3 kg_{sor}.

$$Q_{3\rightarrow 4}=\left(m_{Sor}c_p^{eff}+m_{H_2O}c_p\right)\left(T_H-T_{ads}\right)$$

$$=(\underbrace{\frac{1\text{ kg}_{H_2O}}{0.09\frac{\text{kg}_{H_2O}}{\text{kg}_{sor}}-0.06\frac{\text{kg}_{H_2O}}{\text{kg}_{sor}}}}_{33.3\text{ kg}}\times 1.77\frac{\text{kJ}}{\text{kgK}}+1\text{ kg}_{H_2O}\times 4.19\frac{\text{kJ}}{\text{kgK}})(65-34)\text{K}=1957\text{ kJ}$$

For further heating to the final desorption temperature, an energy quantity of

$$Q_{4\rightarrow 5}=h_{ads}+\left(m_{Sor}c_p^{eff}+m_{H_2O}c_p\right)\left(T_{des}-T_H\right)=2622\text{ kJ}+\left(61.3\frac{\text{kJ}}{\text{K}}\right)(80-65)\text{ K}=3541\text{ kJ}$$

is needed.

The heat given off in the cooling phase corresponds to the amount of heat during the heating phase: $Q_{5\rightarrow 6}=Q_{3\rightarrow 4}$.

This amount of heat can be recovered with a heat recovery efficiency η, around 70% in this case. The result is thus a performance figure for the example calculated of

$$COP = \frac{2492 \frac{\text{kJ}}{\text{kg}}}{1957 \frac{\text{kJ}}{\text{kg}} + 3541 \frac{\text{kJ}}{\text{kg}} - 0.7 \times 1957 \frac{\text{kJ}}{\text{kg}}} = 0.6$$

Better performance figures are achieved if the operating conditions are less extreme. At higher permissible cold water temperatures the adsorption machine can be run at higher evaporator temperatures. Thus the load concentration in the adsorbent rises, and at higher degassing widths the desorption temperature can be lowered. Alternatively the sorbens mass can be increased, and the circulating refrigerant mass flow adsorbed and desorbed at small degassing widths. The machines are then, however, very large and heavy, and for short cycle times the connected heating power must be very high.

4.3 Absorption cooling technology

Absorption technology, known of since the beginning of the 19th century, has attracted increasing interest in recent years, since possibilities of saving primary energy by using waste heat and thermal solar energy exist in heat pump operation for heating purposes and in cooler applications. If solar heat is available in the temperature range of 90–140°C, cold with temperatures of under –30°C can be produced.

While cold-production with absorption facilities has been common for decades, heat pump applications have only in recent years become meaningful in primary-energy terms, due to the improvement in the performance figures; thus small gas-driven absorption heat pumps achieve coefficients of performance of approximately 1.5, i.e. 1 kWh of the assigned primary energy of the gas is converted into 1.5 kWh of heat (Schirp, 1993), which is better than the condensing boilers presently available on the market with maximum COPs of 1.1. These small devices with 3.5 kW heat output are manufactured at present in pilot production and are about to be marketed. Different manufacturers are producing absorption heat pumps with 10–40 kW output which achieve COPs of about 1.3 at advance heating temperatures of around 50°C.

Absorption coolers, on the other hand, have been mass-produced since the 1960s. Arkla Industries alone (today Robur SpA) has produced over 300 000 small coolers with outputs of 10.5–17kW (Lazzarin, 1996). Current coolers cover a large power range from 10 to over 5000 KW.

The most commonly used pairs of working materials are ammonia–water and water–LiBr, with ammonia or water as refrigerants and water or LiBr as solvents. The thermodynamic properties of the refrigerant determine the possible temperature range of the machines: while ammonia at a pressure of 10^5 Pa boils at –33°C and can thus be used for cooling and air conditioning, the refrigerant water is limited to pure air-conditioning with evaporator temperatures over 0°C. In LiBr–water systems, the extremely low refrigerant pressure of approximately 10^3 Pa at +5°C is favourable for small pump power and uncomplicated constructions. However in LiBr systems the refrigerant concentration in the solution must not drop too sharply, since otherwise crystallisation of the solvent occurs. Due to the poorer solubility of water in LiBr, the absorber and condenser are usually water-cooled.

An advantage of water–LiBr systems is in the high boiling point distance between the refrigerant and solvent, so when the refrigerant is expelled from the solution, pure refrigerant vapour develops. The boiling point distance between ammonia and water on the other hand is only 133 K, so when expulsion occurs, water vapour is always produced and must be separated again in a rectifying column.

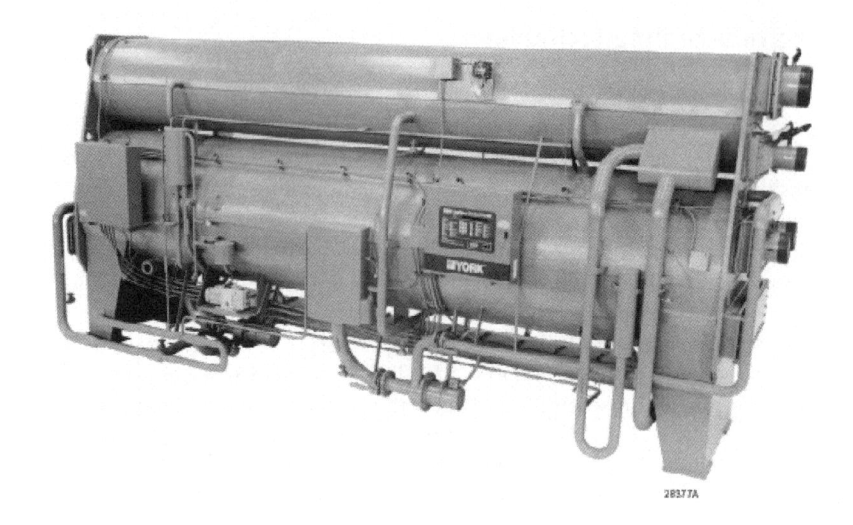

Figure 4.25: Single-lift LiBr–water cooler with hot water or steam operation for power ranges of 420–4840 kW (York International company product information).

High-power coolers are today mainly produced with LiBr technology. One reason is the increasing application of gas-powered double-lift absorption machines, in which first the refrigerant is expelled at high temperatures and the condensation heat is used for further expulsion at lower temperatures and pressures. With such machines COPs of 1.3 or higher are obtained, while single-lift systems are limited to approximately 0.7. Double-lift machines cannot be produced with ammonia refrigerant due to the very high system pressures. In solar energy operation, however, double-lift coolers can only be operated with concentrating thermal collectors, and these are poorly suited to integration in buildings. Since (due to high collector and storage costs) the costs of solar-powered double-lift coolers are not much lower than those of single-lift systems, despite the performance figures, the focus here is on single-lift systems.

4.3.1 The absorption cooling process and its components

Absorption coolers differ from electrically-powered compression refrigerant plants in the substitution of the mechanical compressor by a thermal compressor, and are thus comparable to closed adsorption coolers. The compressor has the function of bringing the evaporated refrigerant to such a high pressure that it can be condensed at high temperatures, and in the cyclic process be supplied again as liquid to the evaporator. In an absorption cooler the compressor process is replaced by absorption of the evaporated refrigerant in a solvent (water or LiBr) and subsequent boiling in the generator at high pressure.

The refrigerant-poor solution from the generator is pumped back into the absorber, where it can take up refrigerant vapour again from the evaporator. A continuous cooling process can be maintained by circulating liquid sorbent, a substantial advantage over adsorption technology with intermittent refrigerant adsorption on the solid silica gel.

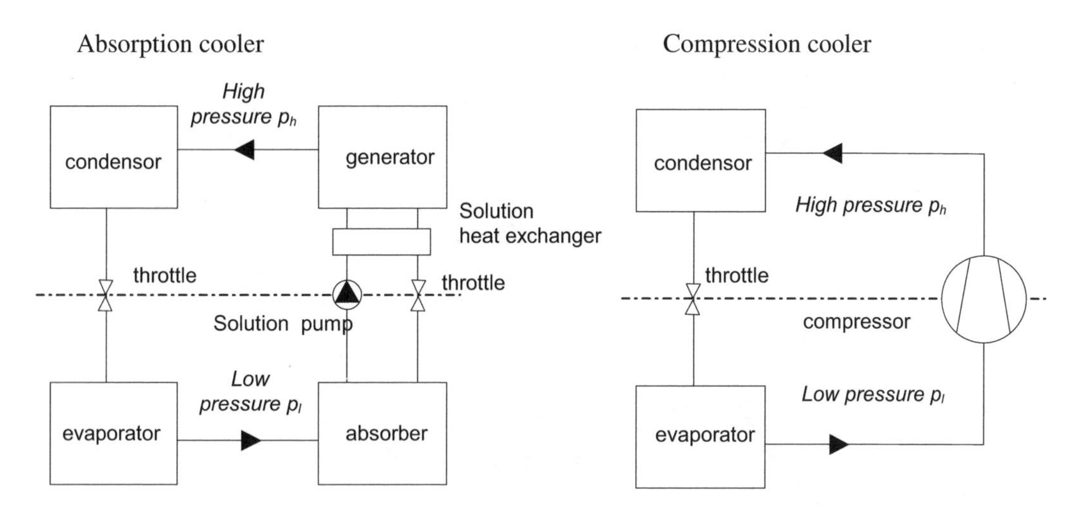

Figure 4.26: Components of the absorption cooler compared with an electrical compression cooler.

Through the representation of the components in the isostere diagram (with the concentration of solution ξ as parameter) the individual process steps can be reconstructed. On the high pressure side with pressure p_h are the condenser and generator; on the low pressure side with pressure level p_l are the evaporator and absorber. In the evaporator and condenser the refrigerant concentration is 100%, which corresponds to a concentration of solution of $\xi = 1.0$. The lowest refrigerant concentration in the solution is produced in the generator (right isostere).

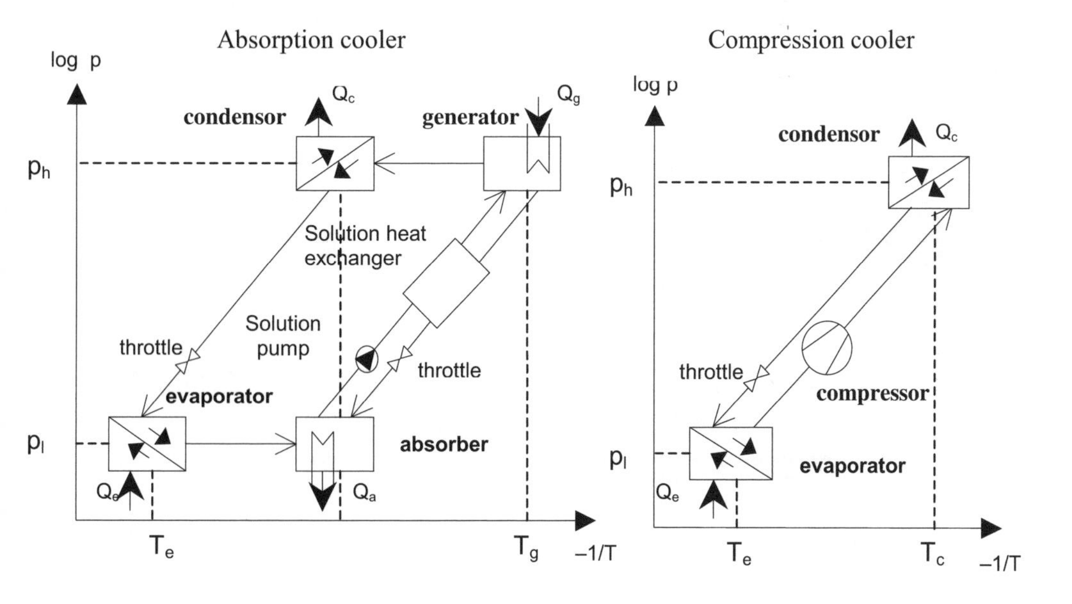

Figure 4.27: Representation of the absorption and compression cooling process in the log p – 1/T diagram.

The arrangement of the components in an absorption cooler is determined by the common pressure level of the generator and condenser on the one hand, and of the evaporator and absorber on the other. The generator and condensor are located in a common upper chamber, while the evaporator and absorber are arranged in the lower area of the machine.

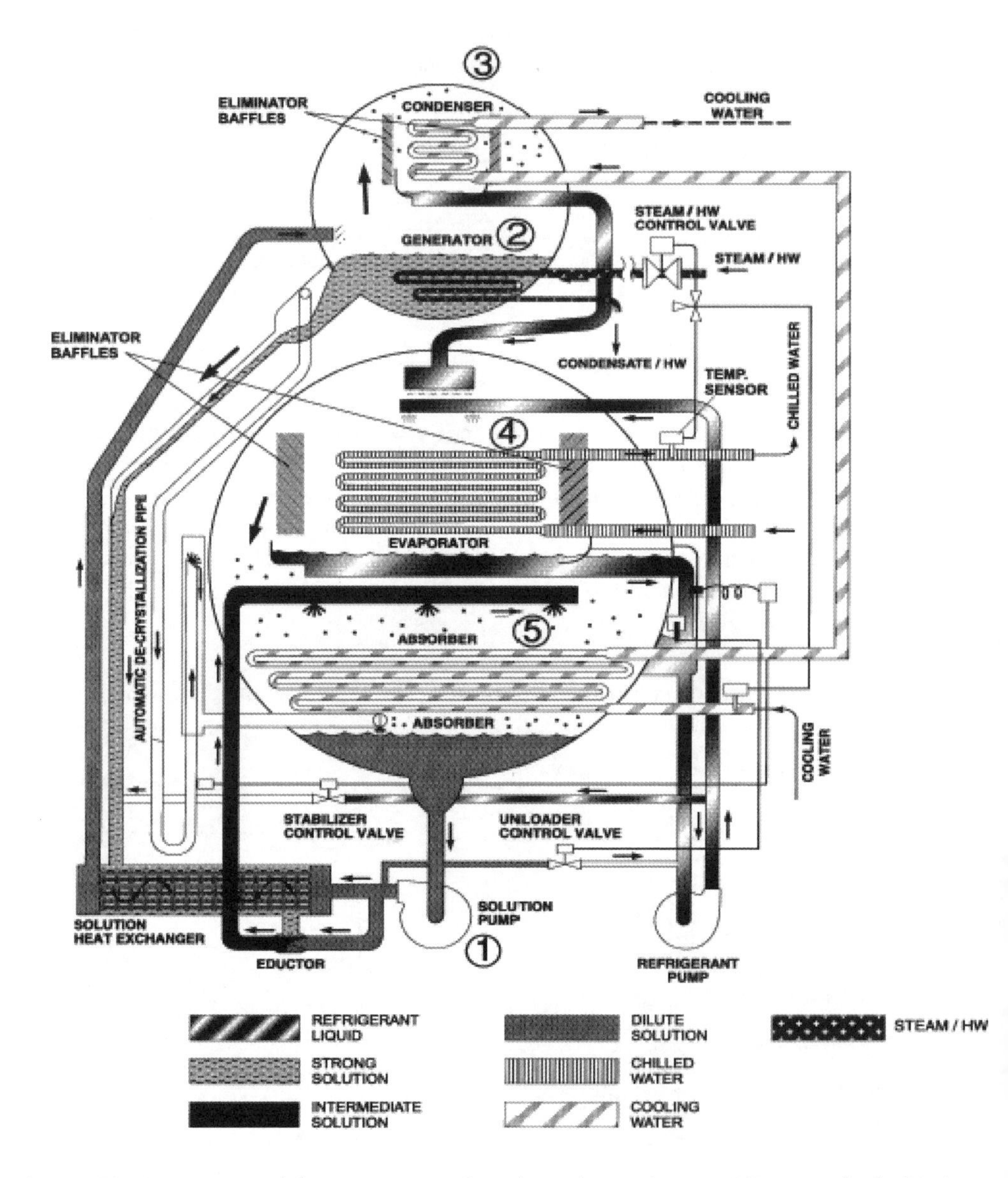

Figure 4.28: Arrangement of the components of an absorption cooler on a LiBr water basis (York company).

4.3.1.1 Double-lift absorption cooling process

In a double-lift absorption process, two generators are operated at different temperature levels. The high-temperature generator with process temperatures of around 150°C produces refrigerant vapour at a high pressure level. This refrigerant vapour condenses at temperatures sufficient to operate a second generator. The refrigerant-rich solution is thus pumped first to the medium temperature generator, where some refrigerant is expelled and then to the high temperature generator, where the solution is further depleted of refrigerant.

The pressure level and thus the temperature in the second generator must be high enough to achieve condensation in the second air- or water-cooled condenser. The condensate from both condensors is expanded into a common evaporator at the low pressure level, where the cooling power is produced.

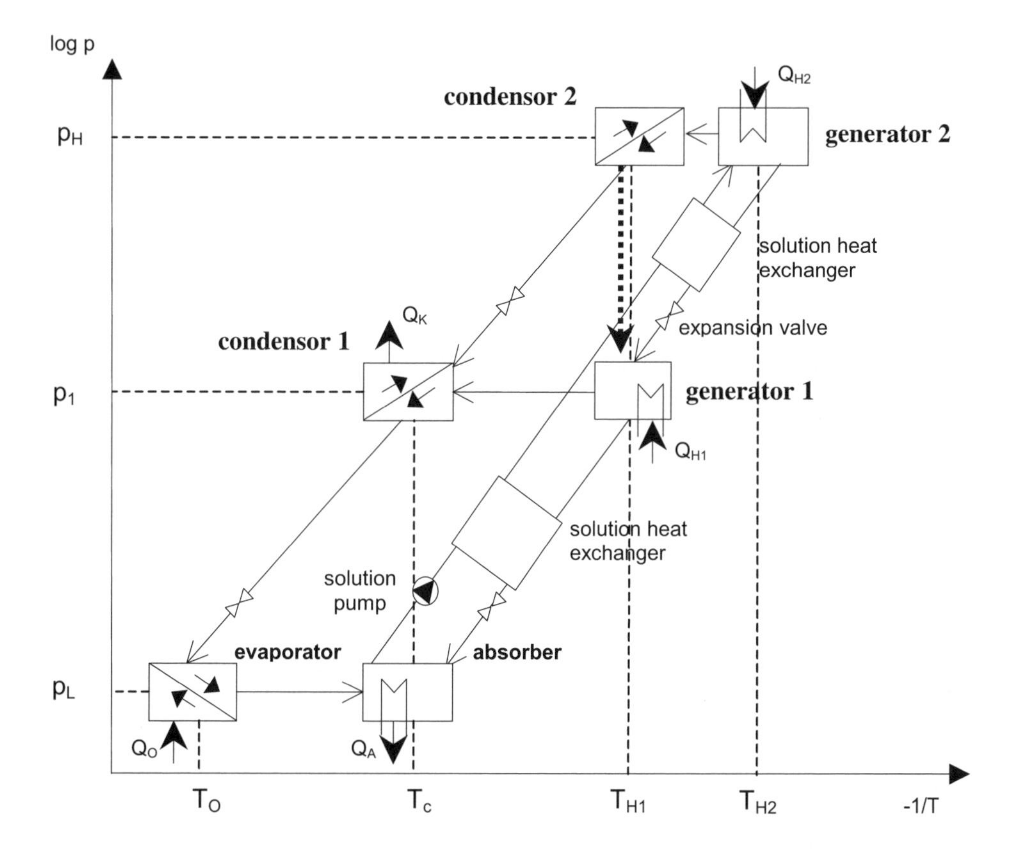

Figure 4.29: Process of double-lift absorption chiller.

The generator heat must be supplied at a high temperature (above 110°). Vacuum tube collectors and parabolic concentrating collectors are very suitable for such applications. Small parabolic concentrators are currently developed for installation on flat building roofs and are attractive alternatives to vacuum tube collectors. These concentrating collectors depend on direct beam irradiance, which varies from about 700 kWh/m²a in Germany (732 in Cologne, 894 in Würzburg) to about 1300 kWh/m²a in southern Europe.

By using condensation heat, the performance figure can clearly be improved from approximately 0.7 for single-effect processes to 1.3 for double-effect absorption chillers and 1.7 for triple effect.

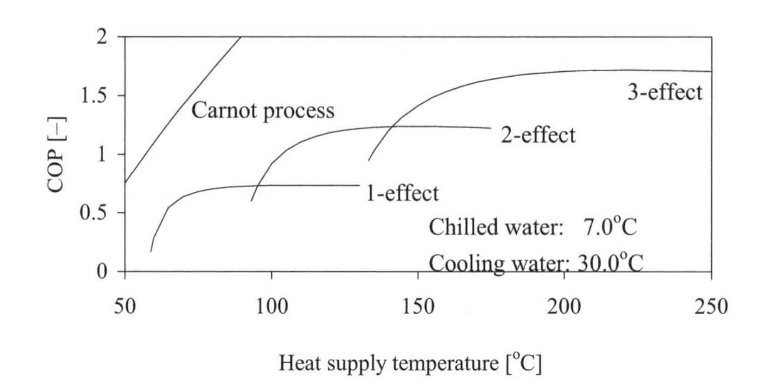

Figure 4.30: Comparison of coefficients of performance for multistage absorption chillers (Grossmann, 2002).

From an energy point of view, the double-lift concept is very interesting and first operating experience was gained in Turkey using a Chinese-produced double-lift absorption chiller with a maximum COP of 1.5, powered by parabolic trough collectors producing steam at 144°C (Lokurlu, 2002).

4.3.1.2 Evaporator and condenser

Evaporator and condenser are conventional components whose heat uptake or release takes place either by air or a liquid circuit. If the refrigerant in the condenser is to condense at high ambient temperatures, the vapour pressure must be correspondingly high. The freed condensation heat is used in heat pump applications for heating purposes, and dissipated in cooling processes to the environment. Before the condensed refrigerant enters the evaporator, the pressure must be reduced to the low evaporator pressure. This is usually done by a throttle valve. Only in diffusion-absorption coolers does an auxiliary gas such as H_2 or He provide the pressure balance between the high and low pressure sides.

Evaporation enthalpy of absorption refrigerants

The evaporation enthalpy h_e (i.e. the amount of heat taken up at constant pressure) of the refrigerant water is almost twice as high as h_e of ammonia, for example at an evaporator temperature of +5°C, 2489 kJ/kg (water) as opposed to 1258 kJ/kg (ammonia). The evaporation enthalpy h_e results from the enthalpy difference between vapour and liquid and is temperature-dependent $h_e = h_v - h_l$. For water vapour the approximation Equations (4.9) and (4.10) are used.

The specific evaporation enthalpy of pure ammonia is calculated with approximation formulae for the high and low pressure range (Sodha, 1983). The equations for the high pressure range are only indicated in the litreature starting from 17.23×10^5 Pa, but they can however be applied with sufficient accuracy from the upper boundary of the low-pressure range at 5.52×10^5 Pa (temperature T in [°C]).

High pressure range: $5.52 \times 10^5 < p \leq 24.13 \times 10^5$ Pa

$$h_l = 6.7702 + 4.7182T \tag{4.68}$$

$$h_v = 1290.28542 + 19.4669 \times 10^{-9}(1.8T + 32)^4 \tag{4.69}$$

Low-pressure range: $3.45 \times 10^5\text{Pa} < p \leq 5.52 \times 10^5$ Pa

$$h_l = -1.4356 + 4.5705\,T \tag{4.70}$$

$$\begin{aligned} h_v &= 1234.944 + (1.8T + 32) \\ &\times \left(0.9672 + (1.8T + 32)\left(-11.5081 \times 10^{-6}(1.8T + 32) + 3.4775 \times 10^{-3}\right)\right) \end{aligned} \tag{4.71}$$

Example 4.12

Calculation of the refrigerant mass flow $\dot{m}_r$ of ammonia or water for a cooling output of the evaporator $\dot{Q}_e$ of 100 kW and an evaporator temperature of 5°C:

In the evaporator the refrigerant is evaporated in the low-pressure range. For water a mass flow of 0.04 kg/s = 145 kg/h results in an evaporation enthalpy h_e = 2510 kJ/kg – 21 kJ/kg = 2489 kJ/kg. For ammonia with an evaporation enthalpy h_e =1279 kJ/kg – 21 kJ/kg = 1258 kJ/kg, a mass flow of 0.079 kg/s = 286 kg/h results.

The refrigerant evaporates only because the pressure in the evaporator is reduced by the constant suction of the compressor or by the absorption in the solvent. Only if the saturation temperature of the respective vapour pressure is under the ambient temperature can heat be taken up from the environment, i.e. can cooling take place. The volume flow rate of the compressor or absorber is regulated in such a way that the evaporator pressure remains constant.

4.3.1.3 Absorber

The refrigerant-poor solution flows back into the absorber from the generator. The refrigerant vapour produced in the evaporator is absorbed there as a function of the absorber temperature and solvent concentration. The evaporator and absorber are at the same refrigerant pressure level. At low evaporator temperatures and correspondingly low vapour pressures, the absorber temperature and the concentration ξ of the refrigerant in the solution must not become too high, since otherwise no more absorption takes place.

The refrigerant-poor solution in the absorber must constantly take up the refrigerant produced in the evaporator, since otherwise the evaporator pressure would rise. Through refrigerant absorption the concentration of refrigerant vapour in the solution rises. The concentration modification between the rich and the poor solution ξ_r and ξ_p is called the degassing width. It is usually between 10 and 25% in the NH_3/H_2O system, and about 4–6%

in the $H_2O/LiBr$ system. The necessary solvent mass flows are determined from a mass balance at the absorber. The mass flow of the rich solution $\dot{m}_r$ consists of the sum of supplied refrigerant vapour $\dot{m}_v$ and the poor solution $\dot{m}_p$ back-pumped from the generator:

$$\dot{m}_r = \dot{m}_v + \dot{m}_p \tag{4.72}$$

The associated refrigerant mass flows depend on the concentration of the refrigerant in the rich and poor solution ξ_r and ξ_p, and on the concentration of the refrigerant in the gaseous state, i.e. the purity of the vapour ξ_v.

$$\dot{m}_r \xi_r = \dot{m}_v \xi_v + \dot{m}_p \xi_p \tag{4.73}$$

The mass flow relation between the rich solution and the refrigerant vapour is also defined as the specific backflow relation *f*, and for ammonia systems is between 10 and 30.

$$f = \frac{\dot{m}_r}{\dot{m}_v} = \frac{\xi_v - \xi_p}{\xi_r - \xi_p} \tag{4.74}$$

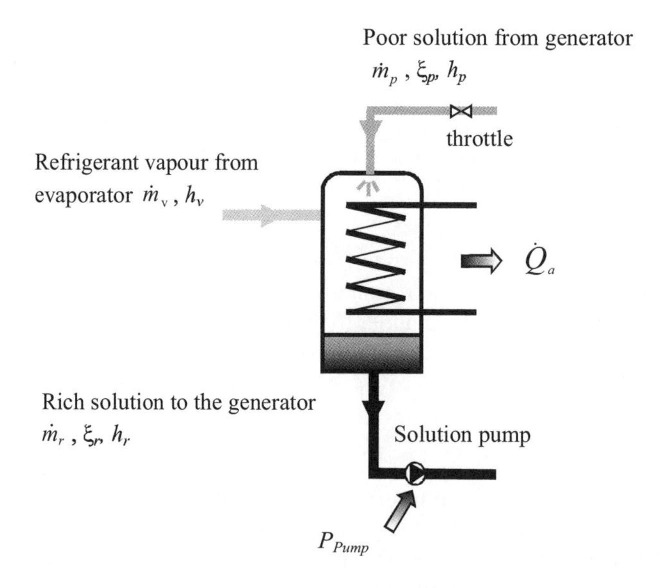

Figure 4.31: Mass balances at the absorber. In the absorber the solution heat $\dot{Q}_a$ is released.

Example 4.13

Calculation of the backflow relation and the necessary solvent circulation for an evaporator output of 100 kW, if the poor solution comes from the generator at a concentration of ξ_p = 0.3 and the degassing width is limited to 10%.

The backflow relation f is only a function of the degassing width. Assuming pure refrigerant vapour ($\xi_v = 1$) results in a backflow relation $f = \frac{1-0.3}{0.4-0.3} = 7$, i.e. seven times the quantity of solvent must be circulated in relation to the refrigerant vapour. For water as refrigerant, the LiBr solvent mass flow is thus 7 × 144.6 kg/h = 1012 kg/h. For ammonia as refrigerant, 7 × 286kg/h = 2003 kg/h must be circulated.

4.3.1.4 Generator

The absorbed refrigerant is expelled again from the generator by heating the refrigerant-rich solution. Conventional heating by gas or other fossil sources of energy can be replaced by thermal solar energy. For operation with solar energy the necessary temperature level in the generator is crucial; it can be well over 100°C with unfavourable boundary conditions, in particular a low solvent concentration. Although today's vacuum collectors can certainly supply such temperature levels with acceptable efficiency, solar use is better at lower temperatures. The physical principles for determining the necessary temperatures and performance figures are discussed in the following sections.

4.3.2 Physical principles of the absorption process

4.3.2.1 Vapour pressure curves of material pairs

In thermal equilibrium, a saturation vapour pressure p_s arises over a pure liquid, depending solely on the temperature. As a function of the evaporation enthalpy of the pure working material, an exponential rise of the saturation vapour pressure with the negative reciprocal value of the temperature $-1/T$ results, based on Clausius–Clapeyron. For pure ammonia, this results in an approximation solution of Bourseau (1986) with the logarithmic function

$$\log_{10} p_s = a - \frac{b}{T} \tag{4.75}$$

indicated by the coefficients $a = 10.018$ and b = 1204.3 for pressures up to 25×10^5 Pa (T in Kelvin).

The low pressure level p_l of the cooler is determined by the saturation vapour pressure at the desired evaporator temperature.

Example 4.14

Vapour pressure of NH_3 at evaporator temperatures T_e of −10°C and +5°C, and at condenser temperatures T_c of +30°C and 50°C.

Component	*Temperature* [K]	*Vapour pressure NH_3* [Pa]
evaporator	263	2.76×10^5
evaporator	278	4.88×10^5
condensor	303	11.1×10^5
condensor	323	19.6×10^5

Due to the high vapour pressures of ammonia, the flow rates of the evaporated refrigerant are clearly lower than the flow rates of the water vapour in LiBr systems, despite higher vapour mass flows. From the pressure levels in the evaporator p_e and the vapour mass flow $\dot{m}_v$, the vapour volume flow $\dot{V}_v$ can be calculated at a given temperature [K] from the ideal gas equation: $p_v \dot{V}_v = \dot{m}_v R_s T$. The specific gas constants R_s of water vapour and ammonia are calculated from the general gas constant R = 8.314 J/molK using the mol mass M: $R_s = R/M$ and are, with mol masses of 18 g/mol for water and 17 g/mol for NH_3, $R_s(H_2O)$ = 462 J/kgK or R_s(NH3) = 489 J/kgK.

Example 4.15

Calculation of the flow rate of refrigerant vapour in the evaporator from water and ammonia from Example 4.12 for 100 kW evaporator cooling capacity.

The pressure level at 5°C evaporator temperature is, using Equation (4.75), 4.88×10^5 Pa for ammonia and from Equation (4.11) 872 Pa for water.

Water: from the mass flow of 145 kg/h or 0.04 kg/s, a very high flow rate of

$$\dot{V}_v = \frac{\dot{m}_v R_s T}{p_e} = \frac{0.04\frac{\text{kg}}{\text{s}} 462 \frac{J}{\text{kgK}} 278K}{872 \text{ Pa}} = 5.9\frac{\text{m}^3}{\text{s}} = 21298\frac{\text{m}^3}{\text{h}}$$

is due to the very low evaporator pressure.

Ammonia: from the almost twice as high mass flow of 286 kg/h a flow rate of only 80 m^3/h results, due to the very high evaporator pressure of 4.85×10^5 Pa.

The temperature level of the condenser determines the high pressure level p_h in the absorption cooler. If the expelled refrigerant is to be liquefied even at high ambient temperatures in an air-cooled condenser, the pressure level of the system rises sharply (typically $p_h > 15 \times 10^5$ Pa for ammonia coolers).

The operational principle of the thermal compressor is based on the fact that the vapour pressure of the pure refrigerant drops if it is absorbed in a more highly boiling liquid. While pure ammonia at +5°C already produces a vapour pressure of about 5×10^5 Pa, in a solution with 50% water it achieves a vapour pressure of 1.3×10^5 Pa at the same temperature; the absorbent fulfils the suction function of the compressor.

Conversely, high temperatures must be produced in the generator to remove ammonia from the solution. The smaller the ammonia concentration in the rich solution, the higher the generator temperature required to produce the vapour pressure for liquefaction.

As a function of the refrigerant concentration ξ in the solution, coefficients a and b of the Clausius–Clapeyron vapour pressure equation are modified:

$$\begin{aligned} &\log_{10} p = a - b/T \\ &a = 10.44 - 1.767\xi + 0.9823\xi^2 + 0.3627\xi^3 \\ &b = 2013.8 - 2155.7\xi + 1540.9\xi^2 - 194.7\xi^3 \end{aligned} \quad (4.76)$$

From the vapour pressure curve of the pure refrigerant, the high and low pressure levels p_h and p_l can thus be determined at condenser and evaporator temperatures T_c and T_e. On the low pressure side, the maximum concentration of solution at which absorption can still occur is determined by the absorber temperature T_a. If the absorber is re-cooled at low temperatures, absorption can occur at higher concentrations of solution. The generator temperature T_g on the high-pressure side determines the minimum solution concentration and so determines the degassing width as the concentration difference between the rich solution in the absorber and the poor solution in the generator.

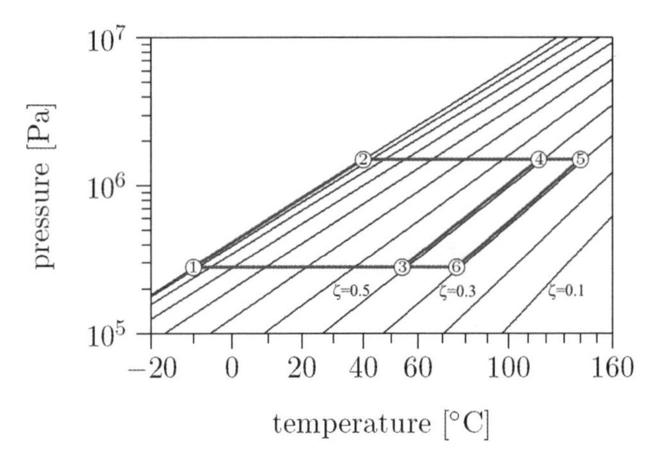

Figure 4.32: Vapour pressure curves over ammonia–water solutions in the log p – 1/T diagram.

The concentration of solution varies in 10% steps from 0.1 to 1 (pure ammonia corresponds to $\xi = 1$, left curve). The process runs via the evaporator (status point 1), condenser (2), the absorption of the refrigerant in the solution (3), the entry of the rich solution into the generator (4), the poor solution at the end of the expulsion process (5) and the poor solution cooled by the solution heat exchanger before re-entry into the absorber (6). The generator temperature can be calculated for given concentrations of solution from the modified Clausius equation:

$$T = \left(\frac{1}{b(\xi)} \times \left(a(\xi) - \log_{10} p \right) \right)^{-1} \tag{4.77}$$

The maximum generator temperature results when refrigerant vapour has already been expelled from the solution, i.e. at the end of the expulsion process. The refrigerant vapour in the generator is then in equilibrium with the refrigerant-poor solution leaving the generator.

Example 4.16

Calculation of the generator temperatures required to produce a vapour pressure of 20×10^5 Pa, if the concentration of the poor solution leaving the generator ξ_p is 0.2, 0.35 and 0.5.

Solution concentration ξ_p [–]	*Generator temperature* T_g [°C]
0.2	156
0.35	120.6
0.5	92.5

High concentrations of solution are therefore favourable for lowering the generator temperature and adjusting it to solar operation. High concentrations of solution, on the other hand, also mean high vapour pressures in the absorber and thus a restriction in suction efficiency of the absorber. If the evaporator temperatures are to be very low, i.e. produce low vapour pressures, the concentration of solution must be correspondingly low to maintain the evaporator pressure at a given absorber temperature.

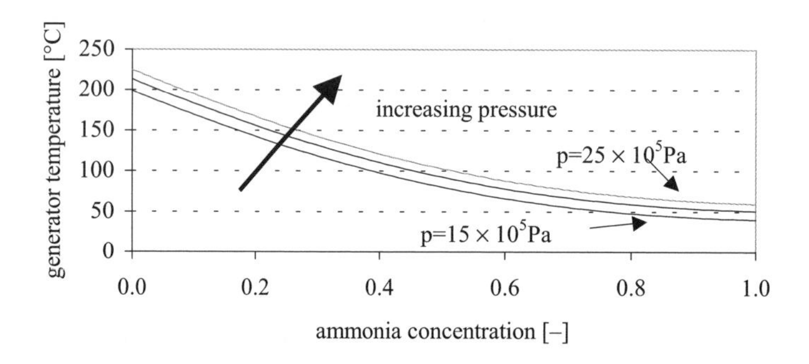

Figure 4.33: Necessary generator temperature as a function of the concentration of the refrigerant ammonia in the solution.

The pressure level is given by the condenser temperature and varies in 5×10^5 Pa steps. With rising condenser pressure, the required generator temperatures also rise.

Example 4.17

In an absorber coolable to 40°C, the refrigerant vapour must still be taken up at an evaporator temperature of –10°C. The maximum concentration of solution is determined by resolving the modified Clausius equation for ξ.

Vapour pressure at –10°C: 2.75×10^5 Pa

Maximum concentration of solution: $\xi = 0.38$

The absorption cooling process of water–LiBr systems takes place at very low absolute pressure levels (under 5000 Pa) between the high and low pressure levels of the condenser and evaporator. The minimum concentration of solution is limited to 35%, since otherwise crystallisation of the solvent LiBr occurs.

4.3.3 *Refrigerant vapour concentration*

If the boiling points of the refrigerant and solution are close (as in ammonia-water systems), a not insignificant proportion of the solvent is co-evaporated in the generator (not a problem in LiBr systems). The lower the refrigerant concentration of the rich solution, the higher are the required generator temperatures and the more solvent is evaporated. Two problems result from this co-evaporation: on the one hand, aqueous refrigerant can no longer be completely evaporated, and on the other a substantial energy quantity is used for water evaporation, which reduces the performance. The purity of the ammonia evaporated in the generator is characterised by the vapour concentration ξ_v , with $\xi_v = 1$ representing pure refrigerant. The higher the refrigerant concentration in the liquid solution ξ_l, the higher the concentration of refrigerant in the vapour. The vapour concentration depends not only on the refrigerant concentration in the solution, but also on the temperature or the total pressure p necessary for liquefaction in the condenser. At high temperatures and pressures, more solvent is co-evaporated and the vapour purity falls.

To calculate the vapour purity (and afterwards the enthalpy), for better curve fitting a division is made into two pressure ranges (Sodha, 1983): the low-pressure range to 5.52×10^5 Pa and the high pressure range for $p > 5.52 \times 10^5$ Pa.

$$\xi_v = 1.0 - (1.0 - \xi_l)^R \tag{4.78}$$

The parameter R for the high pressure range $p > 5.52 \times 10^5$ Pa is

$$\begin{aligned} R = 7.1588 - 0.6171\times10^{-6}\,p + ((((10.7490\xi_l - 17.8690)\xi_l \\ + 4.0297)\xi_l - 1.3086)\xi_l + 0.3715\times10^{-6}\,p)\xi_l \end{aligned} \tag{4.79}$$

and for the low-pressure range $p < 5.52 \times 10^5$ Pa:

$$\begin{aligned} R = ((((108.485\xi_l - 229.009)\xi_l + 155.247)\xi_l - 41.0442) \\ \xi_l)\xi_l + 11.2925 - 4.532\times10^{-6}\,p + 3.0934\times10^{-6}\,p\xi_l^{\,2} \end{aligned} \tag{4.80}$$

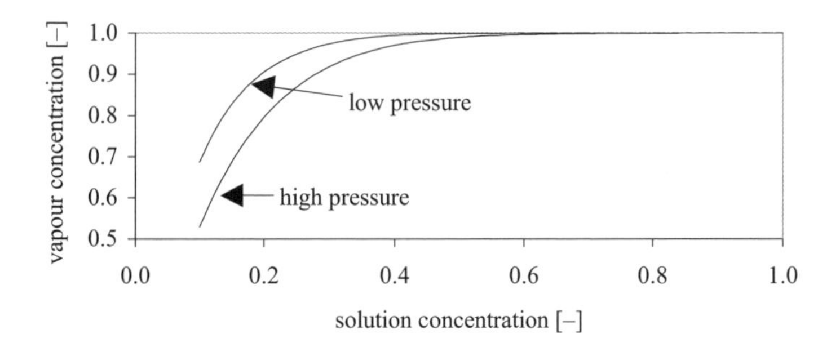

Figure 4.34: Vapour concentration as a function of the concentration of solution.

If the temperature is given instead of the pressure p, the pressure is determined first according to the modified Clausius Equation (4.76).

The pressure curves in the low-pressure range (here represented in 0.5×10^5 Pa steps from $0.5–5 \times 10^5$ Pa) coincide, as do the presure curves in the high-pressure range (here of $10–20 \times 10^5$ Pa in 10^5 Pa steps).

With Equation (4.78) the equilibrium concentration between the solution and vapour can be calculated. Very low concentrations of solution, which are produced at high generator temperatures and which are necessary for very low evaporator temperatures, are in equilibrium with impure vapour, from which high proportions of solvents must be still condensed out.

Example 4.18

Calculation of the equilibrium concentration between liquid and vapour at 20×10^5 Pa total pressure for concentrations of solution of 0.2 and 0.5.

Solution concentration [–]	Vapour purity [–]
0.2	0.7395
0.5	0.9807

The higher concentrations of solution lead directly to a higher vapour purity and thus to lower rectification heat losses.

4.3.4 *Energy balances and performance figures of an absorption cooler*

4.3.4.1 Ideal performance figures

In an ideal absorption process the cyclic process of the refrigerant is regarded as loss-free and thermodynamically reversible. Based on the principle of conservation of energy, the heat taken up in the evaporator and in the generator must equal the delivered heat in the condenser and absorber.

$$Q_e + Q_g = Q_c + Q_a \tag{4.81}$$

Since an ideal process runs reversibly, the entropy must remain constant according to the second law of thermodynamics. The reduced entropy in the condenser corresponds to the entropy increase in the evaporator, and the entropy decrease in the absorber corresponds to the entropy increase in the generator. If instead of the energy balance a power balance is set up and the power is related to the circulating refrigerant mass flow, the result is:

$$\frac{\dot{Q}_c / \dot{m}_v}{T_c} = \frac{\dot{Q}_e / \dot{m}_v}{T_e} \tag{4.82}$$

$$\frac{\dot{Q}_a/\dot{m}_v}{T_a} = \frac{\dot{Q}_g/\dot{m}_v}{T_g} \tag{4.83}$$

The coefficient of performance (COP) of an absorption cooler is defined by the relation of the power taken up in the evaporator to the supplied power in the generator, and can by reformulating the above equations be represented as a temperature relation:

$$COP_{cool} = \frac{\dot{Q}_e}{\dot{Q}_g} = \frac{T_g - T_a}{T_g} \times \frac{T_e}{T_c - T_e} \tag{4.84}$$

The COP of an absorption cooler is thus the product of a right-circulating Carnot thermal engine between the temperatures of the generator and the absorber and a left-circulating Carnot cooling machine between the temperatures of the evaporator and condenser. The ideal COP of an absorption heat pump is defined as the ratio of the heat released in the absorber and condenser to the generator heat.

$$COP_{heat} = \frac{\dot{Q}_a + \dot{Q}_c}{\dot{Q}_g} = 1 + \frac{T_g - T_a}{T_g} \times \frac{T_e}{T_c - T_e} = 1 + COP_{cool} \tag{4.85}$$

Example 4.19

Calculation of the ideal output figures for an absorption cooler with the following temperature conditions:

Temperatures [°C]	*Example 1*	*Example 2*	*Example 3*	*Example 4*	*Example 5*
evaporator T_e	–10	–10	+5	+5	+5
Condensor T_c	30	50	50	30	50
absorber T_a	30	50	50	30	50
generator T_g	140	140	140	100	100
COP_{heat}	1.75	0.95	1.34	2.09	0.83
COP_{cool}	2.75	1.95	2.34	3.09	1.83

Good COPs result if the condenser and absorber temperatures can be kept low, since on the one hand the temperature lifting capacity of the thermal machine rises between the absorber and generator, and on the other hand the temperature difference between the evaporator and condenser for an efficient cooler cyclic process remains small.

4.3.4.2 Real performance figures and enthalpy balances

In an ideal absorption cooling process, Example 4.19 shows that COPs over 1.0 can readily occur. In a real cooler, however, irreversible processes occur during absorption of the refrigerant in the solution. The real COPs depend furthermore on whether the freed amounts of heat in the absorber, condenser and rectifier can be recovered and supplied to the process again.

To calculate the real performance figures, the enthalpies of the liquid (solvent + absorbed refrigerant) and of the vapour (refrigerant plus co-evaporated solvent) for the different equilibrium concentrations in the absorber and generator must be known. For pure ammonia in the evaporator and condenser, the enthalpies are calculated using Equations (4.68) to (4.71). The enthalpy of the solution depends on the refrigerant concentration in the solution ξ_l and on the temperature or pressure. In the following approximation equations, a differentiation is again made between the high and low-pressure ranges.

For the high-pressure range with $p > 5.52 \times 10^5$ Pa, the enthalpy h_l in kJ/kg at the temperature T in °C is calculated as follows:

$$\begin{aligned} h_l = &-94.7974 + 4.7182T + ((((1306.7740\xi_l - 4487.8637)\xi_l \\ &+ 5450.0471)\xi_l - 1926.7160)\xi_l - 240.6738)\xi_l \end{aligned} \tag{4.86}$$

For the low-pressure range $p \le 5.52 \times 10^5$ Pa applies:

$$\begin{aligned} h_l = &-51.7296 + 4.5705T + (((-1526.79\xi_l + 3160.60)\xi_l \\ &- 1158.988)\xi_l - 424.528)\xi_l \end{aligned} \tag{4.87}$$

For the curve fitting of the vapour enthalpy, in the high pressure range an additional difference is made between low and high concentrations of solution. The vapour enthalpy depends not only on the temperature (or pressure) but also on the vapour purity ξ_v, which is calculated using Equation (4.78).

For the high-pressure range with $p > 5.52 \times 10^5$ Pa and a low concentration of solution between $0.1 \le \xi_l \le 0.36$, the vapour enthalpy is given by:

$$\begin{aligned} h_v = &\,2506.6744 + 19.4669\times10^{-9}(1.8T+32)^4 + (((-3122.7354\xi_v + 6871.3435)\xi_D \\ &- 5780.3107)\xi_v + 910.2483)\xi_v + (((-8.7804\times10^{-5}(1.8T+32) \\ &+ 0.0634)(1.8T+32) - 13.8220)(1.8T+32))(1-\xi_v) \\ &+ 1.2714(1.8T+32)(1-\xi_v)^2 \end{aligned} \tag{4.88}$$

For $\xi_l > 0.36$:

$$\begin{aligned} h_v = &\,2113.2189 + 19.4669\times10^{-9}(1.8T+32)^4 + (((0.1599XVT + 4.8363)XVT \\ &+ 57.7705)XVT + 336.3805)XVT + (((-8.7804\times10^{-5}(1.8T+32) \\ &+ 0.0634)(1.8T+32) - 13.8220)(1.8T+32))(1-\xi_v) \\ &+ 1.2714(1.8T+32)(1-\xi_v)^2 \end{aligned} \tag{4.89}$$

with

$$\begin{aligned} XVT &= \ln(1-\xi_v) \quad \text{for } \xi_v < 0.99996 \\ XVT &= \ln(0.00004) \quad \text{for } \xi_v \ge 0.99996 \end{aligned} \tag{4.90}$$

The enthalpy equations in the low-pressure range are valid for the entire concentration range of the liquid.

For $p < 5.52 \times 10^5$ Pa the vapour enthalpy is given by:

$$\begin{aligned} h_v = & 1234.944 + ((-11.5081\times10^{-6}(1.8T+32)+3.4775\times10^{-3})(1.8T+32) \\ & +0.9672)(1.8T+32)+(((9.4324\times10^{-5}(1.8T+32) \\ & -0.0675)(1.8T+32)+15.7951)(1.8T+32))(1-\xi_v) \end{aligned} \quad (4.91)$$

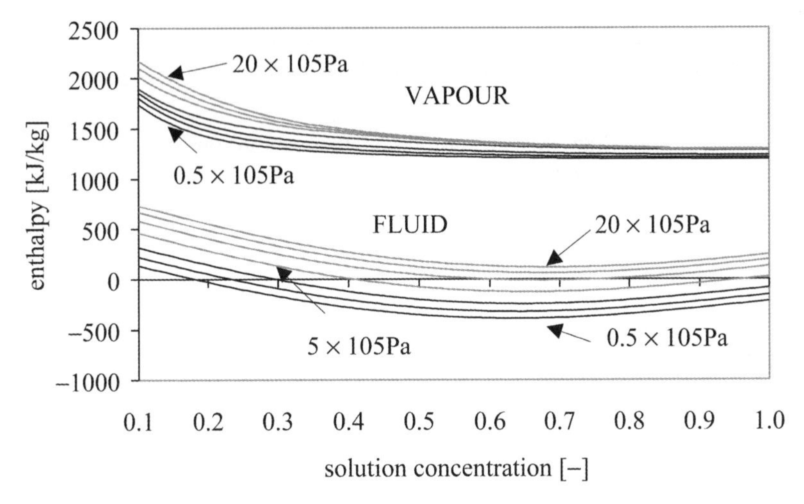

Figure 4.35: Enthalpy of ammonia–water mixtures in liquid and vaporous states as a function of the solution concentration.

In the curves illustrated, the parameter pressure rises from 0.5×10^5 Pa via 1×10^5 Pa, 2×10^5 Pa, 5×10^5 Pa, 10×10^5 Pa, 15×10^5 Pa and 20×10^5 Pa. From the enthalpy diagram as a function of the concentration of solution the enthalpy of the vapour is known (the vapour is in equilibrium with the solution), however, it is not evident what refrigerant concentration is contained in the vapour. To determine this, the vapour enthalpy is plotted not over the concentration of solution, but over the vapour concentration.

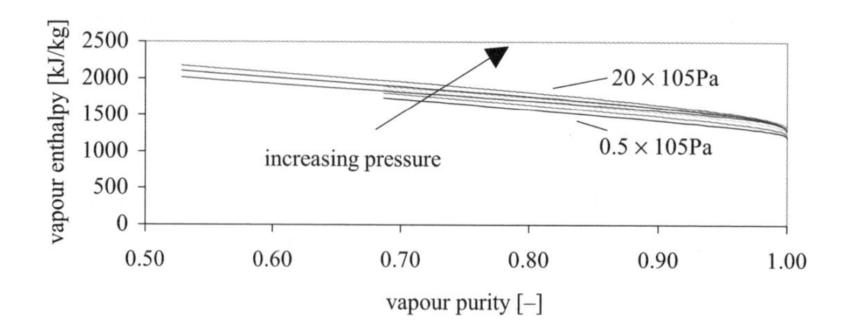

Figure 4.36: Enthalpy of the vapour as a function of the vapour concentration (= vapour purity) of the refrigerant ammonia. The pressure level rises from 0.5×10^5 Pa through 1, 2, 5, 10 to 20×10^5 Pa.

If the two enthalpy diagrams are combined the vapour purity resulting for a given solution concentration can be determined at the same vapour enthalpy.

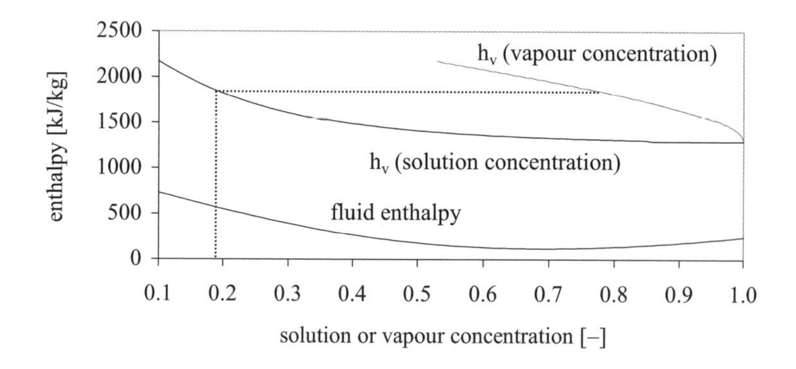

Figure 4.37: Combined enthalpy diagram for the fluid and vapour enthalpy as a function of the concentration of the solution or of the vapour concentration of the refrigerant.

For the concentration of solution of 20% shown, the vapour enthalpy is determined as a function of the concentration of solution at 1867 kJ/kg (for a total pressure of 20×10^5 Pa). This vapour enthalpy also results at a vapour concentration of 74%, so the solution is in equilibrium at exactly this vapour concentration. From the enthalpy differences between vapour and fluid the energy quantity transferred can be calculated for each component of the absorption process.

To calculate the performance figure, a view of the energy flows in the generator is sufficient, since power has only to be supplied there. The heat to be supplied to the generator contains, apart from the enthalpy of the expelled vapour, the heat of absorption released. The coefficient of performance during a single-lift absorption process is therefore always under 1.0, and in commercial machines is typically between 0.5 and 0.8.

The rich solution from the absorber, pre-heated by a heat exchanger, is supplied to the generator (specific enthalpy $h_{r,g}$). The specific heating energy of the generator related to the vapour mass flow is termed q_g. From the generator, heat is removed by the expelled vapour (h_v), which must first be freed in a rectifying column from any solvent still present. The rectification heat (q_r) freed by this condensation is part of the dissipated heat. Furthermore, heat is dissipated from the generator by the discharge of the hot poor solution ($h_{p,g}$).

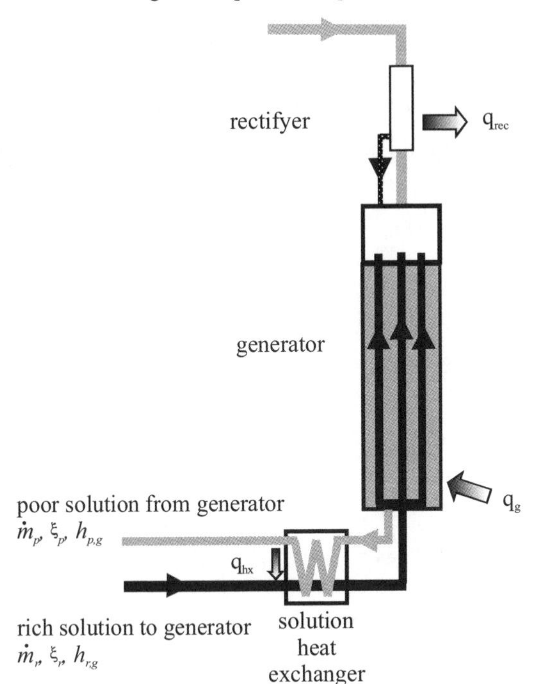

Figure 4.38: Energy balances of the generator with rectifier and heat exchanger between the hot poor solution and cold rich solution.

The energy balance related to the vapour mass flow $\dot{m}_v$ thus reads:

$$\frac{\dot{m}_r}{\dot{m}_v} h_{r,g} + q_g = q_{rec} + h_v + \frac{\dot{m}_a}{\dot{m}_v} h_{a,g} \tag{4.92}$$

Using the specific backflow relation $f = \frac{\dot{m}_r}{\dot{m}_v}$ there follows:

$$\begin{aligned} q_g &= q_{rec} + h_v + f\left(h_{p,g} - h_{r,g}\right) - h_{p,g} \\ &= q_{rec} + h_v + \left(\xi_v - \xi_p\right)\frac{h_{p,g} - h_{r,g}}{\xi_r - \xi_p} - h_{p,g} \end{aligned} \tag{4.93}$$

Instead of the enthalpy difference between $h_{p,g}$ and $h_{r,g}$ at the generator entry and exit, the enthalpy difference of the poor and rich solution at the absorber entry and exit $h_{p,a}$ and $h_{r,a}$ can also be used.

$$q_g = q_{rec} + h_v + (\xi_v - \xi_p)\frac{h_{p,a} - h_{r,a}}{\xi_r - \xi_p} - h_{p,a} \tag{4.94}$$

Rectification heat q_{rec}

The rectification calorific losses depend on the ammonia concentration at which the expelled gas leaves the generator, i.e. which equilibrium between solution and vapour is present. The purest possible vapour mass flow $\dot{m}_v$ leaving the rectifier is the vapour mass flow $\dot{m}_v^1$ produced in the generator minus the liquid quantity condensed in the rectifier $\dot{m}_{fl}$

$$\dot{m}_v = \dot{m}_v^1 - \dot{m}_{fl} \tag{4.95}$$

with the ammonia concentration

$$\dot{m}_v \xi_v = \dot{m}_v^1 \xi_v^1 - \dot{m}_{fl} \xi_{fl} \tag{4.96}$$

From this results for the mass flows

$$\dot{m}_v^1 = \dot{m}_v \frac{\xi_v - \xi_{fl}}{\xi_v^1 - \xi_{fl}} \tag{4.97}$$

The lower the original ammonia vapour concentration ξ_v^1 in the generator, the higher the condensing liquid mass flow $\dot{m}_{fl}$ and the rectification heat become.

$$\dot{m}_{fl} = \dot{m}_v \frac{\xi_v - \xi_v^1}{\xi_v^1 - \xi_{fl}} \tag{4.98}$$

The energy q_{rec} released from the rectifying column, related to the pure vapour mass flow $\dot{m}_v$, therefore results from the difference between the initial vapour enthalpy $\dot{m}_v^1 h_v^1$ and pure vapour $\dot{m}_v h_v$ and the condensed liquid enthalpy $\dot{m}_{fl} h_{fl}$.

$$q_{rec} = \frac{1}{\dot{m}_v}\left(\dot{m}_v^1 h_v^1 - \dot{m}_{fl} h_{fl} - \dot{m}_v h_v\right) \Leftrightarrow$$

$$q_{rec} + h_v = \frac{\xi_v - \xi_{fl}}{\xi_v^1 - \xi_{fl}} h_v^1 - \frac{\xi_v - \xi_v^1}{\xi_v^1 - \xi_{fl}} h_{fl} \tag{4.99}$$

$$= \frac{\xi_v - \xi_{fl}}{\xi_v^1 - \xi_{fl}}\left(h_v^1 - h_{fl}\right) + h_{fl}$$

To the enthalpy of the liquid h_{fl}, therefore, the enthalpy is added which results from the gradient $\frac{h_v^1 - h_{fl}}{\xi_v^1 - \xi_{fl}}$ multiplied by the concentration difference of the pure vapour to the liquid $\xi_v - \xi_{fl}$. From the combined enthalpy–concentration diagram, the enthalpy of the pure vapour and the rectification heat can be read off.

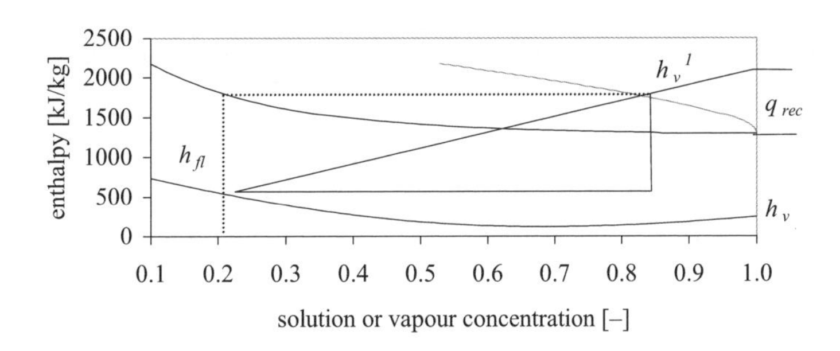

Figure 4.39: Determination of the rectification heat q_{rec} at a concentration of solution of 20% and an initial vapour concentration of 79.5% at 20 × 10^5 Pa.

From the gradient of the enthalpy difference as a function of the concentration difference the result is, at a vapour concentration of 1.0, i.e. pure ammonia, the sum of the vapour and rectification enthalpy q_{rec}.

Example 4.20

Calculation of the rectification heat using Equation (4.99) for the two concentrations of solution 0.2 and 0.5 at 20 × 10^5 Pa total pressure, if the exiting vapour is to have an NH_3 concentration of 100%.

Solution concentration [–]	ξ_{v1} [–]	h_{v1} [kJ/kg]	h_{fl} [kJ/kg]	h_v [kJ/kg]	q_{rec} [kJ/kg]
0.2	0.74	1867	553	1295	1206
0.5	0.98	1439	181	1295	195

At low solution concentrations, it is far more important to supply the rectification heat to the process than at high concentrations with high initial vapour purity.

The temperature in the rectifier results from the desired vapour purity. The pure vapour is in equilibrium with a liquid whose temperature must be all the lower, the purer the vapour has to be.

Absorption heat

To complete the energy balance of the generator using Equation (4.94), the absorber balance $\left(\xi_v - \xi_p\right)\dfrac{h_{p,a} - h_{r,a}}{\xi_r - \xi_p} - h_{p,a} = h_\Lambda$ still has to be created, which results from the enthalpy difference between the poor and rich solution at the absorber entry and exit, and which is also described in the litreature as h_Λ. For the calculation, giving the degassing width ξ_r-ξ_p and the desired vapour purity ξ_v is sufficient. The enthalpy of the solution is calculated for the low-pressure level in the absorber using Equation (4.87).

Example 4.21

Determination of the enthalpy h_Λ for degassing widths $\xi_r - \xi_p = 0.25 - 0.20$ or $0.52 - 0.50$ at a low-pressure level of 5×10^5 Pa.

ξ_r [–]	ξ_p [–]	h_r [kJ/kg]	h_p [kJ/kg]	h_Λ [kJ/kg]
0.25	0.20	206	286	994
0.52	0.50	–85.7	–73.7	372.2

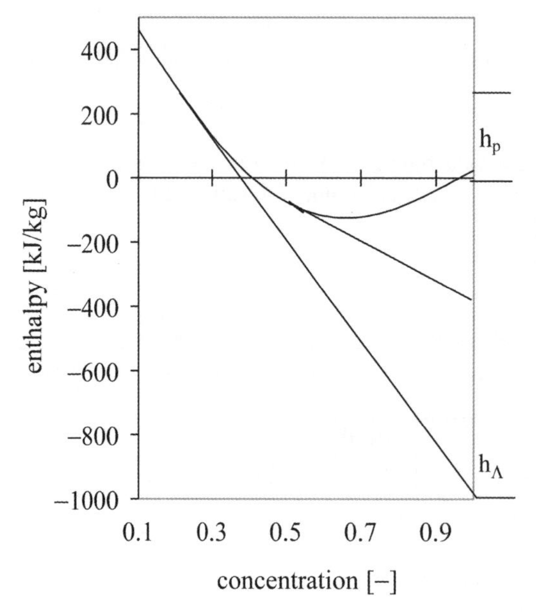

Figure 4.40: Enthalpy h_Λ of the low and high concentrations of solution.

Coefficient of performance
Taking into account the rectification heat q_{rec}, the enthalpy of the pure vapour h_v and the enthalpy balance at the absorber, the coefficient of performance of the absorption cooler can now be calculated for different evaporator temperatures and pressure levels. The concentration of solution in the absorber is adjusted to the evaporator pressure level, i.e. it is so high that at evaporator pressure, refrigerant can still be absorbed. If low evaporator temperatures are to be achieved, the concentration of solution is lowered. With a 20% poor solution, absorption can still take place at a pressure level under 10^5 Pa, i.e. at an evaporator temperature of −30°C. To accelerate the absorption process, the vapour pressure over the solution in the absorber is selected higher than the equilibrium vapour pressure. In the following example, the absorption process is considered at an evaporator temperature of −10°C (corresponding to a pressure level of 2.75×10^5 Pa), and the solution is degassed to 20% refrigerant concentration. Alternatively, a process with a higher evaporator temperature of +5°C is analysed, which enables higher concentrations of solution.

Table 4.6: Boundary conditions for two absorption processes with low and high evaporator temperatures.

No.	ξ_p [–]	ξ_r [–]	*Evaporator temperature* T_e [°C]	*Evaporation enthalpy* h_e [kJ/kg]	*Vapour pressure evaporator* [Pa]
1	0.2	0.25	−10	1296	2.75×10^5
2	0.5	0.52	+5	1258	4.85×10^5

Table 4.7: Enthalpies and performance figures of two absorption processes.

No.	*Rectification heat* q_{rec} [kJ/kg]	*Enthalpy of pure vapour* h_v [kJ/kg]	*Absorption enthalpy* h_A [kJ/kg]	*Generator heat* q_g [kJ/kg]	*COP* [–]
1	966	1290	994	1206 + 1295 + 994 = 3495	0.4
2	195	1295	372	195 + 1295 + 372 = 1860	0.69

The performance figures can be improved if both the rectification heat and the heat of absorption during the process are recovered and used, with the usability of the heat dependent on the temperature level of the freed heat. The heat freed in the rectifier at high temperatures can, for example, be used to pre-heat the cold rich solution flowing from the absorber. The heat freed in the absorber develops especially at high solvent concentrations at low temperatures, and is often released unused to the environment. In single-lift absorption refrigeration systems, real performance figures of approximately 0.5–0.7 can thus be achieved.

4.3.5 Absorption technology and solar plants

The temperature level necessary in the generator results, as already shown, from the operating conditions of the condenser and absorber (water or air-cooled) and from the evaporation temperatures or the concentration of solution of the system. The solar plant often has to supply a temperature level over 100°C via a heat exchanger. Conventional heat storage with buffer stores is necessarily affected by high losses due to the high temperature difference from the environment of around 70–100°C. Cold water storage is more sensible, due to the lower temperature difference from the environment, particularly since energy already converted is then stored. The storage capacity is limited, however, due to the small usable temperature range of a few Kelvin (10 K at most, using cooled ceilings). Of future interest are concepts for storing condensed refrigerant in the cooler itself, which can then be evaporated if there is insufficient solar irradiation.

With the given temperature level of the solar circuit, the efficiency of the solar plant and the surface-related performance are determined. Via the real performance figure of the cooling process the necessary surface of the solar plant can be determined at a given cooling load with a maximum irradiation of 1000 W/m^2.

The buffer store volume on the primary side or the cold-water side should be designed for short-term energy storage of 1–3 days.

5 Grid-connected photovoltaic systems

Photovoltaics (PV) means the direct conversion of short-wave solar irradiance into electricity. Today's market is dominated by semiconductor solar cells on the basis of crystalline silicon, but new technologies based on plastics, organic materials or thin film cells with diverse semiconductor combinations are increasingly achieving marketability.

Solar modules as direct current producers are connected to the 230 V low-voltage electricity grid in systems with about 1 kW to 1 MW of electrical power. The inverters necessary for this are available on the market with a wide input voltage and power range with efficiencies of over 90%.

Photovoltaic systems are characterised by an extremely modular structure, since in principle each module, connected with an inverter, can act as a producer of alternating current. By encapsulating the extremely thin semiconductor cells in a glass–glass or glass–plastic combination, photovoltaic modules are particularly suitable for integration in buildings, since the usual glazing constructions can be used. Only the presence of cables differentiates a photovoltaic module from the use of conventional glazing. The components necessary for the system engineering, such as circuit-breakers, fuses and inverters, can be placed anywhere in the building (or even outside), and take up very little space.

The almost limitless dimensions and designs of photovoltaic modules, the selectable module colour design and the possibility of partial transparency offer special architectural possibilities, in particular for facades. The thermal aspects of integrating photovoltaics in buildings are considered separately in Chapter 6.

5.1 Structure of grid-connected systems

A mains-connected photovoltaic system consists of a solar generator, an inverter, and switching and protection elements. The solar generator consists of PV modules with powers between 10–50 Watts (W) for PV roof tiles, 50–110 W for standard modules, and up to around 600 W for large-scale modules. The connecting of the PV modules depends on the DC (direct current) voltage level of the system, which can be in the low voltage class below 120 volts up to power levels of 5 kW, requiring little effort for safety, but which increasingly covers the voltage range of 200–500 V_{DC}. To reduce assembly time, module wiring boxes with the usual screw connections are increasingly provided with multi-contact plugs, so the PV-generator connection can be performed by people other than electricians. The cables of the modules connected in series are combined into parallel strings in a PV junction box containing over-voltage protection and possibly string diodes. From there the direct current main cable leads to the inverter. If each string is provided with its own inverter the junction box is omitted, a concept which has even been implemented in very large systems in the Megawatt range.

A DC circuit-breaker before the inverter enables disconnection of the system for maintenance work at the inverter, but it is not necessary for safe functioning of the PV system. Monitoring the mains voltage and disconnection of the PV system during mains power loss is usually integrated in the inverter, but the PV system must also be disconnectable manually from the mains after the inverter. Up to 5 kW power, single-phase supply to the mains is normal, and only with more power are three-phase inverters used.

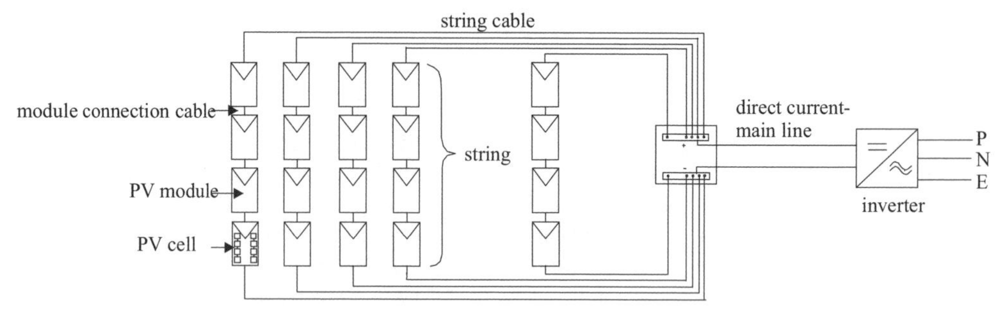

Figure 5.1: Mains-connected photovoltaic system with a common inverter for all module strings.

The inverter converts the photovoltaically produced direct current into alternating current, which in building-integrated systems is fed into the 230 V low-voltage mains. Inverters are available in a wide capacity range for single modules (50–100 W) up to large-scale installations of several 100 kW.

Module-integrated small inverters increase the modularity of PV systems and make possible the usual AC wiring, but they require more material to produce and, at low power, have rather low efficiencies.

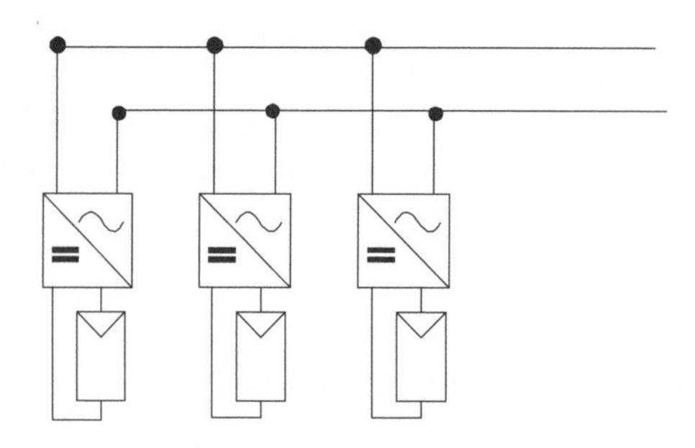

Figure 5.2: Module inverter with one inverter per module.

String inverters with up to 1 kW performance are often used today to simplify DC interconnecting and to decouple the individual module strings from each other. The high DC input voltages, due to the series interconnecting of all modules in the string, result in higher requirements for electrical safety than systems below 120 V_{DC}. In the power range up to 5 kW, a variety of devices are available, with very good transformation efficiencies of over 90%.

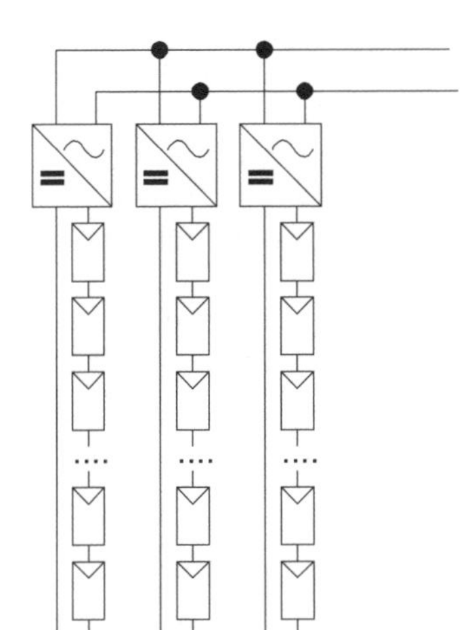

Figure 5.3: PV system with string inverters.

In order to reduce materials and costs and to lower the weight and size of the inverters, transformerless inverters are increasingly being developed, their power electronics supplying a sinusoidal current at a DC voltage level above the mains voltage.

5.2 Solar cell technologies

Today's solar cell market is dominated by crystalline silicon technologies, with a market share of over 80%. While the best small-dimension laboratory cells (2 cm × 2 cm) achieve efficiencies of some 25%, the world-wide highest module efficiency is about 23%. There are mono-crystalline modules with up to 14% efficiency and polycrystalline modules with around 12% efficiency on the market.

Thin-film cells based on amorphous silicon are manufactured both for consumer products of small power and also as modules for power plants or building applications. By using extremely thin double- or three-fold diodes, the problem of light-induced efficiency degradation can be reduced. Amorphous silicon cells are applied both to flexible metal substrates and to coated glass and achieve stabilised efficiencies of 5–8%.

New thin-film technologies on a cadmium telluride (CdTe) or copper-indium-diselenide (CIS) basis are now entering the market. Developments in the field of polycrystalline silicon thin-film cells are promising, due to savings on materials and lower costs at high yields.

5.3 Module technology

Standard modules are encased by polymer layers or cast resin between the front glazing and the back substrate (either glass or a plastic-aluminium material combination). The commonest encapsulation material is a co-polymer of ethylene and vinyl acetate (EVA), which is put on both sides between the cell and substrate, and is polymerised by heating to 140–160°C after evacuation in a vacuum laminator (to prevent air bubbles). The degradation of the polymer by UV radiation can largely be prevented by the addition of stabiliser materials. Large-scale modules are usually encased using cast-resin technology and are manufactured nowadays up to 6 m^2 in size.

The outside glazing of a PV module consists of iron-poor glass, which is pre-stressed for sufficient mechanical stability either thermally (3–4 mm glass strength) or chemically (2 mm glass strength). As a back glass substrate in glass–glass modules, safety glass is normally used. Frameless modules are fastened with profile systems or glued to a frame construction as a structural glazing system. Modules with aluminium frames are often used if building integration is not required, and are mounted on top of the roof or in front of cold facades.

Besides building-glass constructions with photovoltaics, a variety of photovoltaic roof tile systems are available, in which a special plastic frame construction takes on roof tile functions such as impermeability to driving rain and rain run-off, and which are easy to fit on roof slats and connecting to adjoining standard tiles.

5.4 Building integration and costs

The cost allocation of a roof-mounted standard system is dominated by the PV module costs, about 50%. About 5% is spent on the DC wiring, 20% on the inverter and safety engineering, and 25% on assembly. In the last ten years a clear cost reduction in the system price has taken place; while at the beginning of the 1990s the price per installed kilowatt of power was on average about 12 500 €, in 2002 PV standard systems priced around 6000–7500 € were available.

PV module costs tends to rise with sophisticated building integration solutions, since special modules with particular semi-transparency or colouring and no standard dimensions are often selected. Taking as an example the back-ventilated structural glazing construction of a public library in Mataró in Spain, the cost structure can be examined in more detail and compared with other facade constructions. The cost allocation of the back-ventilated PV facade includes 2.5 m^2 glass–glass special modules stuck to an aluminium profile construction with double glazing at the back, the DC cabling, the inverter and the mains-connected safety engineering with total costs of 1167 €/m².

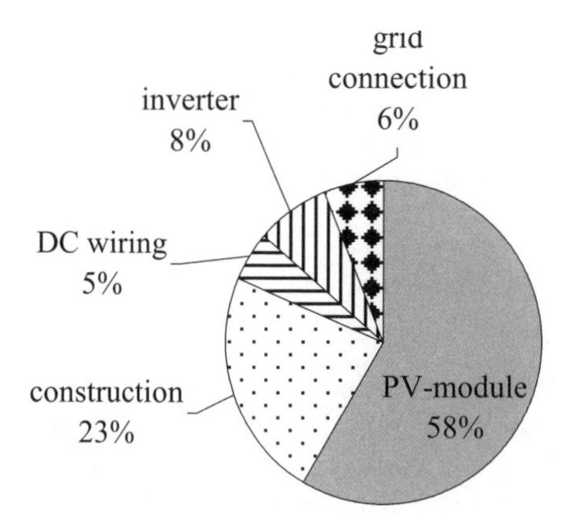

Figure 5.4: Cost allocation of a 245 m^2 PV warm facade as a percentage of the total costs of 1167 €/m^2. The PV facade was constructed in 1995.

In addition to the back-ventilated facade, 325 m^2 PV roof sheds with 50 W frameless standard modules are integrated in the building. The sheds are also back-ventilated and stuck to the profiles as a structural glazing construction, the back double glazing being replaced by an insulated panel. The total cost of the roof shed construction is 1051 €/m^2, with the PV modules dominating the cost here too. For comparison, the cost structure of a conventional curtain facade with laminated glass (6 mm + 6 mm) and of a cold facade with PV standard modules are shown (information from the photovoltaic facade company TFM Barcelona). The surface-related system engineering cost of standard modules are somewhat higher than in the facade special modules, due to higher electrical efficiencies.

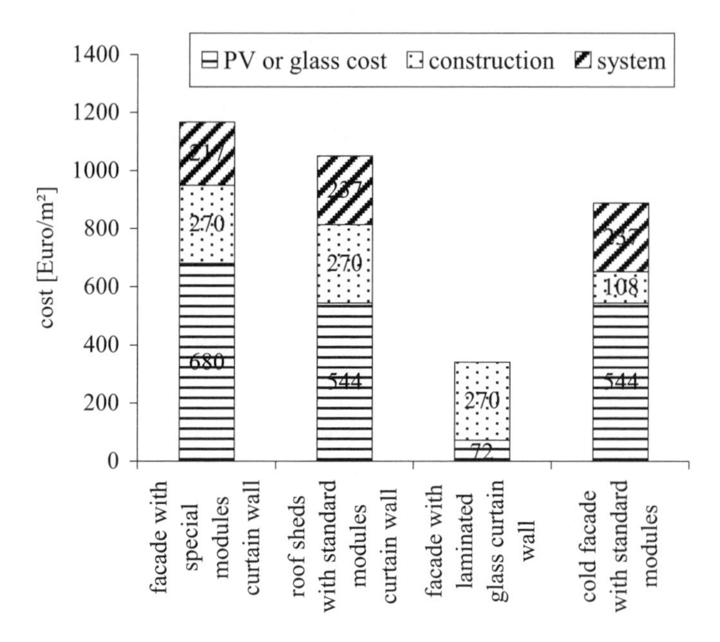

Figure 5.5: Cost allocations for different facade systems with and without photovoltaic modules.

5.5 Energy production and the performance ratio of PV systems

The energy yield of a photovoltaic system is indicated in kWh of electricity supplied, and related to the installed peak power in kW$_p$ under standard test conditions (STC), i.e. 1000 W/m^2 irradiance and a module temperature of 25°C. The measured yields of mains-connected systems under German climatic conditions are between 700–1000 kWh/kW$_p$.

To be able to identify an irradiance-independent characteristic figure of a PV system, the AC energy produced (E_{AC}) is related to the measured irradiance on the module plane G_m resulting in an average system efficiency η_{sys}. This efficiency is compared with the nominal efficiency of the photovoltaic generator under standard test conditions:

$$\eta_{STC} = \frac{P_{DC}}{G_{STC}}.$$

The relation of the two efficiencies is described as the performance ratio *PR* and is typically between 55% and 80%. Instead of the PV system and irradiance energies (in kWh/m²), power ratios can also be used (in kW) to calculate the performance ratio, but the result is then only valid for the one given boundary condition.

$$PR = \frac{E_{AC,measured}}{G_m} \frac{G_{STC}}{P_{DC,STC}} = \frac{E_{AC,measured}}{G_m \times \eta_{STC}} = \frac{\eta_{sys}}{\eta_{STC}} \tag{5.1}$$

The wide value range and the size of the losses, from 20% to as high as 45%, make desirable a more exact loss analysis and performance monitoring of PV systems.

For this, the performance ratio can be subdivided into PV module losses, DC interconnecting losses and AC transformation losses. Like the performance ratio, the actual DC generator energy or power produced is related to the nominal power under standard test conditions and is irradiance-corrected. The Module Ratio *MR* is defined by:

$$MR = \frac{P_{module-DC,measured}}{P_{module-DC,STC}} \frac{G_{STC}}{G_{measured}} \tag{5.2}$$

and contains the power losses of the module due to temperatures above 25°C, deviation from the rated power at low irradiance, and possible shading of individual cells in the module.

The Array Ratio *AR* contains both the module losses and also power losses due to the DC wiring of the generator and adjustment losses by interconnecting non-identical modules.

$$AR = \frac{P_{generator-DC,measured}}{P_{generator-DC,STC}} \frac{G_{STC}}{G_{measured}} \tag{5.3}$$

The Performance Ratio *PR* contains finally the losses of the inverter by the transformation of DC into AC.

5.5.1 Energy amortisation times

The use of photovoltaic systems is only worthwhile in energy economic terms if the energy expenditure in producing the entire plant is clearly lower than the energy quantity produced during its life span.

The energy amortisation time of photovoltaic systems is dominated by the energy expenditure in cell production. The primary energy expenditure to produce a kW of installed photovoltaic power (about 6–8 m^2 of surface depending on module efficiency) is about 2500 kWh/kWp (Palz and Zibetta, 1991). How many years it takes to produce this energy mainly depends on the irradiance on the PV generator, i.e. on the geographic location and module orientation.

With a power production of 800 kWh/kWp, the energy amortisation period is 3.2 years. On a south-facing facade with an energy production of about 70% of the maximum yield, the energy amortisation period increases to 4.5 years.

Manufacturer warranties for PV modules now cover 10–20 years, and the life span can be set at over 25 years. A very long PV module life span is possible, as encapsulation permanently protects the cells from harmful environmental influences, in particular humidity. Initial industrial tests to recycle solar cells from modules have shown that the solar cells themselves display no sign of degradation even after 20 years, and can be encased again as new modules with very small losses in performance.

5.6 Physical fundamentals of solar electricity production

In photovoltaic cells, solar irradiance is converted directly into electricity. The short-wave irradiance is absorbed by the solar cell and produces free electrical charge carriers in the conduction and valence bands.

Suitable materials on a semiconductor basis for solar cells have energy gaps between the valence and conduction bands which are adapted to solar irradiance. A compromise must be found between high current generation with a small gap, where long-wave solar irradiance can be absorbed, and high voltage production with large gaps. The highest efficiencies can be obtained with band gaps between 1.3 and 1.5 electron Volt (eV), e.g. indium phosphide with 1.27 eV, gallium arsenide with 1.35 eV or cadmium telluride with 1.44 eV. The currently most frequently used crystalline silicon has a rather low band gap of 1.124 eV (Goetzberger *et al.*, 1997).

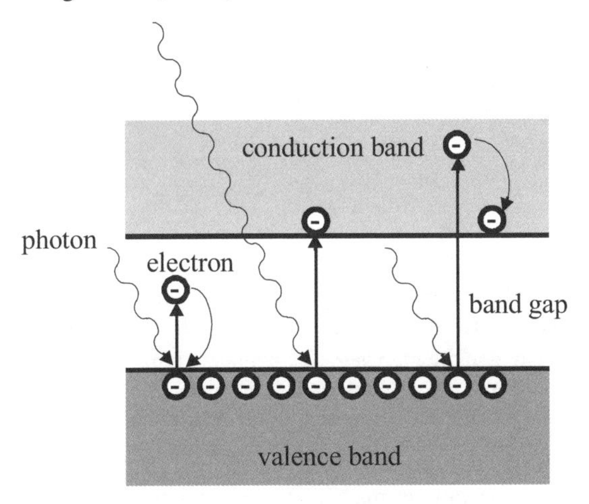

Figure 5.6: Band gaps and charge carrier production in a photovoltaic cell.

The limitation of the theoretical maximum efficiency of a solar cell to 44% is caused mainly by the width of the solar spectrum. Starting from an energy given by the band gap, electrons are lifted from the valence band into the conduction band. Higher-energy photons of the solar spectrum are likewise absorbed, but transfer the surplus energy relative to the gap into thermal energy. Furthermore, there is always a part of the long-wave solar irradiance in the infrared that is not absorbed.

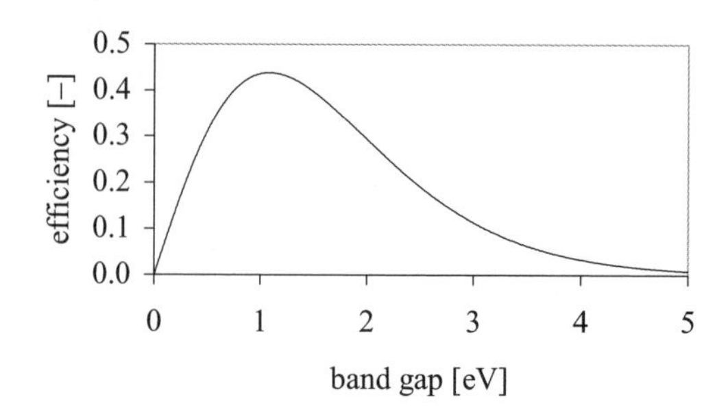

Figure 5.7: Theoretically possible efficiencies of solar cells.

Example 5.1

Calculation of the maximum wavelength of the solar spectrum still absorbed by the following solar cell materials: crystalline silicon with a band gap E_{gap} of 1.124 eV, CdTe with E_{gap} = 1.5 eV and amorphous silicon with E_{gap} = 1.7 eV.

The conversion of the band-gap energy E_{gap} into the wavelength λ by means of Planck's equation results in the longest wavelength of the radiation still absorbed:

$$E_{gap} = h\nu = \frac{hc}{\lambda} \quad [J]$$

with $h = 6.626 \times 10^{-34} Js$, $c = 2.99792 \times 10^{8} \frac{m}{s}$, $q = 1.6021 \times 10^{-19} As$

$$\Rightarrow E_{gap} = \frac{1.24 \times 10^{-6}}{\lambda} [eV] \text{ with } \lambda[m]$$

Photons of greater wavelength are below the band-gap energy and cannot be absorbed. Crystalline silicon therefore absorbs to $\lambda < 1.1 \times 10^{-6}$ m, CdTe to $\lambda < 0.826 \times 10^{-6}$ m and amorphous silicon to 0.729×10^{-6} m only in the visible, due to the high gap. The complete infrared radiation of sunlight is not absorbed by amorphous silicon and is partly responsible for the low efficiencies of this material. The gap of amorphous silicon can be reduced, however, by the addition of germanium.

The electrical field within the photovoltaic diode is produced by endowing the semiconductor material with foreign atoms. If the tetravalent silicon is endowed with pentavalent phosphorus, free electron charge carriers (negative or n-doting) develop. Then if the other side of the silicon wafer is doted with trivalent boron atoms (positive or p-doting), an electron deficiency develops there. The free electrons diffuse into the p-doted area and thus create an electrical field; the p-doted section now displays an electron excess in the boundary layer and is negative, the n-doted section has an electron deficiency and is positively charged. The electric field is in equilibrium with the diffusion process, preventing further electrons of the n-doted layer from moving to the p-layer.

If free charge carriers are now produced by photon absorption, they start to diffuse due to the concentration gradient, which is produced by the exponentially decaying absorption. Electrons diffusing to the n-area are the majority carriers and can leave at the contacts, whereas they remain minority carriers at the p-side and recombine. The increase of electron density in the n-area leads to an increase in potential (so-called quasi-Fermi level), and the increase of holes decreases the hole potential and the difference gives the voltage level

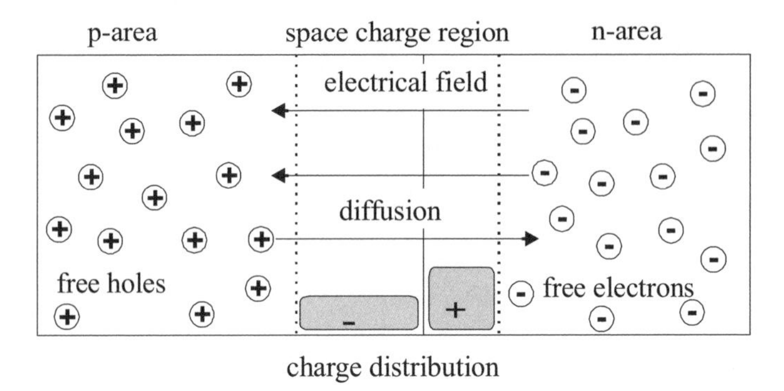

Figure 5.8: Electrical field and movement of the charge carriers.

The photocurrent thus flows in the opposite direction to the forward current of a p/n diode! The photocurrent produced is always a direct current (DC). With AC technology most common in buildings nowadays, an inverter must be introduced between the solar generator and electricity consumer to convert DC into AC.

5.7 Current-voltage characteristics

Photovoltaic cells are photocurrent generators whose electrical characteristic is determined by the superposition of a diode characteristic and a voltage-independent photocurrent in good approximation. The voltage produced by the photodiode is reduced by series resistances of the material and the metallic contacts, and part of the photocurrent produced flows off due to finite bypass resistances. The current usable in the outer electrical circuit results from the photocurrent minus the diode and shunt losses. The voltage across the internal photodiode is obtained from the outside voltage measured at the contacts plus the voltage drop over the series resistances.

5.7.1 Characteristic values and efficiency

In an open outer electrical circuit (total value of current $I = 0$) the maximum voltage is generated; this is called the open circuit voltage, V_{oc}. In a short electrical circuit (voltage $V = 0$) the short circuit current I_{sc} is obtained.

The electrical power P of the solar cell is calculated by the product of the current and voltage.

$$P = I \times V \quad [W] \tag{5.4}$$

The performance curve indicates a sharp maximum near the open circuit voltage; this is called the maximum power point (MPP).

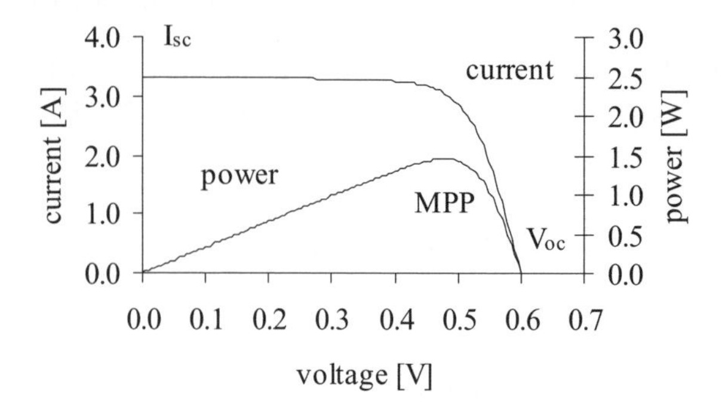

Figure 5.9: Current-voltage characteristic (bold) and power curve (thin) of a single monocrystalline solar cell with 100 cm^2 surface.

All characteristics were calculated using the simulation environment INSEL (Schumacher, 1991). The efficiency of a solar cell is calculated from the relation of the surface-related power produced at the MPP point and the solar irradiance G.

$$\eta = \frac{P_{MPP}}{A \times G} \tag{5.5}$$

Depending on the reference selected for the surface A, the cell efficiency or the module efficiency is calculated.

5.7.2 *Curve fittings to the current-voltage characteristic*

The best curve fitting to measured current-voltage curves of a crystalline solar cell is obtained from the mathematical description of an alternate circuit diagram characterised by the parallel connection of two diodes with the diode saturation currents I_{01} and I_{02}, and the diode factors n_1 and n_2, the so-called two-diode model. In the alternate circuit diagram a current source produces an irradiance-dependent photocurrent I_{ph}, part of which flows off at the diodes due to charge carrier recombination. The current loss caused by low resistance at the edges of the solar cell is characterised by the shunt resistance R_p, which lies parallel to the diodes and the current source. If the voltage at the contacts of the solar cell, i.e. at the consumer, is called V, then the somewhat higher voltage $V + IR_s$ applies to all the parallel components. R_s is the series resistance of the solar cell across which a voltage drop develops proportionally to the current I.

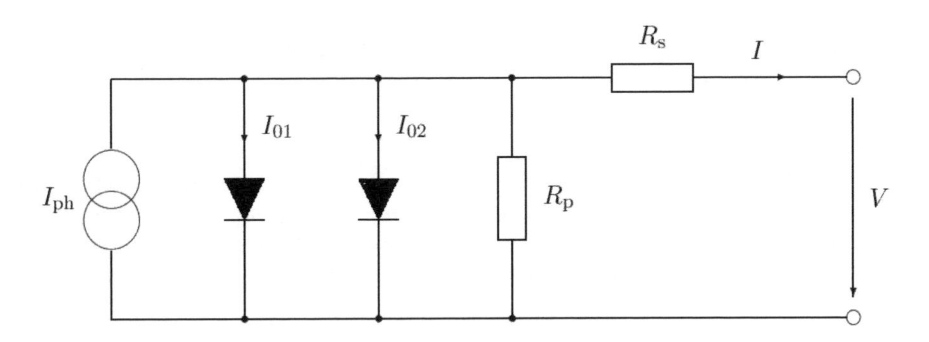

Figure 5.10: Alternate circuit diagram of a solar cell in the two-diode model.

The two-diode model is represented by an implicit equation of the current, which can only be solved iteratively.

$$I = I_{ph} - I_{01}\left(\exp\left(\frac{q(V+IR_s)}{n_1 kT}\right)-1\right) - I_{02}\left(\exp\left(\frac{q(V+IR_s)}{n_2 kT}\right)-1\right) - \frac{V+IR_s}{R_p} \tag{5.6}$$

n_1, n_2: Diode factors [–]
I_{01}: Saturation current from base (p-silicon) and emitter (n-silicon) [A]
I_{02}: Saturation current in the space-charge zone [A]
I_{ph}: Photocurrent of the solar cell [A]
R_s: Series resistance [Ω]
R_p: Bypass resistance [Ω]
k: Boltzmann constant [1.38046×10^{-23} J/K]
q: Elementary charge [1.602×10^{-19} C]
T: Temperature [K]

Diode factors and saturation currents

The two diode equations with the saturation currents I_{01} and I_{02} and with the diode factors n_1 and n_2 describe diffusion and recombination characteristics of the charge carriers in the material itself (index 1) and in the space-charge zone (index 2). Photo-generated charge carriers must first diffuse into the region of the electrical field, to be separated there by the internal electrical field. If they lose their energy during diffusion by dropping back again directly into the valence band or recombining via defect levels in the band gap, the saturation currents rise and the usable photocurrent is reduced.

The saturation current I_{01} depends on the diffusion coefficient and on the life span of the photo-generated charge carriers outside the space-charge zone. From the exponential dependency of the charge carrier concentration on the voltage, the diode factor $n_1 = 1$ results. The temperature dependence of the saturation current I_{01} is non-linear.

$$I_{01} = C_{01} T^3 \exp\left(-\frac{E_{gap}}{kT}\right) \tag{5.7}$$

with

C_{01}: Temperature coefficient obtained from parameter fitting [A K^{-3}]

E_{gap}: Band gap [J or eV], e.g. crystalline silicon:
$1.124\text{eV} = 1.6 \times 10^{-19} \times 1.124J = 1.8 \times 10^{-19}J$

The second diode describes the recombination of charge carriers in the space-charge zone. The saturation current I_{02} rises with charge carrier density and with the recombination rate. The diode factor $n_2 = 2$ can be deduced by recombination of charge carriers at defect levels in the center of the band gap. The temperature dependence of I_{02} is given by the following equation:

$$I_{02} = C_{02} T^{5/2} \exp(-\frac{E_{gap}}{2kT}) \tag{5.8}$$

The values of the temperature coefficients are obtained from parameter fits to the current-voltage characteristics. The value range for a cell with a surface of 100 cm^2 is typically for C_{01} between 150–180 A K^{-3}, and for C_{02} between 1.3–1.7 × 10–2 $AK^{-5/2}$.

Series resistance

The series resistance R_s consists of the internal resistance of the solar cell and the contact resistance. Since with stronger irradiance the current density increases, the power losses at the series resistance also rise. The surface-related series resistance $r_{s,cell}$ of an industrially manufactured solar cell with a surface of 100 cm^2 is typically about 10^{-4}–10^{-5} Ωm^2, approximately half being caused by metallisation at the front and the remainder by the internal cell resistance and other contact resistances. The entire series resistance of a solar module results from the surface-related cell resistance $r_{s,cell}$ [Ωm^2] multiplied by the number of cells n_s connected in series. The total series resistance $R_{s,total}$ [Ωm^2], often given as an absolute value, is obtained by:

$$R_{s,total} = \frac{r_{s,cell} \times n_s}{A_{cell}} \quad [\Omega]$$
$$r_{s,total} = r_{s,cell} \times n_s = R_{s,total} A_{cell} \quad [\Omega m^2] \tag{5.9}$$

In particular, the series resistance influences the gradient of the current-voltage characteristic near the open circuit voltage and can be determined in a simplified model by the gradient of the characteristic at the open circuit voltage V_{oc}.

$$R_s \approx \left.\frac{dV}{dI}\right|_{V=V_{oc}} \tag{5.10}$$

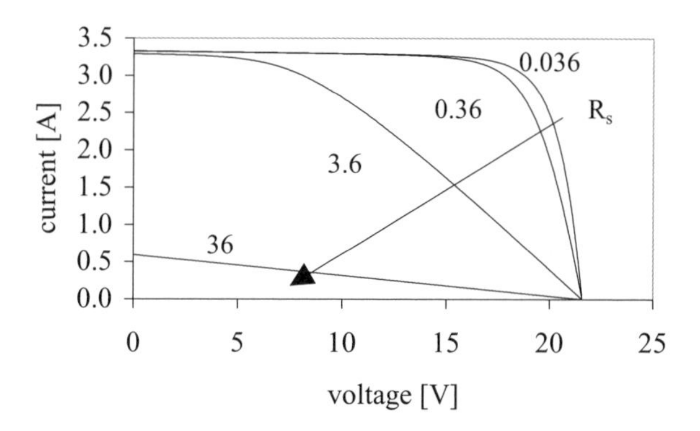

Figure 5.11: Influence of the series resistance [Ω] on the current-voltage characteristic of a module with 36 cells.

With rising series resistance, the gradient near the open circuit voltage decreases and the current-voltage curves flatten. A module series resistance of 0.036 Ω corresponds with 36 cells in series to a cell resistance of 10^{-5} Ωm^2; 36 Ω correspond to a cell resistance of 10^{-2} Ωm^2.

Shunt resistance

The parallel or shunt resistance R_p represents the leakage current, which is lost mainly in the p/n interface of the diode and along the edges. During a series connection of cells, the reciprocal resistances add up, i.e. for n parallel-switched cell strings and the same shunt resistances for each cell ($R_{p,1} = R_{p,2} = .. = R_{p,n_p}$) the total resistance is as follows:

$$\frac{1}{R_{p,total}} = \frac{1}{R_{p,1}} + \frac{1}{R_{p,2}} + \ldots + \frac{1}{R_{p,n_p}} = \frac{n_p}{R_{p,cell}}$$

$$R_{p,total}[\Omega] = \frac{R_{p,total}[\Omega]}{n_p} = \frac{r_{p,total}[\Omega m^2]}{A_{cell}[m^2] \times n_p} \tag{5.11}$$

$$r_{p,total}[\Omega m^2] = \frac{r_{p,total}[\Omega m^2]}{n_p} = R_{p,total}[\Omega] A_{cell}[m^2]$$

Typical values of commercial silicon cells are between 0.1 and 10 Ωm^2. The module manufacturers often give an absolute value in Ω for the shunt resistance.

The shunt resistance influences the gradient of the current-voltage characteristic within the area of the short circuit current and can be determined approximately from the gradient of the characteristic at the point of the short circuit current.

$$R_p \approx \left. \frac{dV}{dI} \right|_{V=0} \tag{5.12}$$

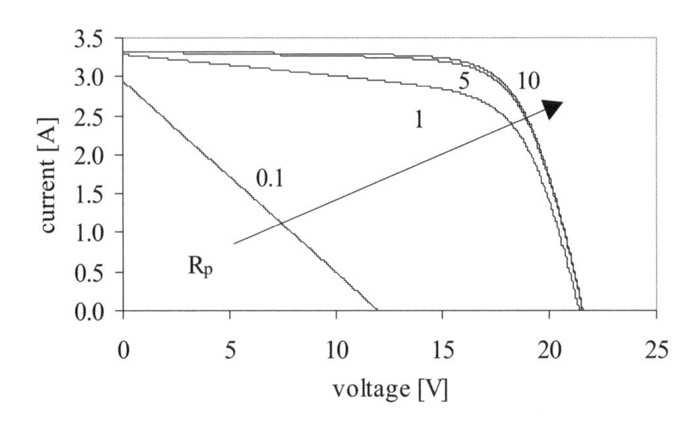

Figure 5.12: Influence of the shunt resistance [Ω] on the current-voltage characteristic for a module with 36 cells in series.

At a cell surface of 100 cm², a module resistance of 0.1 Ω corresponds to a cell resistance of $10^{-3}\ \Omega m^2$, 10 Ω correspond to a cell resistance of $10^{-1}\ \Omega m^2$.

Example 5.2

On the data sheet of a large-scale module with 3 parallel strings of 54 cells each (100 cm² surface), a series resistance of 0.3 Ω and a shunt resistance of 150 Ω are indicated. From these values the surface-related cell resistances are to be calculated.

Series resistance:

$$r_{s,cell}\left[\Omega m^2\right] = \frac{R_{s,total}\left[\Omega\right] \times A_{cell}\left[m^2\right]}{n_s} = \frac{0.3\Omega \times 10^{-2} m^2}{54} = 5.55 \times 10^{-5}\Omega m^2$$

The surface-related total resistance of the module results from the product of the cell resistance and the number of cells in series, and is $3 \times 10^{-3}\ \Omega m^2$.

Bypass resistance:

$$R_{p,total}\left[\Omega\right] = 150\Omega = \frac{r_{p,cell}\left[\Omega m^2\right]}{A_{cell}\left[m^2\right] \times n_p}$$

$$r_{p,cell}\left[\Omega m^2\right] = 150\Omega \times 10^{-2} m^2 \times 3 = 4.5\Omega m^2$$

$$r_{p,total}\left[\Omega m^2\right] = R_{p,total}\left[\Omega\right] A_{cell}\left[m^2\right] = \frac{r_{p,cell}\left[\Omega m^2\right]}{n_p} = 1.5\Omega m^2$$

Photocurrent

The photocurrent of a solar cell depends not only on the wavelength-dependent absorption coefficients but also on the diffusion and recombination characteristics of the material. With sufficient accuracy the photocurrent can be approximated as linear-dependent on irradiance and temperature.

$$I_{Ph} = I_{ph,STC} \frac{G}{G_{STC}} \left(1 + \alpha_I \left(T_{PV} - T_{PV,STC}\right)\right) \tag{5.13}$$

G_{STC}: Irradiance under standard test conditions [1000 W/m^2]
$T_{PV,\,STC}$: Module temperature under standard test conditions [25°C]

In the data sheets of the module manufacturers, the temperature coefficient of the current per Kelvin is indicated as a percentage or in absolute units [AK^{-1}]. The temperature coefficient α_I is usually very small and positive (about 0.03-0.04% of the short circuit current per Kelvin). Neglecting the very small diode current, the photocurrent can be equated to the short circuit current I_{SC}.

Curve fitting of the two-diode model

For an adjustment of the six unknown parameters of the two-diode model to measured data, the error function of the measured currents at a given voltage, irradiance and temperature and of the simulated currents must be minimised.

$$\Phi = \sqrt{\sum_{i=1}^{N} \left(I_{i,meas} - I_{i,calc}\right)^2} \tag{5.14}$$

With little mathematical effort, i.e. a linear regression, only the parameters $I_{ph,\,STC}$ and α_I can be calculated from the linear irradiance-dependence and temperature-dependence of the short circuit current. The remaining parameter set can only be determined by gradient, raster or genetic procedures (Pukrop, 1997).

Example 5.3

Calculation of the parameters $I_{ph,\,STC}$ and α_I of a Siemens M55 module from the following table of values measured.

Irradiance G [W/m²]	*Module temperature* T_{PV} [°C]	*Short circuit current* I_{SC} [A]
220	30	0.74
510	41	1.72
850	54	2.88
1000	60	3.40

To be able to carry out a linear regression of the temperature dependence of the short circuit current using Equation (5.13), the irradiance G must be eliminated from the equation, so that only the temperature difference from standard test conditions remains as an independent parameter. The short circuit current is thus divided by the irradiance, and the regression for I_{SC}/G against $T_{PV}-T_{PV,\,STC}$ is carried out.

Current/irradiance $I_{ph,STC}/G_{STC}$ [$AW^{-1}m^2$]	*Temperature difference* $T_{PV}-T_{PV,STC}$ [K]
3.364×10^{-3}	5
3.373×10^{-3}	16
3.388×10^{-3}	29
3.400×10^{-3}	35

The linear regression produces as axis intercept $I_{ph,STC}/G_{STC} = 0.0034$ $AW^{-1}m^2$, i.e. a short circuit current $I_{ph,STC}$ of 3.4A, and as a gradient $I_{ph,STC}/G_{STC}\,\alpha_I = 1 \times 10^{-6}$ $AW^{-1}m^2K^{-1}$, i.e. $\alpha_I = 3 \times 10^{-4}$ K^{-1} (+0.03%/K).

5.7.2.1 Parameter adjustment from module data sheets

To simplify parameter adjustment, the two-diode model can be reduced to a one-diode model in which, according to the Shockley theory, recombination in the space-charge zone is neglected, so the second diode term is omitted. The current-voltage characteristic is thereby simplified to the following implicit equation:

$$I = I_{ph} - I_0\left(\exp\left(\frac{q(V+IR_s)}{nkT}\right)-1\right) - \frac{V+IR_s}{R_p} \tag{5.15}$$

The irradiance-dependency of the photocurrent is characterised by the current $I_{ph,STC}$ under standard test conditions using Equation (5.13), and the temperature-dependence by the temperature coefficient α_I. The saturation current $I_0 = I_{01}$ is likewise temperature-dependent and depends, based on Equation (5.7), on the fit parameter C_0.

Thus the six parameters $I_{ph,STC}$, α_I , C_0, R_s, n and R_p must be determined for curve adjustment of the one-diode model. In the data sheets of the module manufacturers, however, usually only three operating points are indicated on the current-voltage characteristic (short circuit current at $V = 0$, open circuit voltage for $I = 0$, and the MPP values I_{MPP}, V_{MPP}) plus the temperature coefficients of the voltage α_v and the current α_I, i.e., at most, five unknown parameters can be identified.

The most admissible simplification for crystalline modules is the assumption of an initially infinitely high parallel resistance. After parameter adjustment, a realistic parallel resistance can subsequently be set. The current-voltage equation for the parameter adjustment can be reduced to

$$I = I_{ph}\left(I_{ph,STC},\alpha_I\right) - I_0\left(C_0\right)\left(\exp\left(\frac{q\left(V+IR_s\right)/n_s}{nkT}\right)-1\right) \tag{5.16}$$

with the five parameters $I_{ph,STC}$, α_I, C_0, R_s and n.

The main problem with this simplified approach is that the current-voltage characteristic thus obtained correctly passes through the given three points (I_{sc}, I_{MPP} and $I = 0$ at V_{oc}), but the power so obtained is not necessarily the maximum power at V_{MPP}. A further equation will therefore be derived later, which uses the maximum power condition at V_{MPP}: $d\left(V \times I\right)/dV\big|_{V=V_{MPP}} = 0$. The equation for the parallel resistance can only be solved iteratively, so the simplified method will be presented first.

Since the current-voltage equation can only be solved for one cell, due to the functional characteristics of the exponential function, the module voltages V indicated in the data sheet must be divided by the number of cells n_s connected in series. If several cell strings are switched parallel in the module, all currents used, including the current temperature coefficients, are divided by the number of parallel switched strings n_p and the characteristic is calculated first for one string. The characteristics can then easily be calculated by adding the currents at the same voltage.

Parameter 1: photocurrent $I_{ph,\,STC}$
The first parameter $I_{ph,\,STC}$ is obtained from the data sheet specification of the short circuit current under standard test conditions: $I_{ph,STC} = I_{SC,STC}$.

Parameter 2: temperature coefficient of the photocurrent α_I
The temperature coefficient of the photocurrent α_I is given in the data sheet.

Parameter 3: fit parameter of the temperature-dependence of the saturation current C_0.
Similarly to Equation (5.7), C_0 is given in the one-diode model by

$$C_0 = \frac{I_0(T)}{T^3 \exp\left(-\frac{E_{gap}}{kT}\right)} \tag{5.17}$$

with $E_{gap} = 1.124$ eV $= 1.8 \times 10^{-19}$ J for crystalline silicon.

The saturation current I_0 is calculated from the data sheet specification of the open circuit voltage, which is indicated at the temperature $T = T_{STC}$.

$$I = 0 = I_{ph} - I_0\left(\exp\left(\frac{qV_{OC}/n_s}{nkT}\right) - 1\right) \approx I_{ph} - I_0 \exp\left(\frac{qV_{OC}/n_s}{nkT}\right)$$

$$\Rightarrow I_0 = I_{ph} \exp\left(\frac{-qV_{OC}/n_s}{nkT}\right) \tag{5.18}$$

With this value of the saturation current I_0 at $T = T_{STC}$, the third parameter C_0 is calculated. The fit parameter n, however, still has to be determined.

Parameter 4: series resistance R_s
From the operating point of the data sheet at the point of maximum performance, the parameter R_s can be determined. Here, Equation (5.18) is used for I_0,

$$I_{MPP} = I_{ph} - I_0\left(\exp\left(\frac{q(V_{MPP} + I_{MPP}R_s)/n_s}{nkT}\right) - 1\right) \approx I_{ph} - I_0 \exp\left(\frac{q(V_{MPP} + I_{MPP}R_s)/n_s}{nkT}\right)$$

$$\Leftrightarrow \ln\left(\frac{I_{ph} - I_{MPP}}{I_{ph}}\right) + \frac{qV_{OC}/n_s}{nkT} = \frac{q(V_{MPP} + I_{MPP}R_s)/n_s}{nkT}$$

$$\Rightarrow \frac{R_s}{n_s} = \frac{\frac{nkT}{q} \ln\left(1 - \frac{I_{MPP}}{I_{ph}}\right) + (V_{OC} - V_{MPP}) / n_s}{I_{MPP}} \tag{5.19}$$

with R_s defining the total series resistance of the module.

Parameter 5: diode parameter n

The fifth parameter is determined from the temperature coefficients of the voltage α_V indicated in the data sheet. In addition, the open circuit voltage $V_{oc} = -\frac{n_s nkT}{q} \ln \frac{I_0}{I_{ph}}$ is differentiated by temperature.

$$\begin{aligned}
\frac{dV_{oc}}{dT} = \alpha_V &= \frac{d}{dT}\left(-\frac{n_s nkT}{q} \ln\left(\frac{I_0(T)}{I_{ph}(T)}\right)\right) \\
&= -\left(\frac{n_s nkT}{q}\frac{d}{dT}(\ln I_0(T)) + \frac{n_s nk}{q}\ln I_0(T)\right) + \frac{n_s nkT}{q}\frac{d}{dT}(\ln I_{ph}(T)) + \frac{n_s nk}{q}\ln I_{ph}(T) \\
&= \frac{n_s nk}{q}\ln\left(\frac{I_{ph}(T)}{I_0(T)}\right) + \frac{n_s nkT}{q}\left(\frac{1}{I_{ph}(T)}\frac{d}{dT}I_{ph}(T) - \frac{1}{I_0(T)}\frac{d}{dT}I_0(T)\right)
\end{aligned} \tag{5.20}$$

The photocurrent differentiated by temperature is given by the data sheet figure α_I. The differentiation of the diode saturation current by temperature produces:

$$\frac{dI_0(T)}{dT} = C_0 T \exp\left(\frac{-E_{gap}}{kT}\right)\left(3T + \frac{E_{gap}}{k}\right) \tag{5.21}$$

so

$$\frac{1}{I_0}\frac{dI_0(T)}{dT} = C_0^{-1} T^{-3} \exp\left(\frac{E_{gap}}{kT}\right) C_0 T \exp\left(\frac{-E_{gap}}{kT}\right)\left(3T + \frac{E_{gap}}{k}\right) = \frac{3}{T} + \frac{E_{gap}}{kT^2} \tag{5.22}$$

Inserting Equations (5.21) and (5.22) into Equation (5.20) produces

$$\alpha_V = \frac{n_s nk}{q}\ln\left(\frac{I_{ph}(T)}{I_0(T)}\right) + \frac{n_s nkT}{q}\frac{\alpha_I}{I_{ph}(T)} - \left(3 + \frac{E_{gap}}{kT}\right)\frac{n_s nk}{q} \tag{5.23}$$

The diode fit parameter n is thus calculable as the fifth parameter from data sheet specifications under standard test conditions:

$$n = \frac{\left(\alpha_V - \frac{V_{OC}}{T_{STC}}\right) / n_s}{\frac{kT_{STC}}{q}\frac{\alpha_I}{I_{ph,STC}} - \frac{k}{q}\left(3 + \frac{E_{gap}}{kT_{STC}}\right)} \tag{5.24}$$

Example 5.4

Calculation of the parameters $I_{ph,STC}$, α_I, C_0, R_s and n of the standard modules SM55 and BP585, and of an ASE large-scale module from the following technical specifications of the data sheet:

	Siemens M55	*BP 585*	*ASE*
Nominal power [W]	53	85	202
Short circuit current I_{SC} [A]	3.35	5.0	8.18
MPP-current [A]	3.05	4.72	8.1
Open circuit voltage V_{oc} [V]	21.7	22.3	33
MPP-voltage V_{MPP} [V]	17.4	18.0	25
Temperature coefficient of the voltage α_V [V/K]	−0.074 (−0.34%/K)	−0.086 (−0.39%/K)	−0.1 (−0.3%/K)
Temperature coefficient of the current α_I [A/K]	0.00134 (+0.04%/K)	+0.0025 (+0.05%/K)	+0.006 (−0.074%/K)
Number of cells in series n_s	36 (3 × 12)	36 (4 × 9)	54
Number of parallel strings n_p	1	1	3

With the ASE module, first the current values I_{SC}, I_{MPP} and α_I are calculated for a string; with the values thus obtained the parameters are determined and the characteristic of the cell string calculated using Equation (5.15). Subsequently, at a given voltage the currents are added and so the total characteristic of the module is calculated.

The five parameters result in:

	Siemens M55	*BP 585*	*ASE*
Parameter 1: $I_{ph,STC} = I_{SC,STC}$ (A)	3.35	5.0	8.18/3 = 2.73
Parameter 2: α_I (A/K)	0.00134	0.0025	0.006/3 = 0.002
Parameter 5: n (-)	1.015	1.11	0.97
Parameter 3: C_0 (A/K^3)	114.75	731.96	25.02
Parameter 4: R_s (Ω)	0.66	0.2844	0.648

If the current-voltage or power characteristics of an SM55 module with characteristic values from the parameter identification process are compared with the exact calculations of the two-diode model, then a good agreement of the characteristics can be seen. The one-diode model overrates the performance by 0.6–1.4 W, which at high irradiances amounts to 1.5%, and at low irradiances 15%.

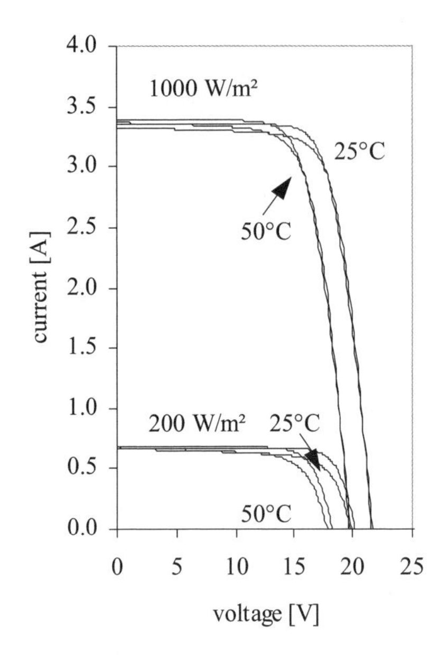

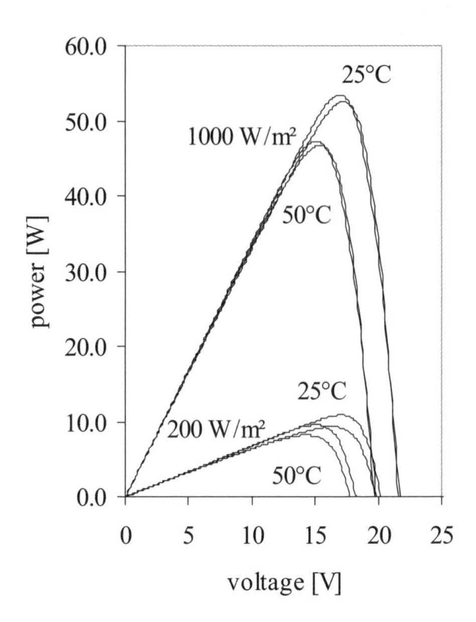

Figure 5.13: Comparison of the current-voltage and of the power characteristic of a SM55 module using the parameter identification procedure (bold curves) and the two-diode model (thin curves).

5.7.2.2 Full parameter set calculation

If all six parameters of the one-diode equation are to be calculated, a further equation has to be derived for the parallel resistance, which uses only manufacturer's data. The equation used is the derivative of the power to the voltage at the MPP point, which has to be zero at this point.

$$d(V \times I)/dV\Big|_{V=V_{MPP}} = 0 \tag{5.25}$$

The shunt resistance thus obtained is given by

$$r_p = \frac{v_{MPP} + i_{MPP} r_s}{\left(i_{SC,STC} - C_0 T_{STC}{}^3 \exp\left(-E_{gap}/kT_{STC}\right) \exp\left(\left(v_{MPP} + i_{MPP} r_s\right)/\left(nkT_{STC}/q\right) - 1\right) - i_{MPP}\right)} \tag{5.26}$$

using normalised current and voltage values for a single cell with surface area A_{cell}, and n_s series connected and n_p parallel connected cells within the module:

$$i = I/\left(n_p A_{cell}\right)$$
$$v = V/n_s$$

As the equation for the series resistance r_s now contains a term with the parallel resistance, both equations have to be solved together in an iterative process.

$$r_s = \frac{\frac{nkT_{STC}}{q}\ln\left(1+\left(i_{MPP}-\frac{v_{MPP}}{r_p}\right)\frac{nkT_{STC}}{q}\frac{T_{STC}^{3}\exp\left(-E_{gap}/kT_{STC}\right)\times\left(\exp\left(qV_{oc}/nkT_{STC}\right)-1\right)}{\left(i_{SC,STC}-v_{oc}/r_p\right)T_{STC}^{3}\exp\left(-E_{gap}/kT_{STC}\right)v_{MPP}}\right)-v_{MPP}}{i_{MPP}} \tag{5.27}$$

5.7.2.3 Simple explicit model for system design

The characteristic obtained from the parameter identification can only be solved iteratively, even if the shunt resistance is not considered. Only by ignoring the series resistances can a simple explicit connection between current and voltage be obtained from Equation (5.15) which enables fast system designs.

$$I = I_{ph}\left(I_{ph,STC},\alpha_I\right)-I_0\left(C_o\right)\left(\exp\left(\frac{q\left(V/n_s\right)}{nkT}\right)-1\right) \tag{5.28}$$

Of the four parameters $I_{ph,STC}$, α_I, C_0 and n, the photocurrent under standard test conditions $I_{ph,STC}$ and the temperature coefficient of the current α_I are indicated in the data sheet. The temperature coefficient of the saturation current C_0 and the diode parameter n can be determined by three different methods:

1. For parameter determination, the temperature-dependence of the voltage α_V and current α_I, as well as two operating points of the characteristic at open circuit voltage and short circuit current are used. The parameters are calculated using the above equations of the identification method and only the series resistance is set at zero. In this case the MPP condition is not used and the MPP performance is overestimated at all irradiances. However, the temperature dependence of the performance and the open circuit voltage correspond well with real characteristics.

2. As above, α_V (for the diode parameter n) and α_I (for the photocurrent) are used. However, the saturation current I_0 and thus C_0 are no longer calculated from the open circuit voltage condition, but at the MPP point:

$$I_0 = \frac{I_{ph}-I_{MPP}}{\exp\left(\frac{qV_{MPP}/n_s}{nkT}\right)} \tag{5.29}$$

This condition results in substantially better agreement for the power values; however, the open circuit voltage does not correspond with the real characteristic.

3. Alternatively, the parameters can be determined from three operating points (with short circuit current, open circuit voltage and the MPP condition) and from the temperature

dependence of the current. Here, however, the condition of the significant temperature dependence of the voltage is not used.

The diode parameter n is now obtained from the MPP condition:

$$I_{MPP} = I_{ph} - \frac{I_{ph}}{\exp\left(\frac{qV_{oc}/n_s}{nkT}\right)} \exp\left(\frac{qV_{MPP}/n_s}{nkT}\right)$$
$$\Rightarrow n = \frac{q(V_{MPP} - V_{oc})/n_s}{kT \ln\left(1 - \frac{I_{MPP}}{I_{ph}}\right)} \quad (5.30)$$

where I_0 has been determined using Equation (5.18) from the operating point at open circuit voltage. If I_0 is known, the temperature factor of the saturation current C_0 can be calculated using Equation (5.17).

The use of the MPP condition results in correct performances at 25°C. At other irradiances and above all different module temperatures, however, the performance is calculated with errors over 20%, so this calculation method is not recommended.

In the following, the comparison of the characteristic calculation is carried out based on the exact two-diode model with series and parallel resistance and the three methods of parameter identification based on the simple explicit model. For this, current-voltage characteristics and power curves have been calculated at 1000 W/m^2 irradiance and a module temperature of 50°C. Method 1 clearly overestimates the MPP performance, with the open circuit voltage resembling the two-diode value. Method 3 underestimates the performance and calculates the voltage level with a large error. The second method uses both the temperature dependence of the voltage and the MPP condition and results in by far the best curve adjustment. At 1000 W/m^2 irradiance and a module temperature of 50°C, the following performance curves are obtained for a standard module (calculated from manufacturer's data based on the simple explicit models and the two-diode model):

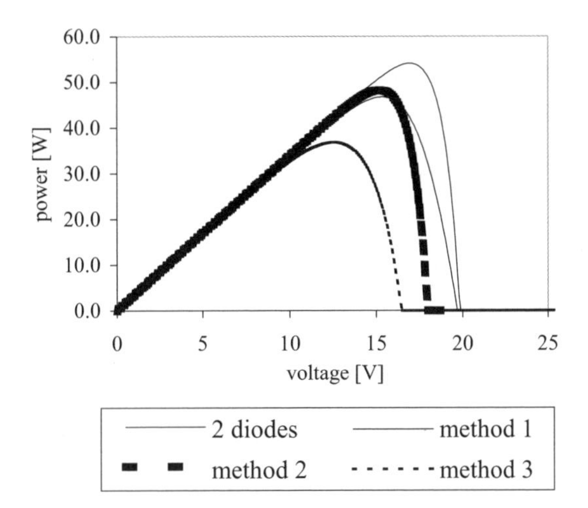

Figure 5.14: Comparison of the power curves of an SM55 module.

5.7.3 *I-V characteristic addition and generator interconnecting*

To calculate the current-voltage characteristic of a photovoltaic generator, the characteristics of the individual modules have to be added. Here the voltage contributions V_{ij} of the modules i of the string j concerned are added in serial connection at a common current. Since the currents through each module are equal in serial interconnecting, the current is limited by the short circuit current of the worst module (the exception being shading with bypass diodes).

In parallel interconnecting of the strings, the generator characteristic results from adding the respective string currents I_j at a given voltage between zero and the string voltage.

Example 5.5

Calculation of PV generator power and voltage in the MPP point for a system with thirty-six modules of 53 W MPP power (17.4 V MPP voltage), which is divided into nine strings, each with four modules connected in series.

Serial interconnecting:

For the four series connected modules the MPP voltages are added at equal current (I_{MPP} = 53 W/ 17.4 V = 3.05 A).

$$V_{MPP,string} = V_{MPP,module1} + V_{MPP,module2} + .. + V_{MPP,module4} = 69.6V$$

Parallel interconnecting:

For the nine strings joined in parallel, the string currents (in each case 3.05 A) are added at equal voltage (V_{MPP}, String = 69.6 V).

$$I_{MPP,Generator} = I_{MPP,String1} + I_{MPP,String2} + ... + I_{MPP,String9} = 27.45A$$

The system thus has an MPP operating point under standard test conditions of 69.6 V and 27.45 A, i.e. a DC performance of 1910 W.

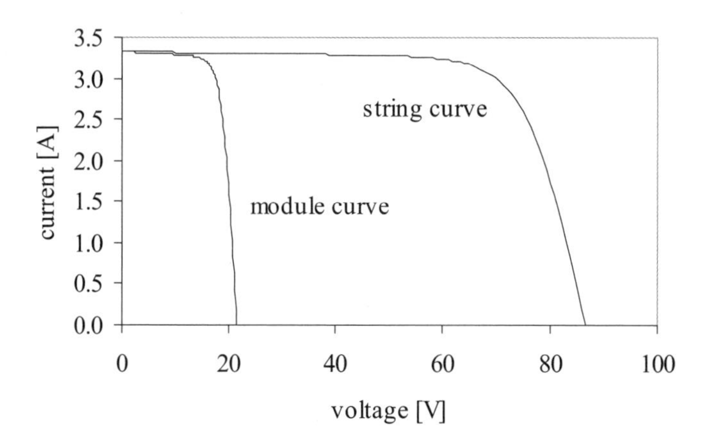

Figure 5.15: I-V characteristic of an individual module with thirty-six cells in series and addition of four series connected module characteristics to the string characteristic.

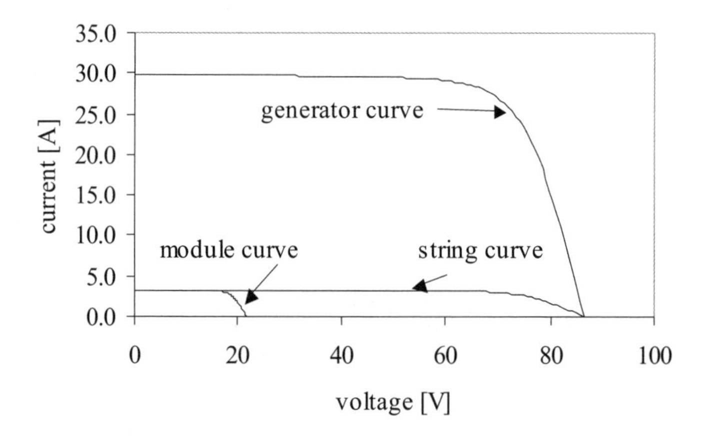

Figure 5.16: Addition of nine parallel-switched PV strings, each with four modules in series, to the generator characteristic.

5.8 PV performance with shading

5.8.1 Bypass diodes and backwards characteristics of solar cells

If individual cells of a module are shaded, they produce a lower photocurrent than the unshaded cells. Since during a series connection equal currents flow through all the cells, starting from zero current at open circuit voltage, the small photocurrent of the shaded cells at first limits the total current. The high photocurrents produced by the unshaded cells can be let through only within the backward voltage area of the shaded cells, with the shaded cells taking up power. Depending on the number of further cells switched in series, the backward voltage can become so high that the breakdown voltage in the shaded cell is achieved and irreversible damage is caused.

The breakdown voltage V_{Br} marks the backward voltage of a diode, at which instead of the extremely low saturation currents an exponential current rise (avalanche breakdown with exponent m) takes place, in commercial solar cells between −10 and −30 V. The power dissipation in the shaded cell can become so high locally that the cell and plastic encapsulation are damaged. The characteristic of a solar cell within the backward voltage area is described by an extension term in the one or two-diode model:

$$I = I_{ph} - I_0\left(\exp\left(\frac{q(V+IR_s)}{nkT}\right)-1\right) - \frac{V+IR_s}{R_p}\left(1+\left(\frac{a}{1-\frac{V+IR_s}{V_{Br}}}\right)^m\right) \quad (5.31)$$

with α as the empirical fit parameter. The parameters for the description of the characteristic in the backward voltage area scatter very strongly and are not a constituent of the usual data sheet specifications.

As an example, the measured breakdown voltage V_{Br} of a polycrystalline cell is –15 V, the exponent m is 3.7 and the fit parameter α is 0.1. The current increase in the backward voltage area resulting from these values is very steep: thus the maximum photocurrent of unshaded cells of a standard module of 3.3 A is let through already at 8.3V, which, as will be shown, puts in question the usual application of one bypass diode per eighteen cells.

First a module characteristic without external bypass diodes will be shown, in order to demonstrate the problems of the power uptake of the shaded cells and the over-proportional power losses of the module. From the total of thirty-six serial connected cells of the module, one cell will be shaded with a remaining diffuse irradiance of 200 W/m^2, while all the other cells are irradiated with 1000 W/m^2. From the characteristic of the shaded cell and from the characteristic of the thirty-five unshaded cells already added, the module characteristic can be constructed.

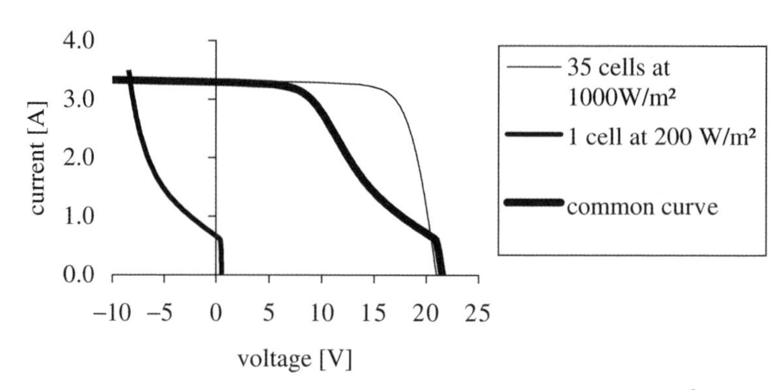

Figure 5.17: Construction of a module characteristic with a shaded cell (200 W/m^2) and thirty-five unshaded cells in series.

At a total current value I = 0, the open circuit voltage of the shaded cell and the unshaded cell are as usual added up to a total voltage, which does not differ from an unshaded module. If the total value of current is now slowly increased, all the cells remain in the positive voltage range up to the short circuit current of the shaded cell. The shaded cell can only let through higher currents than their short circuit current at negative voltages, with the total voltage constantly decreasing. At a total voltage V_{total} = 0, the shaded cell is most strongly negatively polarised (here −8.3 V), while the unshaded cells display further positive voltages (namely +8.3V). The power uptake of the shaded cell results from the product of the current and voltage, and is −8.3V × 3.3A = 27.4 W. The power losses of the total module due to a single shaded cell are 47%! If the backward current of the solar cell did not rise so strongly (as for a "better" diode), the common characteristic would flatten even more strongly.

To reduce the high negative voltages and the power uptake, bypass diodes with reverse current passage direction are switched parallel to the photoelectric cells, ideally one bypass diode per solar cell. If the currents are then added at equal voltages, the characteristic hardly changes for positive voltages, apart from the extremely low saturation current of the diode. As soon as the shaded cell assumes values of around −0.5V to −0.7 V, however, the bypass diode switches and the current can rise exponentially.

If this common characteristic is switched in series with the thirty-five unshaded solar cells, only small performance losses of the module are detected, since only very small negative voltages occur at the shaded cell. However, for technological production problems the concept of cell-integrated bypass diodes diffused into the solar cell has not yet been accepted. Leading out all the cell contact cables to mount external bypass diodes is too complex and expensive. As a compromise, the use of a bypass diode per eighteen cells has become accepted, i.e. in a thirty-six cell standard module a total of three cell access wires must be led out into the external wiring box.

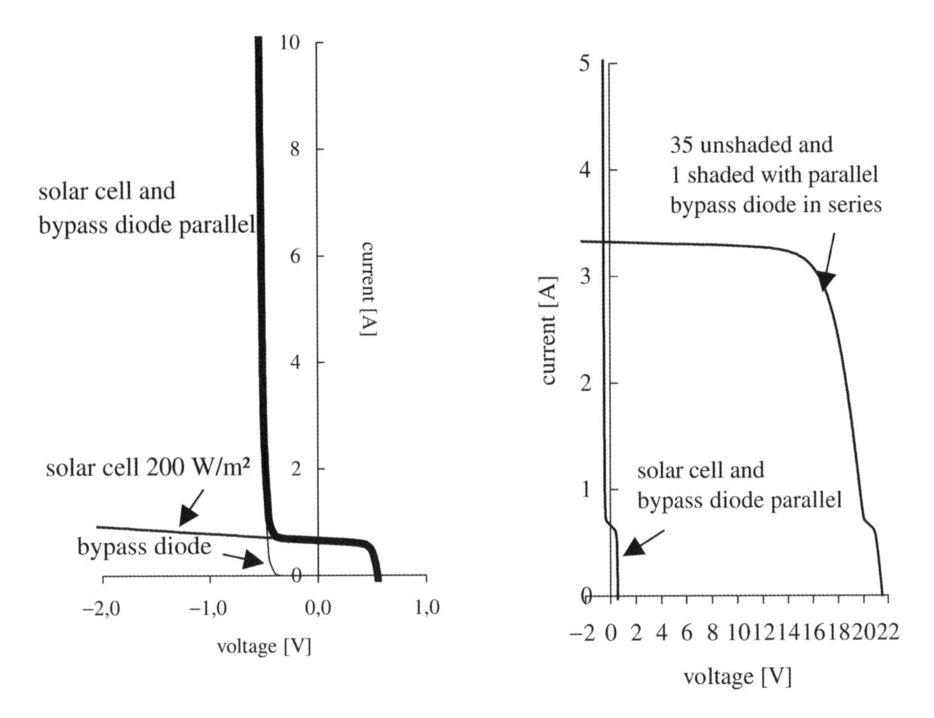

Figure 5.18: Parallel connection of a solar cell with a bypass diode (left) and series connection of thirty-five unshaded cells with the shaded cell with a parallel bypass diode.

In the case of one bypass diode per eighteen cells, a module characteristic with thirty-six cells is constructed in what follows, with one cell shaded.

First the cell string consisting of eighteen cells is considered with the shaded cell and the parallel bypass diode. The characteristic in the positive voltage range is dominated by the backwards characteristic of the shaded cell (similar to Figure 5.17). For a total voltage of zero, there is a high negative voltage of –8.3 V at the shaded cell. The parallel bypass diode would only switch at a negative total voltage, and thus does not yet influence the cell string characteristic. If the shaded cell string is switched in series with the remaining unshaded cell string, the result is a common characteristic which adopts the full current values of the unshaded cells below half the total voltage. The shaded cell string then goes into the backward voltage area and either switches the bypass diode, or the back current is already so high that the photocurrent is allowed through.

With the steep backwards characteristic simulated here the bypass diode does not switch; at half the total voltage the shaded cell string has a voltage level of 0V (+8.3V of the unshaded seventeen cells and –8.3V at the shaded cell), and the full current of the unshaded cells is allowed through. Only with flatter backwards characteristics does the bypass diode contribute to the limitation of the power uptake of the shaded cells.

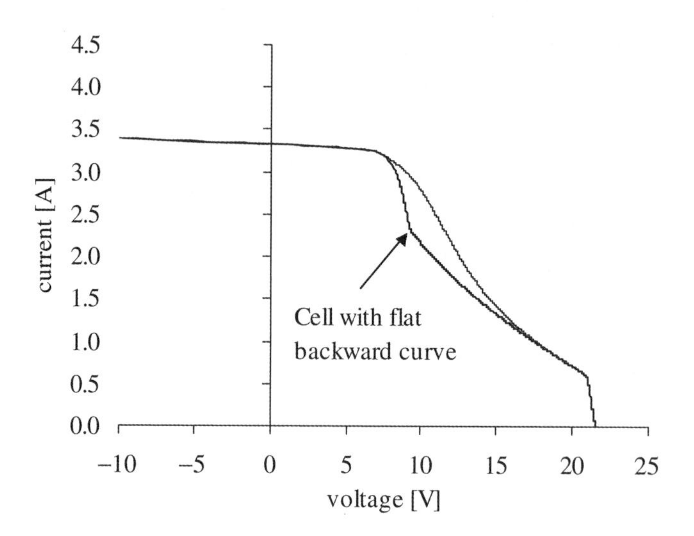

Figure 5.19: Module characteristic with shading of an individual cell with one bypass diode per eighteen cells.

Only if a very flat backwards characteristic is assumed (here with a breakdown voltage of –25 V) does the bypass diode switch below the kink in the characteristic.

5.9 Simple temperature model for PV modules

With high irradiance, the PV module temperature is usually above the STC temperature of 25°C, even at low ambient temperatures. The electrical power P_{MPP} of a PV module decreases linearly with the module temperature, the proportional decrease being calculated from the total of the voltage and current coefficients.

$$\frac{\alpha_P}{P_{MPP}} = \frac{\left.\frac{d(VI)}{dT}\right|_{MPP}}{P_{MPP}} = \frac{\frac{dV}{dT} I_{MPP} + \frac{dI}{dT} V_{MPP}}{V_{MPP} I_{MPP}} = \frac{\alpha_V}{V_{MPP}} + \frac{\alpha_I}{I_{MPP}} \tag{5.32}$$

With average power loss coefficients of 0.3–0.4% per Kelvin, at usual module temperatures of 50°C a performance decrease of as much as 7.5–10% can be expected, which makes a simple model necessary to determine the module temperature. The model applies at first only to the simplest case of free-standing modules, for which simple assumptions can be made for the convective heat transport and the radiation exchange with the sky and the ground. The thermal behaviour of special building integration solutions (double glazing, roof integration etc.) and associated performance reduction is considered in Chapter 6.

In the steady state case, a simple energy balance results in which the absorbed solar radiation $\dot{Q}_G$ minus the electrical power P_{el} and the thermal losses by radiation $\dot{Q}_r$ and convection $\dot{Q}_c$ must equal zero.

$$\dot{Q}_G - P_{el} - \dot{Q}_r - \dot{Q}_c = 0 \tag{5.33}$$

The absorbed solar radiation is a product of PV module surface A, irradiance G and the effective absorption coefficient α, which includes both the ray passage through the glass cover and the absorption of the solar cell material. For this absorption coefficient, values between 0.7–0.9 can be used.

$$\dot{Q}_G = \alpha GA \tag{5.34}$$

For long-wave radiation exchange between the module front and the sky, and the module rear and the ground, a simplified temperature difference is assumed in each case between the module and the ambient temperature T_o, and a simplified form factor between the module surface and an infinitely large confinement surface, so

$$\dot{Q}_r = 2h_r A\left(T_{PV} - T_o\right) \tag{5.35}$$

with

$$h_r = \sigma\varepsilon\left(T_{PV}^2 + T_o^2\right)\left(T_{PV} + T_o\right) \tag{5.36}$$

the emission coefficient of glass ε = 0.88 being assumed and σ desribing the Stefan Boltzmann constant ($\sigma = 5.6697 \times 10^{-8}$ $Wm^{-2}K^{-4}$). The temperatures are to be given in Kelvin.

The convective heat flow is dominated at the module front by forced convection by wind forces ($h_{c,w}$ as a function of the wind velocity v_w); at the module rear, depending on the installation situation, more by free laminar or turbulent convection ($h_{c,\,free}$ as a function of the temperature difference between the module and the environment). For the front and rear, a simplified common heat transmission coefficient h_c is formed for free and forced convection:

$$h_c = \sqrt[3]{h_{c,w}^3 + h_{c,free}^3} \tag{5.37}$$

with the simple approximations

$$\begin{aligned} h_{c,w} &= 4.214 + 3.575\ v_w \\ h_{c,free} &= 1.78\left(T_{PV} - T_o\right)^{1/3} \end{aligned} \tag{5.38}$$

The heat flow by convection is then given by:

$$\dot{Q}_c = 2h_c A\left(T_{PV} - T_o\right) \tag{5.39}$$

The complete energy balance is thus described by

$$\alpha GA - P_{el} - h_r A(T_{PV} - T_o) - h_c A(T_{PV} - T_o) = 0$$
$$\Rightarrow T_{PV} = \frac{(\alpha - \eta_{el}(T_{PV}))G}{h_r(T_{PV}) + h_c(T_{PV})} - T_o \qquad (5.40)$$

with the electrical power $P_{el} = \eta_{el} GA$ used. Since both the heat transmission coefficients and the electrical performance depend on the module temperature, Equation (5.40) must be solved iteratively.

Example 5.6

Calculation of the PV module temperature and electrical power losses at 800 W/m^2 irradiance, an ambient temperature of 10°C and a wind velocity of 3m/s. The electrical efficiency under standard test conditions is 12% and the temperature coefficient of the performance is 0.4%/K. The optical absorption of the module is set at 80%.

First iteration: Assumption of a module temperature of 50°C and calculation of h_r, h_c, η_{el}.

h_r= 5.57 W/m^{2}K

$h_{c\,w}$= 14.9 W/m^{2}K

$h_{c,\,free}$= 6.09 W/m^{2}K

h_c= 15.23 W/m^{2}K

η_{el} =10.8%

From this we can calculate the new module temperature:

T_{PV} = 36°C

Second iteration: Heat transfers and electrical performance at 36°C:

h_r= 5.18 W/m^{2}K

$h_{c,\,w}$= 14.9 W/m^{2}K

$h_{c,\,free}$= 5.27 W/m^{2}K

h_c= 15.11 W/m^{2}K

η_{el} = 11.47%

Module temperature after the second iteration:

T_{PV}= 37°C

Third iteration.

h_r= 5.21 W/m^{2}K

$h_{c\,w}$= 14.9 W/m^{2}K

$h_{c,free}$= 5.34 W/m^{2}K

h_c= 15.31 W/m^{2}K

η_{el} = 11.42%

Module temperature after the third iteration:

T_{PV}= 36.7°C

The electrical efficiency is 11.44%, i.e. the performance reduction in relation to the STC condition is 4.7%.

5.10 System engineering

Photovoltaic system engineering includes on the DC side connecting the PV modules to the PV generator, the sizing of the DC main line and the associated safety engineering (lightning protection and fault current recognition), plus the coupling of the generator via an inverter to the public low-voltage mains.

5.10.1 DC connecting

5.10.1.1 Cable sizing

The DC main line leads the entire generator current from the parallel-switched strings to the inverter. The percentage DC power losses p_R by Ohm resistance R, related to the rated power, should be below 1%. The resistance R is calculated from the product of the specific conductor resistance of copper $\lambda = 0.0178\ \Omega mm^2/m$, the entire line length (back and forth the length l, i.e. a total of $2l$) and the cable diameter A_q [mm²].

$$R = \frac{\lambda 2l}{A_q} \tag{5.41}$$

For a simple line length between the PV branch and inverter l [m], the cable diameter A_q is thus obtained from the admissible DC power loss ΔP_{DC} over the resistance R, related to the DC performance at rated voltage V_n [V] and rated current I_n [A]:

$$\begin{aligned} p_R &= \frac{\Delta P_{DC}}{P_{DC}} = \frac{I_n^{\,2} R}{P_{DC}} = \frac{P_{DC}^2}{V_n^2}\frac{R}{P_{DC}} = \frac{P_{DC}}{V_n^2}\frac{\lambda 2l}{A_q} \\ A_q &= \frac{2l\,\lambda\,P_{DC}}{p_R V_n^2} \end{aligned} \tag{5.42}$$

Example 5.7

Calculation of the cable diameter of a 2 kW system with a rated voltage level of 60 V_{DC} or 240 V_{DC} for a main-line length between roof and cellar of $l = 10$ m. The admissible power loss p_R is to be 1%.

$$60V:\quad A_q = \frac{2l\,\lambda\,P_{DC}}{p_R V_n^2} = \frac{2\times 10\times 0.0178\times 2000}{0.01\times 60^2} = 19.8mm^2$$

$$240V:\quad A_q = \frac{2\times 10\times 0.0178\times 2000}{0.01\times 240^2} = 1.2mm^2$$

Even at the relatively low performance of 2 kW, the cable diameters of DC lines for the small system voltage are very large!

5.10.1.2 System voltage and electrical safety

If the PV open circuit voltage remains under 120 V_{DC} and an inverter with an isolating transformer is used, the PV system conforms to safety category III based on VDE 0100/IEC 364, and additional measures for the protection of individuals are unnecessary. At higher DC voltages the modules must conform to safety category II, i.e. they must be tested for insulation defects at a voltage of 2000 V plus four times the open circuit voltage. In addition, the module access lines must be laid so as to prevent short-circuits and ground faults. This can be most easily achieved by separate positive and negative lines with double insulation.

5.10.1.3 String diodes and short-circuit protection

In short-circuits or low voltages within a string due to shading, a high current can flow into this string during parallel connection from other strings, a current for which the module access lines are not designed. This fault current flows in the opposite direction to the photocurrent and can thus be blocked by a string diode switched in series to the modules of a string. The voltage drop over the string diode of about 1 V leads, however, to a constant power loss, and checking the diode function is maintenance-intensive. In the last few years, therefore, the need to use such a string diode and the level of possible fault currents has been increasingly discussed.

With partial shadings within a string the current can be drastically reduced, but the voltage remains almost unchanged even at low irradiance. In the MPP point, therefore, the shaded string also makes a rather small contribution to the total current value, and no fault currents flow. The common characteristic for four parallel-switched strings, each with four modules in series, is represented with 1000 W/m^2 for a shaded string with an irradiance of 200 W/m^2 and three unshaded strings. The shaded string is first operated within the area of positive power; only near the open circuit voltage of the common characteristic do small negative currents occur.

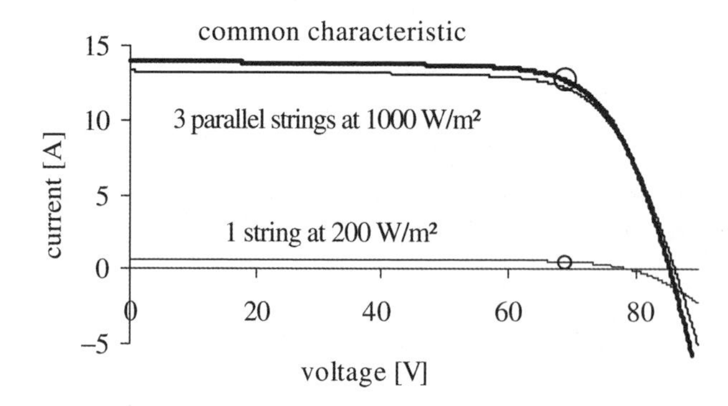

Figure 5.20: Operating points of a PV system with four parallel strings each with four modules in series and one shaded string.

Short-circuits within a string are more serious than shadings. PV generators are usually operated potential-free, i.e. neither the positive nor the negative voltage level is connected with mass, and the inverter has a galvanic separation from the mains. A simple short-circuit therefore causes no fault current at first, since only the potential level at the short-circuit point is brought down to ground level. Only a second short-circuit within a string can short-circuit one or more modules, so the MPP and open circuit voltage of the string now fall significantly.

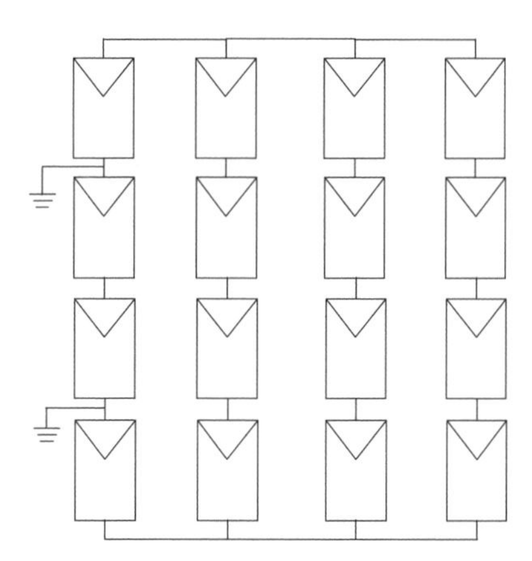

Figure 5.21: Double short-circuit in a module string.

If the generator field is now operated at higher voltages than the open circuit voltage of the partly short-circuited string, the non-short-circuited modules are operated in far forward voltage and take up power. When this happens, the currents can become very high. For example, in a generator with four strings with four modules each in series in a string, two modules are to be short-circuited. In the MPP point, the string with the short-circuit is also operated within the area of positive power, however, the short-circuit string takes up power near the open circuit voltage (at 1000 W/m^2 irradiance about –15 A at 60 V for a standard module).

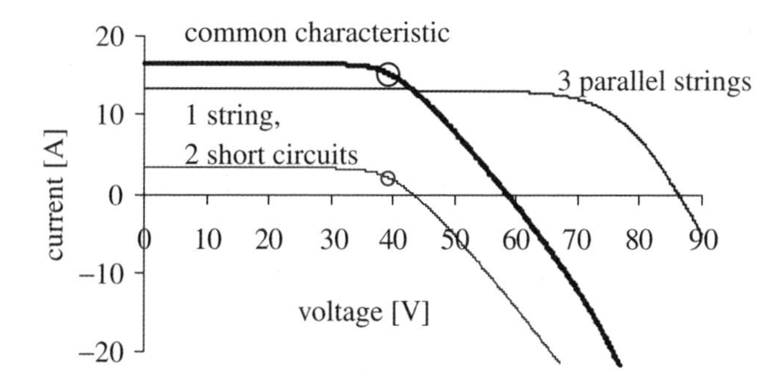

Figure 5.22: Characteristics of a generator with two short-circuits in a string.

The risk of a double short-circuit is, however, negligible with short-circuit and ground fault prevention (single line cables conforming to safety category II). Insulation monitoring can detect a simple short-circuit promptly, so timely elimination of faults is possible without adverse consequences. Protection from high fault currents can also be ensured by fuses, which are preferable to the string diodes due to smaller power losses.

One of the large German electricity suppliers RWE, says that there is no need for string diodes and fuses in systems of up to 5 kW performance. The whole power can be fed in at any place into the house mains, thus enabling simplified and economical installation. In the field of lightning protection too, new investigations by RWE have shown that the grounding of the metallic construction of the PV roof fixation tends to worsen indirect lightning protection. The usual overvoltage protection with resistors can be attached to the DC side of the inverter.

5.10.2 Inverters

5.10.2.1 Operational principle

In the inverter the photovoltaically produced direct current is commuted periodically by controllable power-electronics semiconductor switches between two conductors, and the voltage is converted to the mains voltage by transformers.

In the simplest case, the positive photocurrent per period of the 50 Hz oscillation is switched during the half oscillation period to the one access line of the transformer and during the second oscillation half to the other line, so a rectangular alternating voltage develops. However, a rectangular wave voltage has such a high harmonic content that disturbances to electrical devices can emerge, and at higher inverter power the maximum admissible harmonic currents in the public low-mains system (according to the european standard EN 61000-3-2) are exceeded. The smallest harmonic content of the current to be fed in is obtained by a high frequency commutation of the DC voltage, in which the pulse width of the switching operation is varied in such a way that after smoothing of the current a largely sinusoidal wave develops, a widespread inverter concept known today as pulse-width modulation.

Apart from the maximum values of the harmonic current which have to be taken into account, the quality of the alternating voltage produced is described by the distortion factor k, which is defined as the relation of the effective values of the voltage harmonics to the effective values of the basic and harmonic oscillations.

$$k = \sqrt{\frac{\sum_{n=2}^{\infty} V_n^2}{\sum_{n=1}^{\infty} V_n^2}} \qquad (5.43)$$

The usual distortion factors are between 1.5 and 5%. A further quality factor is the power factor L, which is defined as the relation of the active power to the apparent power and is thus a measure for the reactive power of an inverter.

$$L = \frac{P_{AC,active}}{P_{AC,apparent}} \qquad (5.44)$$

The power factor should be as close as possible to 1. Apart from the adjustment efficiency, which will be discussed later, and the transformation efficiency of an inverter, the quality factors of harmonic content, power factor and electromagnetic compatibility (EMC) should be considered when selecting a product.

5.10.2.2 Electrical safety and mains monitoring

During disconnection from the mains, decentrally feeding photovoltaic systems must be separated immediately from the mains, to be sure of avoiding danger to people or further mains faults. A power failure is detected by under- or over-voltage relays or frequency monitoring integrated in the inverter. A so-called island formation, in which (despite mains failure) photovoltaic systems produce power simultaneously removed by consumers, can nevertheless occur with this passive network monitoring. Active mains-monitoring methods are a more reliable solution, with small disturbances such as pulsed voltages or frequency deviations fed to the mains. In the case of frequency deviation, the inverter constantly tries to modify the feeding frequency and is synchronised with the mains frequency again with each zero crossover of the mains voltage. In the case of power failure, on the other hand, the frequency rises or falls constantly until it is outside a given band and the inverter switches off. The current modifications caused by modifications of the voltage level or the phase are amplified sufficiently by a feedback loop for the inverter to detect the power failure.

5.10.2.3 Inverter efficiencies

MPP Tracking and adjustment efficiency

As well as AC production, an inverter has the function of operating the direct current PV generator at the point of maximum power, i.e. the effective input impedance must be constantly adapted to the irradiance- and temperature-dependent point of maximum power. The MPP regulation is often based on periodic driving through of a voltage ramp until the power measured has achieved the optimum.

The adjustment efficiency η_{adj} is defined as the relation of actual performance at a given MPP regulation and the maximum possible DC performance with ideal MPP operation. and is approximately 97% in commercial inverters irradiance-weighted over a year (Knaupp, 1993).

$$\eta_{adj} = \frac{P_{DC,real}}{P_{DC,ideal}} \tag{5.45}$$

The predicted energy yield of a photovoltaic system should be reduced by this roughly 3% loss of energy.

Transformation efficiency

The most important characteristic of an inverter is its transformation efficiency, which is defined by the relation of AC power output to DC input power. The power output P_{AC} results from the difference between the DC input power P_{DC} and the energy dissipation P_l.

$$\eta = \frac{P_{AC}}{P_{DC}} = \frac{P_{DC} - P_l}{P_{DC}} \tag{5.46}$$

The energy dissipation is made up of the input power-independent losses (internal current supply and magnetisation losses p_{own}), the losses in the semiconductor switches l_{switch} which are linear-dependent on the power output, and the Ohm cable losses r_{ohm} which rise as the square of the AC performance. It can be represented with good accuracy by a second-order polynomial.

To obtain a unit less representation of the coefficients, all absolute performances are related to the DC rated nominal power P_n of the inverter.

$$p_l = \frac{P_l}{P_n} = \frac{P_{own}}{P_n} + v_{schalt}\frac{P_{AC}}{P_n} + r_{ohm}\left(\frac{P_{AC}}{P_n}\right)^2 = p_{own} + v_{switch}p_{AC} + r_{ohm}p_{AC}^2 \tag{5.47}$$

The energy dissipation is represented as a function of AC power output p_{AC}, since for a positive DC input power $p_{DC} = p_{own}$ first the own losses of the inverter are covered (i.e. $p_l = p_{own}$) and no switch or Ohm losses yet occur. The efficiency and thus the power output must be zero.

$$\eta = \frac{(P_{DC} - P_l)/P_n}{P_{DC}/P_n} = \frac{p_{AC}}{p_{DC}} = \frac{p_{DC} - \left(p_{own} + v_{switch}p_{AC} + r_{ohm}p_{AC}^2\right)}{p_{DC}} \tag{5.48}$$

With this description of the energy dissipation, however, both entry and exit powers occur in the efficiency term. The aim must be to calculate the efficiency of the inverter as a function of the input power supplied by the PV generator.

p_{AC} is therefore replaced by $\eta \times p_{DC}$ and Equation (5.48) is solved for η.

$$\begin{aligned} \eta &= 1 - \frac{p_{own}}{p_{DC}} - v_{switch}\eta - r_{ohm}\eta^2 p_{DC} \\ \Rightarrow \eta &= \frac{1 + v_{switch}}{2r_{ohm}p_{DC}} \pm \sqrt{\frac{(1 + v_{switch})^2}{(2r_{ohm}p_{DC})^2} + \frac{p_{DC} - p_{own}}{r_{ohm}p_{DC}^2}} \end{aligned} \tag{5.49}$$

Only the positive term of the quadratic equation supplies physically meaningful values. The three loss coefficients p_{own}, v_{switch} and r_{ohm} can be calculated from three efficiency values η_1, η_2 and η_3 given by most manufacturers in technical data sheets for power ratios of, for example, $p_1 = 10\%$, $p_2 = 50\%$ and $p_3 = 100\%$ of the rated power. The algebraic manipulation of the three equations for the three unknown quantities results in (Schmidt and Sauer, 1996):

$$p_{own} = \frac{p_1 p_2 p_3 \left(\eta_1^2 p_1 (\eta_2 - \eta_3) + \eta_1 (\eta_3^2 p_3 - \eta_2^2 p_2) + \eta_2 \eta_3 (\eta_2 p_2 - \eta_3 p_3)\right)}{\eta_1^2 p_1^2 - \eta_1 p_1 (\eta_2 p_2 + \eta_3 p_3) + \eta_2 \eta_3 p_2 p_3 (\eta_2 p_2 - \eta_3 p_3)} \tag{5.50}$$

$$v_{switch} = \frac{\begin{array}{c}\eta_1^2 p_1^2 (\eta_2 p_2 - \eta_3 p_3 - p_2 - p_3) + \eta_1 p_1 (\eta_3^2 p_3^2 - \eta_2^2 p_2^2) + \ldots \\ \ldots \eta_2^2 p_2^2 (\eta_3 p_3 + p_1 - p_3) - \eta_2 \eta_3^2 p_2 p_3^2 + \eta_3^2 p_3^2 (p_2 - p_1)\end{array}}{(\eta_1 p_1 - \eta_2 p_2)(\eta_1 p_1 - \eta_3 p_3)(\eta_3 p_3 - \eta_2 p_2)} \tag{5.51}$$

$$r_{ohm} = \frac{\eta_1 p_1 (p_2 - p_3) + \eta_2 p_2 (p_3 - p_1) + \eta_3 p_3 (p_1 - p_2)}{\eta_1^2 p_1^2 - \eta_1 p_1 (\eta_2 p_2 + \eta_3 p_3) + \eta_2 \eta_3 p_2 p_3 (\eta_3 p_3 - \eta_2 p_2)} \tag{5.52}$$

Example 5.8

Calculation of the inverter characteristic values for two commercial inverters with the following manufacturer data:

Inverter	*SMA 1800*	*NEG 1400*
η_1: $P_{DC}/P_n = 0.1$	79.4	83.0
η_2: $P_{DC}/P_n = 0.5$	89.9	91.9
η_3: $P_{DC}/P_n = 1.0$	88.9	89.8

The characteristic values calculated using the above equations result in:

Inverter	*SMA 1800*	*NEG 1400*
P_{own}	0.016575	0.015505
v_{switch}	0.045513	0.010553
r_{ohm}	0.067941	0.095879

With these three characteristic values the efficiency can be calculated as a function of the power ratio p_{DC} using Equation (5.48)

At low DC input power, the efficiency is dominated by the consumption of the power electronics; at high input power Ohm losses account for the decrease in efficiency. The maximum inverter efficiency is thus often around half the rated power.

At DC input power above the rated power of the inverter, the efficiency depends on the overload behaviour of the inverter. Many inverters can be used for some time when overloaded, and regulate only on overheating of the electronics away from the MPP point. In a simple approximation, in the case of overload, i.e. $p_{DC} > 1$, the AC power output is held constant at the level of the rated power output given at $p_{DC} = 1$:

$$p_{AC}\big|_{p_{DC}>1} = (p_{DC}\eta)\big|_{p_{DC}=1} = \eta\big|_{p_{DC}=1} = const \tag{5.53}$$

The transformation efficiency thus drops sharply with increasing overload.

$$\eta\big|_{p_{DC}>1} = \frac{p_{AC}\big|_{p_{DC}=1}}{p_{DC}} = \frac{\eta\big|_{p_{DC}=1}}{p_{DC}} \tag{5.54}$$

The following efficiency characteristics of the two inverters were calculated up to the DC rated power ($p_{DC} = 1$) with the parameters from Example 5.8, and above the rated power with Equation (5.54).

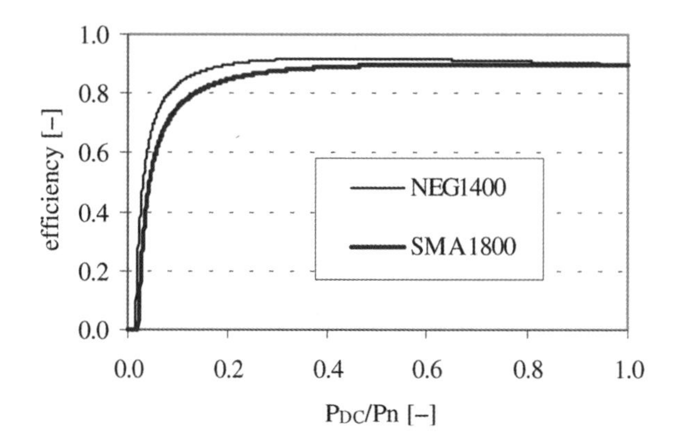

Figure 5.23: Transformation efficiency of two commercial inverters as a function of the DC input power normalised on rated power.

To enable a standardised comparison between inverters, the so-called Euro-efficiency has been introduced, which weights the partial load efficiencies with "average radiation conditions in Central Europe".

$$\eta = 0.03\eta_{p_{DC}=5\%} + 0.06\eta_{p_{DC}=10\%} + 0.13\eta_{p_{DC}=20\%} + 0.1\eta_{p_{DC}=30\%} + 0.48\eta_{p_{DC}=50\%} + 0.2\eta_{p_{DC}=100\%} \tag{5.55}$$

The Euro-efficiency cannot, however, take into account either the influence of module orientation under different radiation conditions or a different performance interpretation of a PV generator and inverter. To determine the real annual efficiency of the inverter, these influences must be analysed.

5.10.2.4 Power sizing of inverters

To determine the annual efficiency of an inverter, we have to know with what energy-weighted frequency the PV generator produces certain relative DC performances p_{DC}. Then the different partial load efficiencies of the inverter, multiplied by the energy-weighted frequency and standardised on the entire DC annual energy, produce the annual efficiency.

Given that under northern European climatic conditions, irradiances over 900 W/m^2 seldom occur and the module temperature at high irradiance values is usually over 25°C and thus causes power losses, it is not necessary even with south-facing roofs to design the inverter for the nominal power of the PV generator under standard test conditions. On south-facing facades, which often occur within the building integration area, irradiance values over 700 W/m^2 play no role in energy terms, so an inverter designed for a PV generator rated power is clearly oversized. Since undersizing the inverter reduces the system costs, in the next section a calculation will be made of how the rated power can be reduced without large efficiency losses for a given module orientation.

For the performance sizing of an inverter for a mains-connected PV system, the following procedure is suggested:

1. First the energy-weighted frequency of the irradiance is determined for the respective module orientation: hourly irradiance values are divided into irradiance classes with mean irradiance G_i and class width ΔG_i, the absolute frequency of the irradiance per class in hours per year ($n_{h,i}$) is determined, and finally the frequency is energy-weighted with the mean irradiance value of the respective class. Thus the energy irradiated annually on the surface in kWh/m^2 in each irradiance interval ΔG_i is obtained. It is sufficient to divide the irradiance of 0–1000 W/m^2 into 10 classes with 100 W/m^2 class width.

2. The irradiated energy per irradiance interval $G_I \times n_{h,i}$ is then converted into electricity $P_{DC} \times n_{h,i}$ using the efficiency of the PV generator η_{PV} and the generator surface A_{PV}, with the frequency distribution unchanged.

$$P_{DC,i} n_{h,i} = \eta_{PV} G_i n_{h,i} A_{PV} \quad [\text{kWh}] \tag{5.56}$$

   With sufficient accuracy for the inverter annual efficiency, the influence of the module temperature is negligible.

3. For an inverter with a given rated DC power P_n, the standardised DC input power $p_{DC,i} = P_{DC,i}/P_n$ is then determined for each irradiance class, and the inverter efficiency η_{inv} ($p_{DC,i}$) is calculated using Equation (5.48). This results in the AC energy:

$$P_{AC,i} n_{h,i} = P_{DC,i} n_{h,i} \, \eta_{inv} \left(p_{Dc,i} \right) \tag{5.57}$$

4. The mean annual efficiency of the inverter is then obtained from the totalled AC energy divided by the total of the DC energy.

$$\eta_{inv,year} = \frac{\sum_{i=1}^{n} \eta_{inv,i} P_{DC,i} n_{h,i}}{\sum_{i=1}^{n} P_{DC,i} n_{h,i}} = \frac{\sum_{i=1}^{n} P_{AC,i} n_{h,i}}{\sum_{i=1}^{n} P_{DC,i} n_{h,i}} \tag{5.58}$$

As an example, for a south-facing facade and a south-facing roof inclined at 45°, the mean annual efficiency of an SMA1800 inverter for a PV generator in Stuttgart with 1.8 kW of rated DC power (12% module efficiency and 15m^2 surface) is to be calculated, and compared with the Euro efficiency.

1. First the energy-weighted frequency distribution is calculated from hourly time series of the irradiance. Although the low irradiance values occur most frequently, the most energy-

relevant irradiance intervals are situated in the middle irradiance area between 400–600 W/m^2 for a south-facing roof or 300–400 W/m^2 for a south-facing facade.

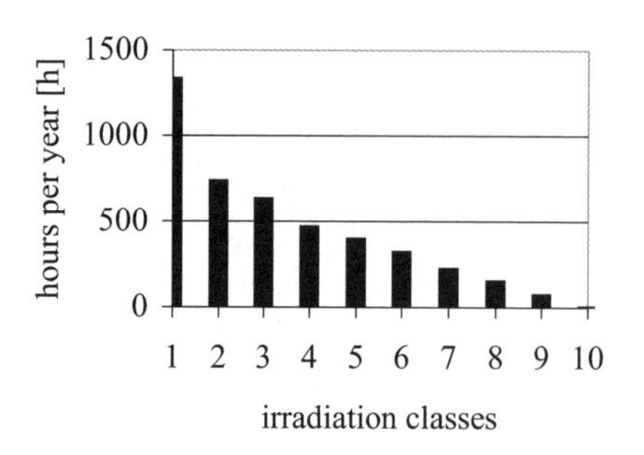

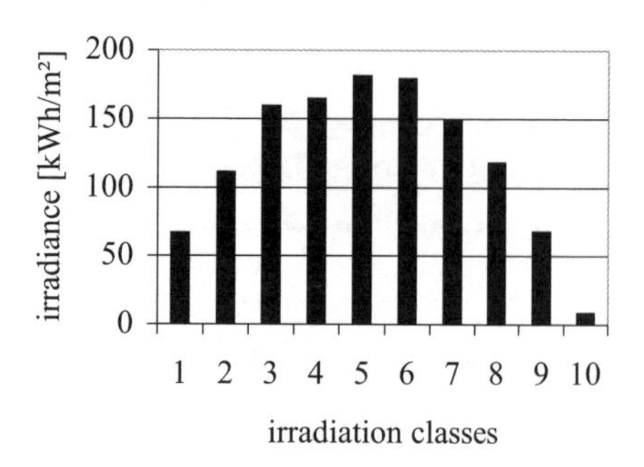

Figure 5.24: Frequency of the irradiance $n_{h,i}$ and energy-weighted irradiance for a south-facing roof in Stuttgart inclined at 45°.

2. In the second step the PV energy is calculated per irradiance class using the constant efficiency of 12% and the surface of 15m^2. Here class 4 is the most energy-relevant class at the south roof with irradiance levels between 400–500 W/m^2, i.e. a mean irradiance of 450 W/m^2. The result is (with 404 h absolute frequency) a DC energy of

$$P_{DC} n_{h,4} = G_4 n_{h,4} \eta_{PV} A_{PV} = 0.45 \frac{kW}{m^2} \times 404h \times 0.12 \times 15m^2 = 327kWh$$

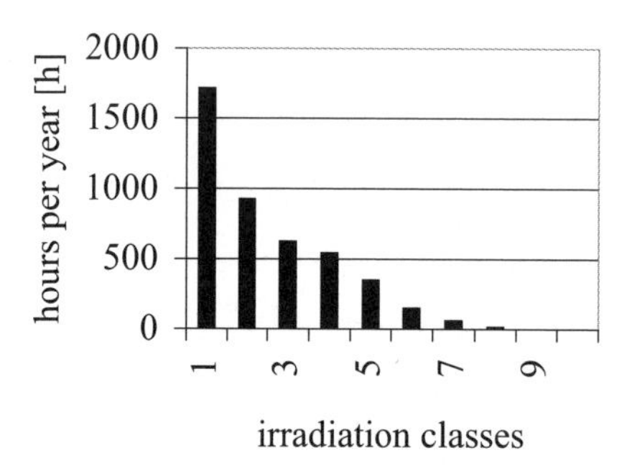

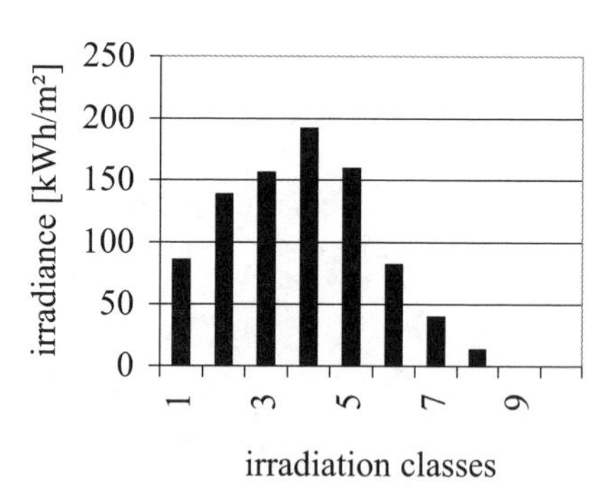

Figure 5.25: Frequency of the irradiance $n_{h,i}$ and energy-weighted irradiance for a south-facing façade.

3. From the DC energies, the mean PV performance for each class is calculated, related to the inverter rated power and converted by the inverter efficiency into AC energy. For class 4 above of the south-facing roof, the mean PV performance is given by 327 kWh/ 404 h = 0.63 kW, i.e. related to the inverter with 1800 W nominal power the result is a relative p_{DC} power of 0.35. The associated efficiency at this partial load is 0.88, so the AC energy of this class is given by

$$P_{AC} n_{h,4} = 327\text{kWh} \times 0.88 = 266\,\text{kWh}$$

4. The mean yearly efficiency results from the total of the AC energy divided by the totalled DC energy. It is 87.5% for the south-facing roof and 85.9% for the south-facing facade. The Euro-efficiency is 86%, independent of orientation.

If the inverter rated power is now selected in any relation to the PV generator power, the mean yearly efficiency can be calculated using this procedure, with only steps 3 and 4 changing. For the south-facing facade, the south-facing roof and a horizontal PV generator orientation, the inverter rated power varied from 30% of the PV generator power, which is extremely undersized, to 200% of the PV generator performance, i.e. greatly oversized. The efficiency characteristic of the inverter takes into account the case of overload by limitation of the power output to the nominal inverter power (by means of regulation away from the MPP point). The results show that the optimal sizing of the inverter for module angles of inclination between the horizontal surface and 45° for a south-facing roof is 80% of the PV rated power (yearly efficiency of 88%, location Stuttgart).

For a south-facing facade the optimal sizing is 60% of the PV generator performance (88% yearly efficiency). Also evident are the relatively flat maxima particularly in the overload area, i.e. oversizing is of no consequence, while undersizing below the optimum leads to relatively large performance losses.

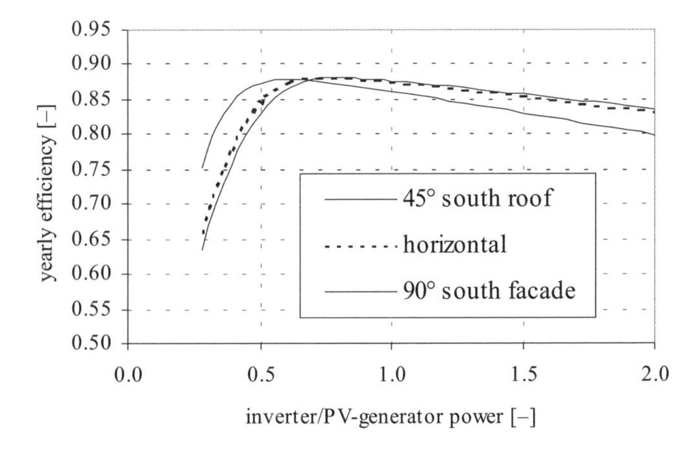

Figure 5.26: Yearly efficiencies of an inverter as a function of the power ratios of the inverter and PV generator in a German climate.

For the inverter nominal power, therefore, a relation to the PV generator power of 60–130% with less than 2% efficiency losses is possible for south-facing roofs with random inclinations, undersized inverters being recommended. For south-facing facades, inverter power ratios of 45–100% of the PV generator power are recommended.

6 Thermal analysis of building-integrated solar components

The substitution of classical building materials by active solar components represents a particularly interesting multi-functional use of solar technology in buildings. Active solar components produce electricity or heat, which is transferred to a fluid such as air or water in a controllable way. In addition, with building integration, heat flows to the building occur which can be described by heat transfer coefficients and total energy transmission factors. By using ventilated double facades with photovoltaics as an example, a methodology can be developed for thermal characterisation which enables computations of the heating energy and cooling load of a building with integrated solar components.

Previous thermal analysis of active solar components (air and water collectors, and photovoltaics) has been based on the assumption of thermal separation from the building, i.e. the calorific losses of the solar radiation absorber were computed on both sides against the ambient air temperature T_o. With building-integrated solar components, in particular warm facades, the assumption of a solar element surrounded by outside air is no longer applicable. While most thermal flat plate collectors have sufficient back insulation (with insulating thickness > 6 cm), partly transparent photovoltaic modules are, for example, often only separated by further glazings from the room (with room air temperature T_i) for architectural reasons.

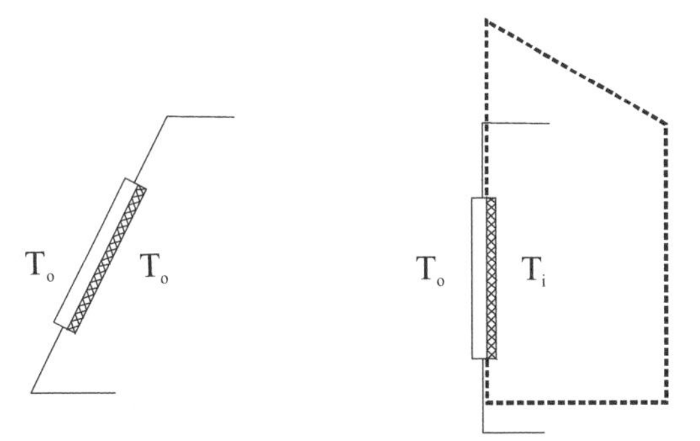

Figure 6.1: Free-standing and building-integrated collectors.

Due to this thermal coupling, heat gains for the space occur which in winter contribute to heating energy cover, but which in summer can cause overheating problems. With photovoltaic modules in a double-glazing construction, of special interest are the surface temperatures of the module (to determine the electrical power) and of the glazing on the room side (to determine the effective total energy transmittance characterised by the g-value). With back-ventilated PV double facades, the heat supplied by the modules can serve as useful thermal energy for pre-heating outside air. At the same time, transmission heat losses of the room can be recovered by the heated gap air.

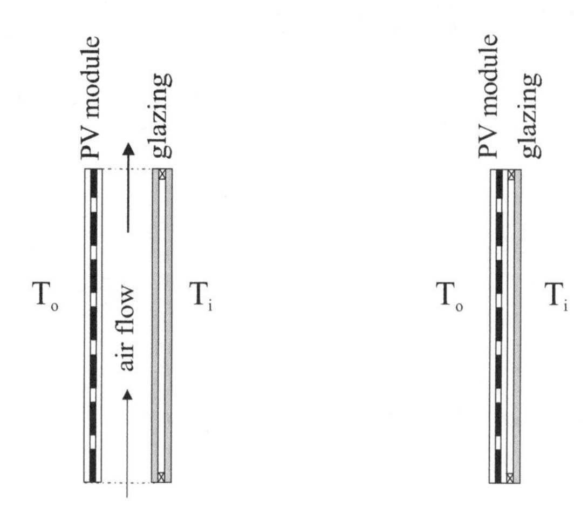

Figure 6.2: PV module in a back-ventilated cavity facade with double glazing on the room side and a PV module integrated in an double glazing construction.

First an overview is given of typical temperature conditions of building-integrated solar components, mainly photovoltaic modules, in different building-integration versions. The empirical equations obtained from measured data enable fast analysis of the temperature influence on the annual electrical efficiency and on the thermal behaviour of the building. Subsequently, a more detailed thermal model of a back-ventilated photovoltaic facade is developed which describes heat flows and temperature conditions in conventional double facades. With this model, the monthly thermal energy gains of a PV cavity facade can be calculated.

6.1 Empirical thermal model of building-integrated photovoltaics

Solar modules are generally characterised by high optical absorption coefficients in the wavelength area of short-wave solar radiation. In current-generating photovoltaic modules, however, only a small part of the absorbed irradiance, about 10–15%, is converted into electricity, with mainly heat being produced. This heat can be used for heating and actively cooling buildings, but it reduces the electrical power of the PV generator due to the module's rise in temperature. Crucial for the temperature levels are, at a given solar irradiance, the convective heat transfer mechanisms at the front and rear of the module, which depend mainly on the wind velocity.

The convective and radiant heat transition of the module rear are especially influenced by the installation situation. For detailed calculations of the temperatures, the respective relevant Nußelt correlations must be determined in each case; they depend on geometry, heat flow density, degree of turbulence etc.

For a rough estimate of the temperature conditions in different installation situations, it is sufficient to use linear regressions of temperature rises against the irradiance derived from measurements. The linear connections between the module temperature and irradiance neglect the strong dispersion of the measured values, in particular due to wind influences, but they lead to a sufficiently exact estimate of the electrical power losses and the mean temperature rise at a given irradiance. A thermal model has been developed by Sauer (1995), validated at building-integrated components, and regression analyses covering all relevant integration possibilities have been performed for twelve German climatic test reference years for different installation situations (see Table 6.1):

Table 6.1: Installation situations of building-integrated solar elements.

No.	*Module mounting*	*Ventilation*
1	free-standing module	optimum ventilation
2	roof-mounted module, large distance between module and roof tiles	optimum ventilation
3	roof-mounted, mean distance between module and roof tiles	good ventilation
4	roof-mounted, small distance between module and roof tiles	limited ventilation
5	roof-integrated	without ventilation
6	cold facade with large air gap	good ventilation
7	cold facade with small air gap	limited ventilation
8	facade integrated	without ventilation

The gradient of the linear regression curves shows the rise in the temperature difference between the module and environment $\Delta T = T_{Module} - T_o$ per W/m^2 of irradiance increase ΔG. From this the temperature rise at 1000 W/m^2 irradiance can then be calculated, to compare the installation situations.

$$\left(T_{Module} - T_o\right)\Big|_{1000W/m^2} = \frac{\Delta T}{\Delta G} \times 1000 \frac{W}{m^2} \tag{6.1}$$

The gradient obtained as an average of the test reference years varies from a minimum of 0.019 K/(W/m^2) for a free-standing module, up to 0.052 K/(W/m^2) for a facade without back-ventilation, so at 1000 W/m^2 irradiance, module temperatures of 19–52 K above the ambient temperature appear.

The mean temperature rises at 1000 W/m^2 irradiance are represented together with the minima and maxima of the twelve test reference years for all eight installation situations. From this the regression coefficient for each installation situation can be read off. The fluctuations for a given installation situation are mainly caused by different wind velocities at the locations. In addition, the relative electrical power losses are represented compared to those of the free-standing module. In the most unfavourable version – the non back-ventilated facade – 7.5–10% less electricity is produced annually than by the free-standing module, due to temperature effects alone. The electrical energy losses are calculated relative to the annually produced energy of a free-standing module.

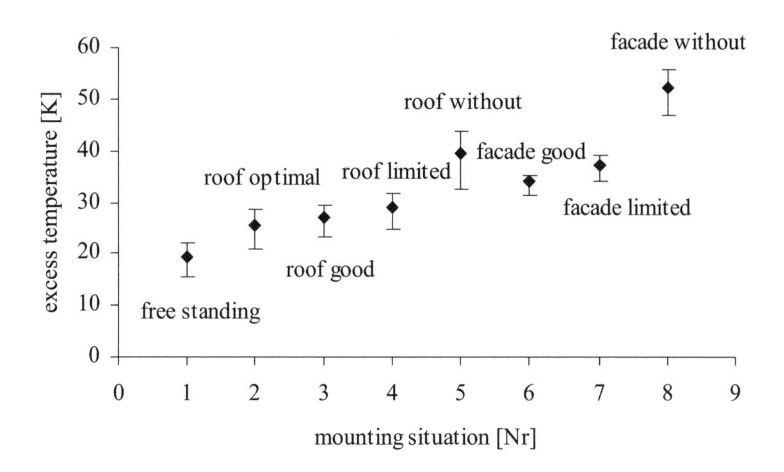

Figure 6.3: Temperature rise of a PV module in different building integration solutions with optimal, good and limited ventilation or without ventilation.

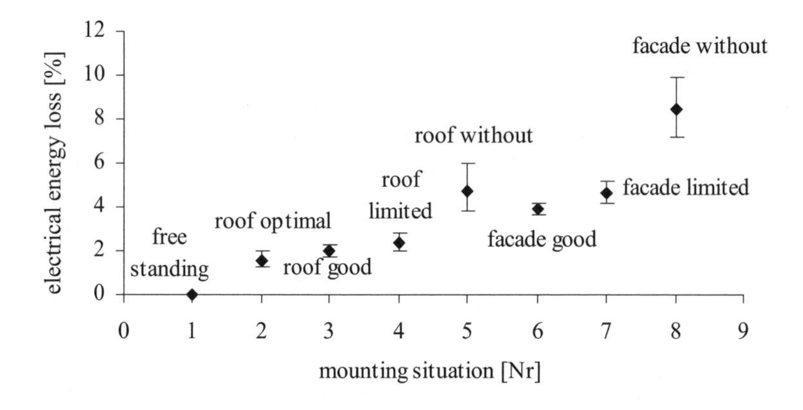

Figure 6.4: Yearly electrical energy loss of a PV module in different building integration solutions.

6.2 Energy balance and stationary thermal model of ventilated double facades

While the empirical regression equations for building-integrated photovoltaic elements provide sufficient accuracy for module temperature calculation and electrical performance analysis, a more exact model has to be created to analyse the effective heat transfer and thermal use of the module waste heat. This model must permit different temperatures as a boundary condition of the integrated solar component (ambient temperature and room temperature), and take into account the heat flows from the absorbing solar element or from the room into a ventilation gap.

Energy balances for each temperature node are set up to calculate the temperatures, and the set of equations produced thus is solved. Since most solar components have only small thermal masses, a stationary energy balance is sufficiently exact. The calculation methodology, which can be generalised for different installation situations, is explained by the example of a back-ventilated photovoltaic heated facade (Vollmer, 1999).

The structure of the facade corresponds to a typical double facade construction, with the photovoltaic module constituting the outer shell, back-ventilated with outside air. The back-ventilation can be via free convection or fan-operated. In cavity facades, the back-ventilation gap has dimensions of between about 0.1–1 m, so in contrast to commercial air collectors, low flow velocities can generally be expected. The gap dimensions and flow rates influence in particular the convective heat transfer in the air gap and thus the thermal efficiency, which in general is far lower than in turbulent-throughflowed air collectors.

For the energy balance, three temperature nodes are considered: Node *a* for the absorber (here the PV module), node *f* for the fluid (here air) and node *b* for the gap-closing glazing to the room. Due to the thinness of the PV laminates, typically 4 mm + 6 mm glass, only one temperature node is used for the photovoltaic module. A stationary energy balance is set up for the three temperature nodes.

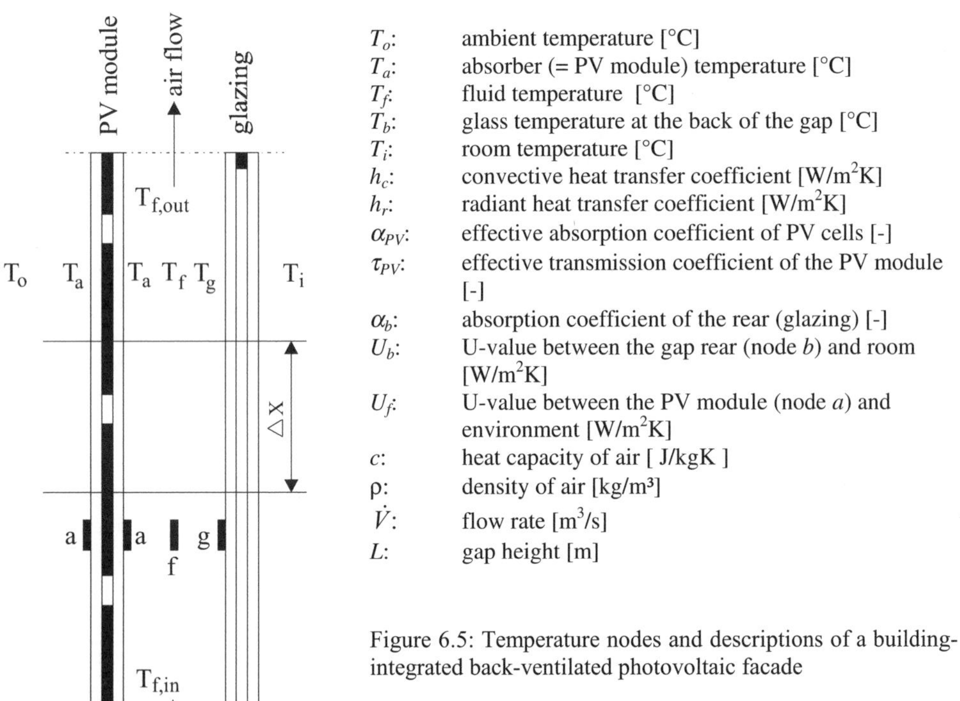

Figure 6.5: Temperature nodes and descriptions of a building-integrated back-ventilated photovoltaic facade

Node *a*:

The PV module as a thermal absorber absorbs solar irradiance G with the effective absorption coefficient α_{PV}, which includes absorption by the solar cell, reflection losses and remaining transmittance of the PV glass pane, and is typically about 80%. The calorific losses of the absorber are divided into a heat flow to the environment via the facade front heat transfer coefficient U_f, a convective heat flow from the absorber to the gap air with the heat transfer coefficient h_{ca}, and into a radiation exchange with the gap rear, the heat transfer coefficient being h_r. The electrical power $\dot{Q}_{el}$ is deducted in the balance from the absorbed irradiance.

$$G\alpha_{PV} - U_f\left(T_a - T_o\right) - h_{ca}\left(T_a - T_f\right) - h_r\left(T_a - T_b\right) - \dot{Q}_{el} = 0 \tag{6.2}$$

Node *f*:
By the convective heat transfer of the two gap confinement surfaces (absorber and gap rear) with temperatures T_a and T_b and gap width b, the fluid temperature is increased over the distance dx.

$$c\rho\dot{V}\frac{dT_f}{dx} = h_{ca}b\left(T_a - T_f\right) + h_{cb}b\left(T_b - T_f\right) \tag{6.3}$$

For the gap width b, the unit length $b = 1$ m is used here.

Node *b*:
The solar irradiance transmitted by the PV module, with the transmission coefficient τ_{PV}, is absorbed with the absorption coefficient of the back (glazing) α_b. The calorific losses consist of the radiation exchange with the PV module, the convective heat transfer to the fluid with the heat transfer coefficient h_{cb} and of the losses to the room, with the heat transfer coefficient U_b.

$$G\left(\tau_{PV}\alpha_b\right) - h_r\left(T_b - T_a\right) - h_{cb}\left(T_b - T_f\right) - U_b\left(T_b - T_i\right) = 0 \tag{6.4}$$

The front heat transfer coefficient of the facade U_f is calculated as usual from the thermal resistance of the PV module with layer thickness s_{PV} [m] and heat conductivity λ_{PV} [W/mK], and from the outside heat transfer coefficient h_a [W/m^{2}K]. The U-value of the rear U_b consists of the thermal resistance of the gap rear R_b [m^{2}K/W] (e.g. 0.3 m^{2}K/W for a non-coated double glazing), and of the internal thermal resistance between the surface and room $1/h_i$:

$$U_f = \frac{1}{\frac{s_{PV}}{\lambda_{PV}} + \frac{1}{h_a}} \tag{6.5}$$

$$U_b = \frac{1}{R_b + \frac{1}{h_i}} \tag{6.6}$$

Equations (6.2) and (6.4) are used to represent the absorber temperature T_a and the glass temperature T_b as a function of T_f, T_i and T_o, and then to use them in Equation (6.3).

$$T_a = \frac{(B\alpha_{PV} + h_r\tau_{PV}\alpha_b)G + h_rU_bT_i + BU_fT_o + (Bh_{ca} + h_rh_{cb})T_f - B\dot{Q}_{el}}{AB - h_r^2} \tag{6.7}$$

$$T_b = \frac{(A\tau_{PV}\alpha_b + h_r\alpha_{PV})G + AU_bT_i + h_rU_fT_o + (Ah_{cb} + h_rh_{ca})T_f - h_r\dot{Q}_{el}}{AB - h_r^2} \tag{6.8}$$

From this results a differential equation for the fluid temperature T_f, which can be solved by separation of the variables.

$$c\rho\dot{V}\frac{dT_f}{dx} = D_1T_i + D_2T_o - D_3\dot{Q}_{el} + D_4G - D_5T_f \tag{6.9}$$

The general solution is

$$T_f(x) = \frac{1}{D_5}\left(D_1T_i + D_2T_o - D_3\dot{Q}_{el} + D_4G - Ce^{-Zx}\right) \tag{6.10}$$

with the integration constant C dependent on the boundary condition and the constants D_1 to D_5 given by

$$D_1 = \frac{h_{ca}h_rU_b + h_{cb}AU_b}{AB - h_r^2} \qquad D_2 = \frac{h_{cb}h_rU_f + h_{ca}BU_f}{AB - h_r^2}$$

$$D_3 = \frac{h_{ca}B + h_{cb}h_r}{AB - h_r^2}$$

$$D_4 = \frac{h_{ca}h_r\tau_{PV}\alpha_b + h_{ca}B\alpha_{PV} + h_{cb}h_r\alpha_{PV} + h_{cb}A\tau_{PV}\alpha_b}{AB - h_r^2}$$

$$D_5 = h_{ca} + h_{cb} - \frac{\left(2h_{ca}h_{cb}h_r + h_{ca}^2B + h_{cb}^2A\right)}{AB - h_r^2}$$

with $A = U_f + h_{ca} + h_r$ $\quad B = U_b + h_{cb} + h_r$ and $\quad Z = \dfrac{D_5}{c\rho\dot{V}}$

The integration constant C results from the boundary condition at the gap inlet; here the ambient temperature T_o:

Boundary condition: with x = 0 $\quad T_f = T_o$

$$C = D_1T_i + (D_2 - D_5)\,T_o - D_3\dot{Q}_{el} + D_4G \tag{6.11}$$

Thus the special solution results

$$T_f(x) = \left(1 - e^{-Zx}\right)\frac{D_1 T_i + D_2 T_o - D_3 \dot{Q}_{el} + D_4 G}{D_5} + T_o e^{-Zx} \tag{6.12}$$

The mean fluid temperature of the entire flow channel is obtained by integrating Equation (6.12) over the entire gap length L.

$$\begin{aligned} \bar{T}_f &= \frac{1}{L}\int_0^L T_f(x)\,dx \\ &= \frac{D_1 T_i + D_2 T_o - D_3 \dot{Q}_{el} + D_4 G}{D_5}\left(1 + \frac{1}{LZ}\left(e^{-ZL} - 1\right)\right) + \frac{T_o}{LZ}\left(1 - e^{-ZL}\right) \end{aligned} \tag{6.13}$$

With the mean fluid temperature, the mean absorber and rear temperatures of the gap can also be calculated, using Equations (6.7) and (6.8).

The coefficients D_1 to D_5 depend on the heat transfer coefficients for convection and radiation in the gap, and on the heat transfer coefficients to the environment or to the interior. Since these are also temperature-dependent, the gap temperatures can only be determined iteratively.

Before the effective heat transfer coefficients (U- and g-values) are determined from the temperatures, a brief examination will be made of the necessary heat transfer coefficients and relevant Nußelt correlations for the heat transfer in double facades with large gap dimensions.

6.2.1 Heat transfer coefficients for the interior and facade air gap

The heat transfer coefficient inside h_i depends on the room geometry, the heating system, the type of ventilation and other parametres. For most applications, the assumption of a constant heat transfer coefficient is sufficiently exact. Euro standard EN 832 states that a mean thermal resistance of $1/h_i = 0.13$ m^{2}K/W can be used inside.

The convective heat transmission coefficients in the air gap h_{ca} and h_{cb} depend on the facade geometry and on the flow type (free or forced). With cavity facade geometries with gap dimensions over 10 cm, the flow rates are generally small (under 0.5 m/s), so the free convection proportion cannot be neglected even with fan-driven forced ventilation. The derivation of heat transfer coefficients from first principles is extremely complex, so that mostly empirical correlations for Nußelt numbers as a function of flow velocity and fluid properties are used. Experiments by Schwab (2002) on asymmetrically heated double facades with a height to gap width ratio of 50 showed a steady increase of the heat flux density with rising Reynolds number. If one of the plates were cooled below the inlet air temperature, the heat flux densities on the hot plate even decreased with rising Reynold number. The correlations used below fit experiments on a 6.5 m high double facade with 0.14 m gap distance, where the "hot" plate was a photovoltaic module and the back of the air gap was a double glazing.

Under mixed convection conditions the Reynolds number consists of a free and a forced convection part.

$$\mathrm{Re} = \sqrt{\mathrm{Re}_{free}^2 + \mathrm{Re}_{for}^2} \tag{6.14}$$

The free convection proportion results from the temperature-induced density variation over the gap height *L*.

$$\mathrm{Re}_{free} = \sqrt{\frac{Gr}{2.5}} \tag{6.15}$$

$$Gr = \frac{g\beta' L^3 \left|\left(\bar{T}_f - \bar{T}_{a,b}\right)\right|}{\nu^2} \tag{6.16}$$

$$\mathrm{Re}_{for} = \frac{\mathrm{v}L}{\nu} \tag{6.17}$$

With flow conditions over an individual plate, the boundary layer flow at the lower panel edge can begin laminarly and after a certain length become turbulent, the transition point being about $\mathrm{Re} = 2 \times 10^5$ (Merker and Eiglmeier, 1999). The mean Nußelt number is formed from a laminar and a turbulent proportion.

$$Nu = \sqrt{Nu_{lam}^2 + Nu_{turb}^2} \tag{6.18}$$

For Prandtl numbers between $0.6 < \mathrm{Pr} < 10$, integration over the local Nußelt number produces the laminar proportion:

$$Nu_{lam} = 0.664\sqrt{\mathrm{Re}}\sqrt[3]{\mathrm{Pr}} \tag{6.19}$$

The turbulent proportion is calculated from an empirical correlation derived from numerical integration of the boundary layer equation:

$$Nu_{turb} = \frac{0.037\,\mathrm{Re}^{0.8}\,\mathrm{Pr}}{1 + 2.443\,\mathrm{Re}^{-0.1}\left(\mathrm{Pr}^{\frac{2}{3}} - 1\right)} \tag{6.20}$$

with $\mathrm{Pr} = \dfrac{\nu c \rho}{\lambda}$

From the mean Nußelt number, the heat transmission coefficients result as usual in $h_c = \dfrac{Nu\,\lambda}{L}$.

Example 6.1

Calculation of the two convective and of the radiation heat transfer coefficient of a 6.5 m high back-ventilated PV facade with 14 cm gap depth, for PV module temperatures of 50°C and a back gap temperature (e.g. glazing) of 30°C. The flow velocity of the gap air is 0.3 m/s and the fluid temperature is 40°C.

The material values for the left, warm PV side are calculated with the mean temperature between the surface and the gap air, i.e. here 45°C; for the right side similarly from the surface temperature of the glazing and the mean gap air temperature (i.e. 35°C).

Material values of the gap air:

Mean temperature of	*45°C (PV)*	*35°C (glass)*
Kinematic viscosity ν:	17.546×10^{-6} m²/s	16.60×10^{-6}
Heat conductivity of air λ:	0.02758 W/mK	0.02554
Density ρ:	1.095 kg/m³	1.130
Heat capacity c:	1008.25 J/kgK	1007.75
heat expansion coefficient β':	0.00314 K^{-1}	0.0032

Re_{for}:	111 134	117 486
Gr:	2.8×10^{11}	3.123×10^{11}
Re_{free}:	334 329	353 435
Pr:	0.75	0.74
Re:	352 316	372 450
Nu_{turb}:	863	894
Nu_{lam}:	325	332
Nu:	922	954

From these come the heat transfer coefficients h_{ca} = 3.66 W/m^{2}K, h_{cb} = 3.75 W/m^{2}K and h_r = 3.07 W/m^{2}K.

If the temperatures are not given, the heat transfer coefficients and temperatures must be determined iteratively. Firstly temperatures at the gap confinement surfaces are assumed, so that the heat transfer coefficients can be calculated and then the mean temperatures recalculated. With the new temperatures, heat transfer coefficients are again calculated etc.

Example 6.2

Calculation of the mean temperatures and air outlet temperature of a 6.5 m high, 1 m wide and 0.14 m deep back-ventilated photovoltaic facade under the following boundary conditions:

Irradiance on the facade	G = 800 W/m²
Ambient temperature	T_o = 10°C
Room air temperature	T_i = 20°C
Wind velocity	v_w = 3 m/s
Flow velocity in the gap	v = 0.3 m/s
Absorption coefficient of the PV module	α_{PV} = 0.8
Transmission coefficient of the PV module	τ_{PV} = 0.1

Absorption coefficient of the back glazing	$\alpha_b = 0.05$
Layer thickness of the PV module	$s_{PV} = 0.01$ m
Heat conductivity of the PV module	$\lambda_{PV} = 0.8$ W/mK
Heat resistance of the back glazing	$R_b = 0.18$ m²K/W
Heat transfer coefficient inside	$h_i = 8$ W/m²K
Electrical efficiency	$\eta_{el} = 0.12$
Emission coefficient	$\varepsilon = 0.88$

Solution:

The temperatures from Example 6.1 are used as initial values, resulting in the heat transfer coefficients of the first example. With these heat transfer coefficients the mean fluid temperature is calculated and inserted into the equations of the heat transfer coefficients. After four iterations, the error in the temperatures is less than 0.1°C and the coefficients are:

Heat transfer coefficient absorber fluid	$h_{ca} = 5.1$ W/m^{2}K
Heat transfer coefficient glazing fluid	$h_{cb} = 3.5$ W/m^{2}K
Heat transfer coefficient for radiation	$h_r = 2.9$ W/m^{2}K
Thus a mean fluid temperature results	$T_f = 19.6$°C.

If the mean fluid temperature is inserted into Equations (6.7) and (6.8), this produces for the

mean PV module temperature	$T_a = 40.9$°C
glazing temperature on the gap side	$T_b = 26.5$°C.

The fluid outlet temperature after 6.5 m height is $T_{f,out} = 27.2$°C,
i.e. the PV facade has warmed up the ambient air by 17.2°C.

The temperature distribution over the facade height for the above boundary conditions clarify the temperature rise of the fluid and of the gap confinement surfaces. The surface temperature of the back glazing on the room side is also raised slightly above the room temperature despite low outside temperatures of 10°C.

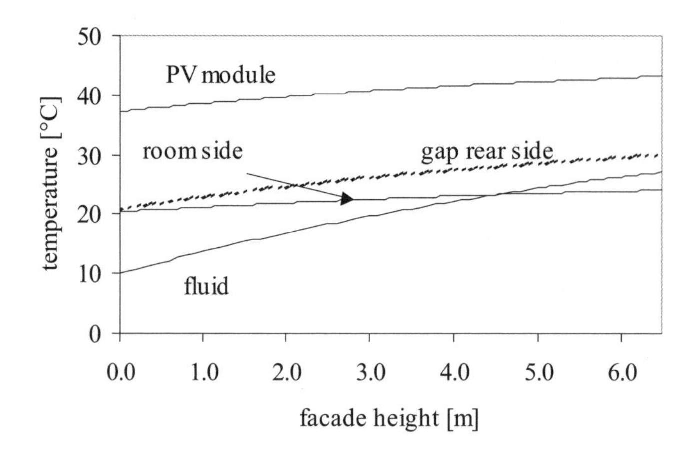

Figure 6.6: Rise in temperature of the ambient air entering through a back-ventilated PV facade. The boundary conditions are 800 W/m^2 irradiance, 3 m/s wind velocity, 10°C outside temperature, 20°C room air temperature and a flow velocity in the gap of 0.3 m/s.

6.3 Building-integrated solar components (U- and g-values)

To determine the influence of building-integrated solar components on the thermal behaviour of the building, it is a good idea to use the usual component characteristic values such as the heat transfer coefficient U and the total energy transmission factor g. These characteristic values are constant in conventional components. Since, however, the energy flows and temperature levels of absorbing solar elements depend greatly on irradiance and ambient temperature, the U and g-values must be determined time-dependently.

Since energy gains from warm-air use by back-ventilated facades are also to be considered, the U and g-values are divided into two components, which differentiate between transmission and ventilation heat flows (Bloem *et al.*, 1997). U_{trans} characterises the entire calorific losses from the interior, U_{vent} the calorific losses from the inside to the air gap, which can be recovered by the back-ventilation, g_{trans} the solar radiation gains transmitted directly into the interior, and g_{vent} the absorbed radiation gains contributing to the heating up of the air volume flow (Versluis *et al.*, 1997).

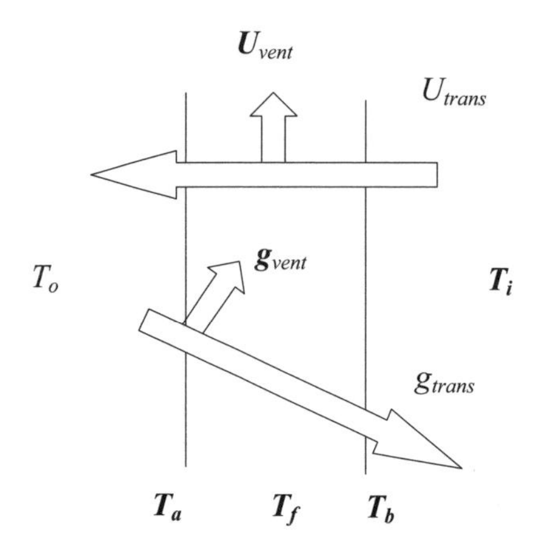

Figure 6.7: Characteristic values for the thermal characterisation of a back-ventilated facade.

The transmission heat flow $\dot{Q}_{trans}$ describes the effective heat losses from the room as the difference of transmission heat losses and direct solar gains, calculated similarly to the procedure for the effective U-values of windows. The ventilation heat flow $\dot{Q}_{vent}$ contains that part of the transmission heat losses which can be recovered by the back-ventilation (related to the temperature difference between the room and environment) plus the heat gains of the air gap due to absorption of solar radiation.

$$\dot{Q}_{trans} = U_{trans}\left(T_i - T_o\right) - g_{trans} G \tag{6.21}$$

$$\dot{Q}_{vent} = U_{vent}\left(T_i - T_o\right) + g_{vent} G \tag{6.22}$$

The heat transfer coefficient U_{trans} is defined via the heat transfer coefficient of the rear U_b, which is known and calculated from the mean temperature $\bar{T}_b$. For this definition, the heat flow from the room $U_b(T_i - T_b)$ is normalised to the temperature difference between the room and the environment.

$$U_b(T_i - \bar{T}_b) = U_{trans}(T_i - T_o) \tag{6.23}$$

$$U_{trans} = \frac{U_b(T_i - \bar{T}_b)}{(T_i - T_o)} \tag{6.24}$$

The direct energy transmission coefficient g_{trans} contains the optical transmission of the facade and consists of the transmission of the PV module and the back glazing. For example g_{trans} results, for a facade with 15% glass proportion of the PV module and a back double-glazing, from the product of the glass proportion of 15%, the single glazed PV module transmission of approximately $\tau_{PV\text{-}glass}$ = 90% and the transmission of the double glazing of approximately $\tau_{gap,\ rear}$ = 80%.

$$g_{trans} = \frac{A_{glass}}{A_{module}} \tau_{PV-glass} \tau_{gap\,rear} = 0.15 \times 0.9 \times 0.8 = 0.108 \tag{6.25}$$

The useful energy $\dot{Q}_{vent}$ of the heated air within the gap is calculated directly from the mean surface temperatures of the gap confinement surfaces and from the mean fluid temperature.

$$\dot{Q}_{vent} = h_{ca}(\bar{T}_a - \bar{T}_f) + h_{cb}(\bar{T}_b - \bar{T}_f) \tag{6.26}$$

If the mean temperatures from the energy balance Equations are used, $\dot{Q}_{vent}$ can be calculated as a function of the room and ambient temperatures, irradiance and the produced electricity,

$$\dot{Q}_{vent} = \left(\frac{1-e^{-ZL}}{ZL}\right)(T_i D_1 - T_o (D_5 - D_2)) + \left(\frac{1-e^{-ZL}}{ZL}\right)\left(GD_4 - \dot{Q}_{el} D_3\right) \quad (6.27)$$

with L describing the height of the facade.

In agreement with Equation (6.22), $\dot{Q}_{vent}$ is divided into a solely temperature-dependent term, described by U_{vent}, and an irradiance-dependent term, characterised by g_{vent}. Since the PV electrical efficiency is essentially irradiance-dependent, it is integrated into the g_{vent} term. The normalisation to the temperature difference between the room and environment thus produces, for U_{vent}:

$$U_{vent} = \frac{\left(\frac{1-e^{-ZL}}{ZL}\right)\left(T_i D_1 - T_o (D_5 - D_2)\right)}{T_i - T_o} \quad (6.28)$$

The irradiance-dependent term is normalised to the irradiance G. A sufficiently exact approximation in thermal terms results from assuming a constant electrical efficiency η_{el} for the PV module.

$$g_{vent} = \left(\frac{1-e^{-ZL}}{ZL}\right)\left(D_4 - \frac{\dot{Q}_{el} D_3}{G}\right) = \left(\frac{1-e^{-ZL}}{ZL}\right)\left(D_4 - \eta_{el} D_3\right) \quad (6.29)$$

Example 6.3

Calculation of the component characteristic values U_{trans}, U_{vent} and g_{vent} of the photovoltaic double facade, with the boundary conditions of the last example.

$U_{trans} = -2.14\ W/m^2K$

$U_{vent} = 1.07\ W/m^2K$

$g_{vent} = 0.178$

Thus the ventilation gains at an irradiance of 800 W/m^2 amount to

$$\dot{Q}_{vent} = U_{vent}(T_i - T_0) + g_{vent} G = \underbrace{1.07 \frac{W}{m^2K}(20-10)K}_{10.7} + \underbrace{0.178 \times 800 \frac{W}{m^2}}_{142.4} = 153.1 \frac{W}{m^2}$$

The gains from the absorption of solar radiation at a solar efficiency g_{vent} of 17.8% dominate.

Since the back-glazing temperature is higher than the room temperature T_i, the transmission coefficient U_{trans} is negative and in effect heat gains are supplied to the room:

$\dot{Q}_{trans} = -97.4\ W/m^2$.

6.4 Warm-air generation by photovoltaic facades

With the above method, hourly U and g-values can be calculated, and hence hourly energy balances for the facade system can be drawn up. Weighted monthly average values can then be used for heating-energy calculations based on the monthly balance procedure of EN832. The thermal gains of the back-ventilated facade consist of direct solar gains (described by the constant g_{trans}-value), indirect solar gains from solar radiation absorption and subsequent heat transfer to the gap air (g_{vent}), plus the heat recovered from the interior (U_{vent}). The irradiance-weighted g_{vent} value is given by

$$\bar{g}_{vent} = \frac{\sum_{j=1}^{hours\ per\ month} g_{vent,j} G_j}{\sum_{j=1}^{hours\ per\ month} G_j} \tag{6.30}$$

The useful energy from heat recovery is calculated by the temperature difference-weighted $\bar{U}_{vent}$ -value.

$$\bar{U}_{vent} = \frac{\sum_{j=1}^{hours\ per\ month} U_{vent,j} \left(T_i - T_{o,j}\right)}{\sum_{j=1}^{hours\ per\ month} \left(T_i - T_{o,j}\right)} \tag{6.31}$$

The entire transmission heat loss of the room is expressed by a mean heat transfer coefficient $\bar{U}_{trans}$.

$$\bar{U}_{trans} = \frac{\sum_{j=1}^{hours\ per\ month} U_{trans,j} \left(T_i - T_{o,j}\right)}{\sum_{j=1}^{hours\ per\ month} \left(T_i - T_{o,j}\right)} \tag{6.32}$$

Since in the monthly energy balance procedure only solar gains are taken into account, $\bar{U}_{vent}$ can also be deducted from the total transmission heat loss coefficient $\bar{U}_{trans}$, so an effective transmission heat loss $\bar{U}_{trans,eff} = \bar{U}_{trans} - \bar{U}_{vent}$ remains.

With these characteristic values, the contribution of back-ventilated facades to building heating energy can be calculated and compared with conventional facade systems. In summer the thermal load of the PV facade for the interior is easily calculated from the total of direct solar gain (g_{trans}) and indirect gain (U_{trans}). When active solar cooling is used, the useful energy is calculated as usual from the rise in temperature of the gap air via g_{vent} and U_{vent}.

With the above procedure, monthly component characteristic values for a back-ventilated PV facade for a Mediterranean (Barcelona) and a German (Stuttgart) climate have been calculated. The total ventilation gains Q_{vent} result from the sum of the ventilation gains by irradiance and the heat flow from the room into the air gap, multiplied by the number of hours in the month n_h. The monthly irradiance G_m is given by the sum of hourly values in kWh/m^2.

$$Q_{vent} = \bar{g}_{vent} G_m + \bar{U}_{vent} \left(\bar{T}_i - \bar{T}_o\right) n_h \tag{6.33}$$

The mean ventilation g-value $\bar{g}_{vent}$ can be interpreted directly as the solar thermal efficiency of the back-ventilated facade. The monthly transmission heat loss Q_{trans} is calculated from the total heat flow from the room, minus the direct solar gains.

$$Q_{trans} = \bar{U}_{trans} \left(\bar{T}_i - \bar{T}_o\right) n_h - \bar{g}_{trans} G_m \tag{6.34}$$

The g_{trans} value only takes into account the optical transmission of the glazing system and is constant here at 0.108. The thermal efficiency at the low flow velocity of 0.3 m/s is 13% on average. From the total transmission heat losses of 50 kWh/m^2 in the heating season, 40 kWh/m^2 can be recovered by feeding back the heated gap air (interior temperature constant at 20°C).

Table 6.2: Climatic boundary conditions, ventilation gains and transmission heat losses of the back-ventilated south-facing facade in Barcelona.

Month	G_m [kWh/m²]	T_o [°C]	n_h [-]	$\bar{g}_{vent} = \eta_{th}$ [-]	$\bar{U}_{vent}\left(\bar{T}_i - \bar{T}_o\right)n_h$ [kWh/m²]	$\bar{U}_{trans}\left(\bar{T}_i - \bar{T}_o\right)n_h$ [kWh/m²]	Q_{trans} [kWh/m²]	Q_{vent} [kWh/m²]
January	82	9.75	744	0.142	8.05	11.00	2.16	19.67
February	88	9.95	672	0.130	7.03	9.30	–0.20	18.50
March	106	11.26	744	0.128	6.98	8.08	–3.37	20.58
April	94	12.92	720	0.124	5.18	5.78	–4.32	16.80
May	80	16.16	744	0.126	2.91	1.80	–6.88	13.02
June	80	20.06	720	0.122	0.17	–3.69	–12.32	9.90
July	88	23.65	744	0.118	–2.29	–9.06	–18.61	8.15
August	100	23.46	744	0.129	–2.20	–9.60	–20.36	10.64
September	108	21.26	720	0.134	–0.71	–6.84	–18.48	13.75
October	107	17.01	744	0.130	2.26	–0.52	–12.12	16.25
November	90	12.70	720	0.137	5.42	6.16	–3.57	17.73
December	84	10.75	744	0.146	7.10	9.25	0.12	19.45
Sum/ Mean value	1107	15.7	8760	0.13	40	22	–98	184

In Stuttgart the average thermal efficiency is somewhat higher at 15%. Despite the high U-value of the non-coated double glazing, 4 W/m^{2}K, the transmission heat loss of the PV facade system is in effect clearly smaller due to the thermal energy gains, and the real U_{trans} value varies between 1.5-1.8 W/m^{2}K. Of the total transmission heat losses in the heating season, 142 kWh per m^2 of facade system, 99 kWh/m^2 can be recovered.

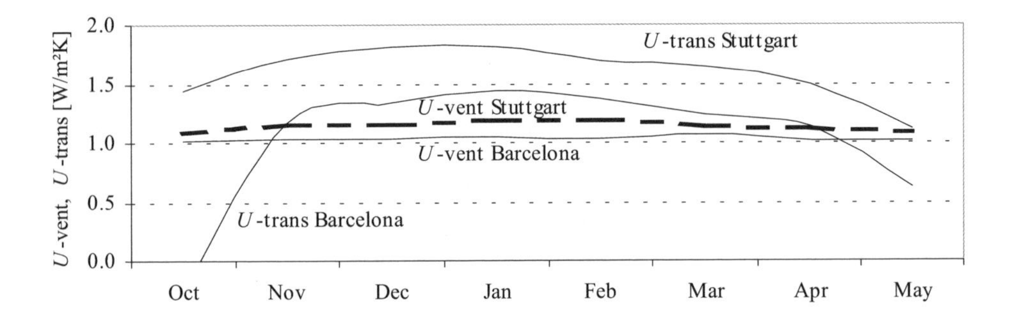

Figure 6.8: Mean monthly component characteristic values of the back-ventilated photovoltaic facade with conventional double glazing as the gap rear (U = 4 W/m^{2}K) in Stuttgart and Barcelona.

With the methodology described, different facade types can be compared in energy terms, with the thermal gains depending greatly on the thermal separation of the gap rear from the building. If the thermal separation is improved, the transmission heat losses to the gap fall, but so too do the ventilation heat gains. At the same time, however, the summer load of the room is reduced, so the best possible thermal separation is always to be recommended.

While a back-ventilated PV facade produces solar ventilation gains of between 83–93 kWh/m^{2}a in Stuttgart, depending on the gap rear construction, the solar ventilation gains of a vitreous cavity facade are only 15 kWh/m^{2}a. The heat recovered from the room depends only on the quality of the thermal separation between the room and the back-ventilation gap, and is 27 kWh/m^2 for cavity facades with and without PV when heat-protecting glass is used. The direct solar gains are very high in fully vitreous facade types, between 250–285 kWh/m^2, and must in all cases be controlled with shading devices. With commercialg-values for external sun protection of about 20%, overheating problems in the summer must be expected in such all-glass facades.

7 Passive solar energy

Passive solar energy use supplies a significant contribution to the energy demand of every building, mainly by the short-wave solar irradiance transmitted by glazing, which is converted into heat by absorption on wall surfaces and provides daylight. Here that form of energy transfer is described as passive which takes place solely by thermal conduction, solar irradiance, long-wave radiation and free convection, i.e. is not line-bound and requires no auxiliary mechanical energy for moving a heat carrier.

Solar irradiance is absorbed without transport losses directly by the building shell or internal storage masses. Besides windows and the associated internal storage mass, the possibilities of passive use also include transparent thermal insulation on a heat-conducting external wall. Despite lower efficiency compared to windows, transparent thermal insulation in connection with a massive building component enables a temporal phase shift between irradiance and utilisation of heat, and so reduces the overheating problems of large glazings.

Unheated conservatories rank among the classical forms of passive solar use. As elements placed in front of the building shell they reduce, however, the direct solar irradiance through windows lying behind. Thus both the daylight and the direct heat entry into the adjoining heated rooms are clearly reduced. Furthermore, the indirect heating of glazed conservatories by adjoining rooms often leads to an increase in the heating requirement of buildings. Only with a very energy-conscious use of an unheated conservatory can energy gains for the building actually be achieved.

7.1 Passive solar use by glazings

Glazings are characterised by the fact that they display high transmittances for short-wave solar irradiance up to around 2.5 μm, but are impermeable to long-wave radiant heat emitted from building components with a maximum intensity of around 10 μm. The greenhouse effect results from this transmission and conversion of the solar irradiance by absorption in structural elements into heat, whose long-wave radiation proportion is not transmitted through the glazing.

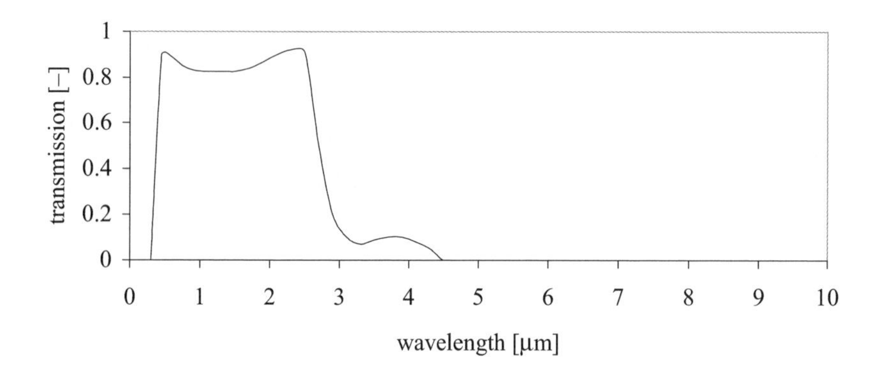

Figure 7.1: Wavelength-dependent transmittance of a single glazing.

7.1.1 *Total energy transmittance of glazings*

Apart from the direct transmission of the short-wave solar radiation with transmittance τ, part of the irradiance is absorbed in the panes, and by heating them causes a heat flow towards the room which contributes to the total energy transmission factor (g-value). The absorption coefficient of single glazing can be up to 30% for special sun-protection glazings. The secondary heat emission degree q_i is defined according to DIN EN 410 as the relation of the heat flow on the room side $\dot{Q}_i$ per square metre of window area A_w to the impacting solar radiation G, and calculated by solving heat balance equations. For each pane of a multi-pane system the degrees of transmission, absorption and reflection must be known. The total energy transmission factor then results from the relation of the total heat flow $\dot{Q}_{total}/A_w$ into the room to the irradiance:

$$g = \frac{\dot{Q}_{total} / A_w}{G} = \frac{\tau G + \dot{Q}_i / A_w}{G} = \tau + q_i \tag{7.1}$$

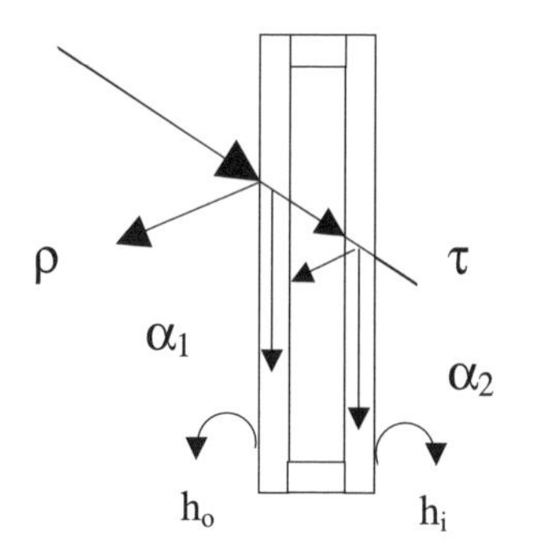

Figure 7.2: Transmission τ, reflection ρ, absorption α and heat transfer coefficients at the inside (h_i) and outside (h_o) of a double glazing.

The transmittance τ is calculated on the basis of DIN EN 410 by integration of the wavelength-dependent transmission over the solar spectrum. With perpendicular incidence it is approximately 90% for an uncoated simple float glass, and about 80% for a two-pane system. Due to today's commonly used metallic coating of the pane on the room side in thermally insulating glazing, the transmittance clearly falls, so the total energy transmission factor of the two-pane system is rarely over 65%. The absorption factor α of the short-wave solar radiation is likewise calculated for each pane by integration over the spectrum, and inter-reflections in multi-pane systems are taken into account.

The solutions of the heat balance equations for single, double and triple glazing are indicated in DIN EN 410. Using the example of single glazing, a heat balance can be described; the intensity αG absorbed by the pane of surface A_w is divided into a heat flow inward $\dot{Q}_i/A_w$ and outward $\dot{Q}_o/A_w$.

$$\alpha G = \frac{\dot{Q}_i}{A_w} + \frac{\dot{Q}_o}{A_w} \tag{7.2}$$

These heat flows can be calculated with the help of the heat transfer coefficients h_i (standard value 7.7 W/m^{2}K) and h_o (standard value 25 W/m^{2}K), and the temperature difference between pane surface T_s and room air T_i or outside air T_o. Temperature differences between the outside and inside surface are neglected.

$$\frac{\dot{Q}_i}{A_w} = h_i\left(T_s - T_i\right) \qquad \frac{\dot{Q}_o}{A_w} = h_o\left(T_s - T_o\right) \tag{7.3}$$

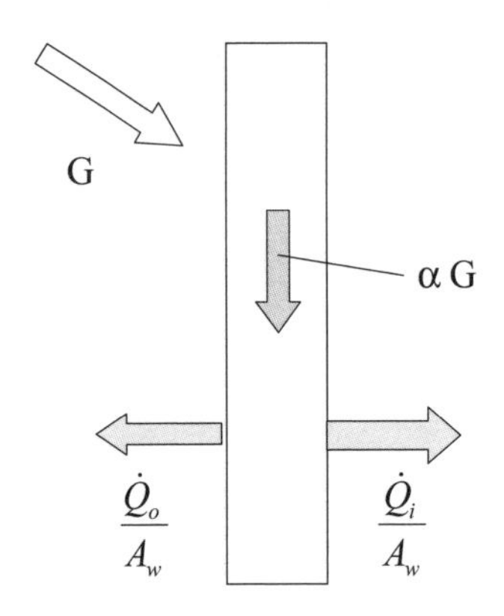

Figure 7.3: Irradiance G, absorption αG and secondary heat flows of single glazing outward and inward.

From the heat flow balance, first determine the pane surface temperature T_s:

$$\alpha G = \left(h_i + h_o\right)T_s - h_i T_i - h_o T_o$$
$$\Rightarrow T_s = \frac{\alpha G + h_i T_i + h_o T_o}{h_i + h_o} \tag{7.4}$$

With this pane temperature T_s, the heat flow inward can be calculated:

$$\frac{\dot{Q}_i}{A_w} = h_i\left(T_s - T_i\right) = \frac{h_i}{h_i + h_o}\left(\alpha G - h_o\left(T_i - T_o\right)\right) = \underbrace{\frac{h_i}{h_i + h_o}\alpha G}_{secondary\ heat\ flow} - \underbrace{\frac{1}{\frac{1}{h_i} + \frac{1}{h_o}}\left(T_i - T_o\right)}_{Transmission\ losses} \quad (7.5)$$

Since the transmission heat losses of the glazing are calculated separately via the U-value, for the definition of the secondary heat emission degree q_i the ambient temperature can be set equal to the outside temperature. The result for q_i is:

$$q_i = \frac{\dot{Q}_i / A_w}{G} = \alpha \frac{h_i}{h_i + h_o} \quad \textit{für } T_i = T_o \quad (7.6)$$

For double glazing the characteristic values are calculated accordingly, though for the outside pane a further heat balance must now be created. Defining the absorption coefficient for the outside pane as α_1 and for the internal pane as α_2, the secondary heat emission degree also depends on the layer thicknesses s_1, s_2 and heat conductivities λ_1, λ_2 of the two panes and on the thermal resistance R_{air} of the standing air layer between the panes:

$$q_i = \frac{\left(\frac{\alpha_1 + \alpha_2}{h_o} + \frac{\alpha_2}{\frac{s_1}{\lambda_1} + R_{air} + \frac{s_2}{\lambda_2}}\right)}{\frac{1}{h_i} + \frac{1}{h_o} + \frac{s_1}{\lambda_1} + R_{air} + \frac{s_2}{\lambda_2}} \quad (7.7)$$

The solar radiation let through (transmitted by) the glazing into the room $\dot{Q}_{trans}$ results directly from the product of the g-value and the solar irradiance:

$$\dot{Q}_{trans} = g\,G \quad (7.8)$$

7.1.2 Heat transfer coefficients of windows

The calorific losses through the window are deducted from the transmitted power, which are characterised by the heat transfer coefficient of the glazing U_g, or of the entire window including the frame U_w. Double glazing coated and filled with heavy noble gases achieve a minimum U_g value of 1.0 W/m^{2}K, triple glazing at best a U_g value of 0.4 W/m^{2}K. Even at a glazing U_g value of 1.3 W/m^{2}K, a wooden or plastic frame increases the window's U_w value slightly. For passive house concepts, specially insulated expanded polystyrene frameworks must be used, so that the low glazing values of triple glazing are not worsened by the frame proportion.

The passive solar gain $\dot{Q}_u$ usable in the room results from the balance of losses and gains. The losses are calculated from the U_w value of the window of surface A_w and from the temperature difference between the room air T_i and the outside air T_o:

$$\frac{\dot{Q}_u}{A_w} = U_w (T_i - T_o) - g\,G \tag{7.9}$$

From the available energy balance, an effective U-value U_{eff} can be defined, which is often used for monthly or annual balance calculations with mean temperature differences and irradiances.

$$U_{eff} = \frac{\dot{Q}_u}{A_w (T_i - T_o)} = U_w - g\frac{G}{T_i - T_o} \tag{7.10}$$

Balanced over a heating season, about 400 kWh/m^{2}a of solar irradiance is available on a south-facing facade in Germany. The mean temperature difference between the inside and outside of about 17°C, multiplied by the number of days in the heating season, results in the so-called heating degree day number, which on average in Germany is about 3500 Kelvin-days per year. The maximum usable energy per square metre of glazing surface for two-pane low-e coated glazing with a U_w value of 1 W/m^{2}K and $g = 0.65$ is thus:

$$\begin{aligned}\frac{Q_u}{A_w} &= 0.65 \times 400 \times 10^3 \frac{Wh}{m^2 a} - 1.0 \frac{W}{m^2 K} \times 3500 \frac{K\ days}{a} \times 24 \frac{h}{day} \\ &= 260 \frac{kWh}{m^2} - 84 \frac{kWh}{m^2} = 176 \frac{kWh}{m^2}\end{aligned} \tag{7.11}$$

The amount of heat effectively usable in the room depends greatly, however, on the storage capability of the structural elements on the inside, since high passive solar heat gains can easily lead to overheating of the interior and thus do not contribute to covering the heating requirement. A detailed analysis of the dynamic storage behaviour of building elements can be found in section 7.3.

In the monthly balance procedure based on EN 832 for calculating the heating requirement, the efficiency of the solar irradiance transmitted by windows is indicated as a function of the relation of the monthly gains to the transmission and ventilation heat losses. For low energy buildings with an annual heating requirement between about 30 and 70 kWh/m^{2}a the result is, for a heat-storing heavy building construction, a flat minimum of the heating requirement for a window area proportion on the south-facing facade of approximately 25%. In administrative buildings with mostly higher internal loads, the window area proportion should be lower still, to avoid overheating in summer. With a light building method with a small storage capacity, the minimum heating requirement is obtained for 0–20% of the window area proportion.

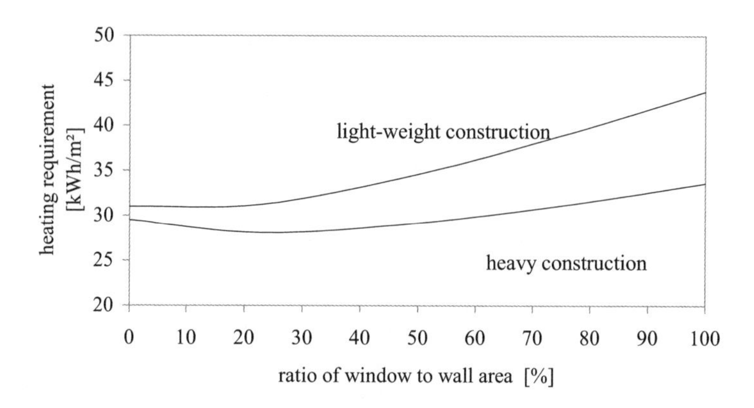

Figure 7.4: Influence of the window area proportion on the heating requirement.

7.1.3 New glazing systems

For flexible control of total energy transmission, glazing systems are being developed which modify their transparency degree temperature-dependently (thermo-tropic) or electrically controlled (electro-chromic).

Thermotropic layers are polymer mixtures or hydrogels which are inserted in a homogeneous mixture between two window panes, and which with rising temperature reduce the light permeability by up to 75% (Hartwig, 2000).

Electrochrome thin films, for example from tungsten oxide, are evaporated on window panes with conductive oxide coatings. On accumulation of cat-ions (e.g. Li+) from the counter-electrode to the tungsten oxide by an external electrical field, the transmittance falls wavelength-dependently to 10-20%. The two thin film electrodes are connected by a polymer ion conductor. A tight sealing at the edges is very important for long-term stability. With the first commercially available glasses with a maximum surface of 0.9 m × 2.0 m, a reduction of the total energy transmission factor from 44% in the bright status to 15% in the dark status is achieved (U-value = 1.6 W/m^2K) (Wittkopf *et al.*, 1999); in systems with a lower U-value of 1.1 W/m^2K, the g-value falls from 36% to 12%.

7.2 Transparent thermal insulation

Since the early 1980s, several thousand square metres of transparent thermally insulated facade systems have been installed in Germany. Compared to conventional thermal insulation in buildings, transparently insulated external walls can use the incoming solar radiation to a far greater extent. The energy potential for the application of such solar systems is high; if a fifth of all existing facades were equipped with transparent insulating systems, approximately 15% of the heat needed for room heating could be supplied (Braun *et al.*, 1992). The technology is particularly interesting for the renovation of old buildings with heavy, very heat-conducting walls (Eicker, 1996). Transparent insulated panes can be stuck directly on the external wall; a transparent plaster protects the material from the weather. By foregoing complex frame constructions, the costs of glass composite structures, high so far, can be significantly reduced. A further application of transparent thermal insulation is daylighting. The light scattering and light directing by the elements can be used for an even illumination of the room, and the very good heat-insulating characteristics allow large-scale application on external walls.

7.2.1 *Operational principle*

If short-wave solar radiation hits an external wall, the radiation is absorbed and converted into heat. The external surface warms up, but most of the heat produced is mainly transferred to the outside air. Only a small part of the heat reaches the building interior. If a transparent insulating layer is attached in front of the wall (in the simplest case a window pane), heat emission to the outside is made more difficult.

The main parametres influencing the extent of the useful heat gain are the transmission coefficient for solar radiation and the heat resistance of the transparent thermal insulation on the one hand, and the absorption coefficient, heat conductivity and storage capability of the adjoining wall on the other. A time delay in the heat flow takes place due to the wall, so the maximum values of the solar-induced heat flow reach the inside when the direct solar gains through the windows have already decreased and outside temperatures are falling. In addition, the thermal characteristics of the entire building play a role, above all the heat-storing capability of the interior structural elements in avoiding overheating.

An external wall which even with 10 cm external insulation still has calorific losses of over 30 kWh per square metre and heating season, becomes a solar collector due to a transparent thermal insulation and produces around 50–100 kWh/m^2 of useful heat for the building.

The heat transition coefficient of an external wall insulated with TWD results as usual from the total of the thermal resistances of the existing external wall and the transparent insulation,

$$U_{eff} = U_{wall+TWD} - \eta_0 \frac{G}{T_i - T_o} \tag{7.12}$$

with layer thickness s_{TWD} and heat conductivity λ_{TWD} of the transparent insulating material, layer thickness s_{wall} and heat conductivity λ_{wall} of the external wall and also the heat transfer coefficients inside h_i and outside h_o.

The absorber is characterised by the absorption coefficient α, the transparent insulation by the diffuse total energy transmission factor g_d. The U-values of 10 cm TWD are typically about 0.8 W/m^{2}K, lower than the best double-glazed heat-protection glass, but still twice as high as the heat transfer coefficients of 10 cm of conventional insulating material with U-values of 0.4 W/m^{2}K.

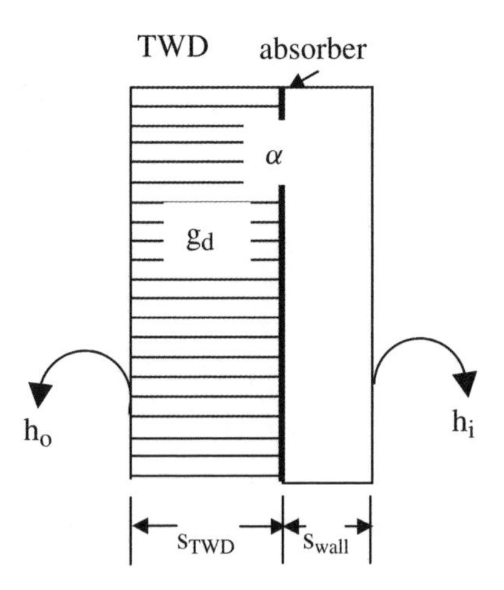

Figure 7.5: System structure of a transparently insulated wall.

Just as with the energy balance of windows, an effective U-value can also be defined for transparent thermal insulation as the difference between losses and solar gains with efficiency η_0.

$$U_{eff} = U_{wall+TWD} - \eta_0 \frac{G}{T_i - T_o} \tag{7.13}$$

Solar efficiencies η_0 up to 50% with a simultaneous low heat transfer coefficient lead to effective U-values which, with a favourable wall orientation, are negative on the annual average and thus are heat gains for the building. Measurements of a 10 cm transparently insulated building in Freiburg/Germany resulted in weekly averaged effective U-values between 0 and –3.5 W/m²K.

The solar efficiency η_0 corresponds to the total energy transmission factor of a glazing and consists of the g_d-value of the transparent insulating material, the absorption factor of the absorber α and the proportion of the heat flow inward to the total heat flow. The heat flow from the absorber inward is calculated from the temperature node of the absorber at temperature T_a to the room air temperature T_i via the heat transfer coefficients of the wall $U_{wall} = \left(\frac{s_{wall}}{\lambda_{wall}} + \frac{1}{h_i} \right)^{-1}$:

$$\frac{\dot{Q}_i}{A} = U_{wall} \left(T_a - T_i \right) \tag{7.14}$$

The total heat flow results from the total heat flow inward (Equation (7.14)) and the heat flow from the absorber outward $\dot{Q}_o$, which is calculated via the heat transfer coefficient U_{TWD} of the TWD material:

$$\frac{\dot{Q}_o}{A} = U_{TWD}\left(T_a - T_o\right) \tag{7.15}$$

$$\eta_0 = \alpha g_d \frac{U_{wall}\left(T_a - T_i\right)}{U_{wall}\left(T_a - T_i\right) + U_{TWD}\left(T_a - T_o\right)} \tag{7.16}$$

Assuming identical temperatures inside and outside, a constant solar efficiency can be defined which is very suitable for material comparisons and estimates of the energy yield.

$$\eta_0 = \alpha g_d \frac{U_{wall}}{U_{wall} + U_{TWD}} \tag{7.17}$$

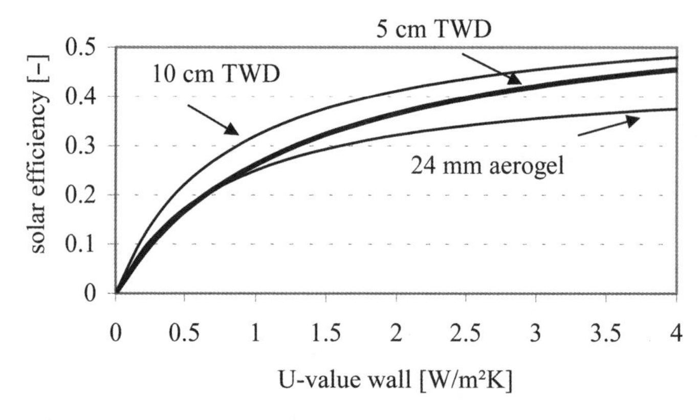

Figure 7.6: Solar efficiency as a function of the heat transfer coefficient of the opaque wall.

With 5 cm TWD capillaries, the U_{TWD} value is 1.3 W/m^{2}K at a g-value of 0.67; with 10 cm it is 0.8 W/m^{2}K and g = 0.64. The aerogel material shows a U_{TWD} value of 0.8 W/m^{2}K with a very small layer thickness of 2.4 cm at a g-value of 0.5.

For optimal use of solar heat, one must ensure that the transparently insulated room is not overheated. Conventional window areas usually bring sufficiently high solar gains into the room during the day; additional gains from the TWD wall must be stored and then used in the evening hours. The temporal shift between heat production at the absorber and maximum heat flow into the room rises with external wall thickness and also depends on the density, thermal capacity and heat conductivity of the wall. With a 24 cm brick or lime-sandstone wall, phase shifts of 6 to 8 hours are achieved; with concrete walls, little more than 5 hours are possible. With sufficiently strong wall constructions, the result is typically 100% efficiencies of the heat produced by the TWD wall within the core months of the heating season; in the transition months around 30% (Wagner, 1998).

The high absorber temperatures of the external wall, which can reach peak values between 70° and 80°C, are effectively dampened at wall thicknesses over 20 cm and are, at the interior surface, rarely higher than 30°C even in the summer. The absorber temperatures are lower with heavy, heat-conducting components than with light walls with densities of around 800 kg/m³. The heat which develops can penetrate quickly into the heavy external wall and be led into the interior. Any thermal tensions are thus correspondingly low.

At an experimental house in Stuttgart the thermal deformations of the external wall were measured on a long-term basis. Compression stresses and slight swelling of the wall due to the high temperature difference did not pose a problem. Fine cracks of about 1 to 2 mm in the plaster resulted from accelerated drying of the new building's brickwork dampness around the absorber; they did not, however, influence the load-carrying capacity of the wall. With new buildings it is worth planning for defined joints at the edges of the TWD surfaces.

The use of shading systems such as blinds or shutters prevents heating of the external wall in the transition period and in the summer months, but it is complex in terms of construction and maintenance issues. Constructional shadings such as balconies or roof projections must be planned very carefully, in order not to obstruct exposure to the sun in the transition period. Foregoing shading mechanisms is possible if the transmittance of the transparent insulation is strongly angle-dependent, so that with a high sun position in the summer with angles of incidence over 60° at the south-facing facade, less than 20% of the irradiance reaches the absorber wall. The transmittance of a transparent heat insulation system falls, for example, from approximately 50% with perpendicular incidence to 15% with a sun elevation angle of 60°.

Simulations for highly insulated low-energy buildings have shown that even when the south-facing facade is largely covered with a TWD heat insulation system, the excess heat in summer can be expelled by night ventilation (Meyer, 1995). The number of hours with ambient temperatures over 26°C is fewer than 50 over the reference building with conventional insulation and is altogether below 300 hours per year.

On the other hand, at angles of incidence of 60°, capillary material with a glass covering displays a very high total energy transmission factor of 35%. In this case, sun protection is unavoidable, when the facade is largely covered. If nocturnal ventilation is not possible, e.g. in office buildings, shading of the TWD surfaces in the summer should likewise be provided for.

The orientation of the transparently insulated facade is crucial both for energy gain in the winter and for protection against overheating in the summer. Facade orientations between south-east and south-west are suitable. Twice as much energy, some 400 kWh/m^2, falls on a south-facing facade in the heating season as on an east or west-facing one. In addition, the low winter sun position leads to good light transmission by the TWD material. In summer, on the other hand, only the south-facing facade offers a certain natural sun protection with low transmittance.

7.2.2 Materials used and construction

TWD capillary or honeycomb structures are assembled from thin-walled plastic tubes and welded by a hot wire section or manufactured into strips of any width from extruder nozzles with an almost square cell cross-section. The typical cell diametre is 3 mm.

Two polymer types are in use today: Polymethyl methacrylates (PMMA) and polycarbonates (PC). PMMA is characterised by high transmittance and by good UV stability. Due to the brittleness of the material and its poor fire-retardance (class B3), PMMA is bound between window panes. For this, a qcomplex mullion-transom construction is necessary, leading to system costs between 400–750 €/m^2. The cost of the 10–12 cm transparent insulation is typically only around 50 €/m^2; it is the glazing and attachment, at approximately 250 €/m^2, plus shading items such as blinds at around 150 €/m^2, which drives up the costs.

Polycarbonates are mechanically more stable and can also be processed without glass covering; they are, however, not very UV-resistant. Their fire retardance is better (class B1) and the material is temperature-resistant to about 125°C. Polycarbonate materials can be used in heat insulating compound systems. The covering plaster is an acryl adhesive mixed with 2.5–3 mm diametre glass balls, which is applied in the factory directly onto the capillary material. Additional UV absorbers can be likewise brought into the cover plaster. Such heat insulating compound systems can be manufactured with substantially reduced costs of around 150 €/m^2, since there are no complex glazing and shading systems. The weight of capillary materials is around 30 kg/m^3.

Capillaries made of glass are manufactured like the polymer structures, but are complicated to produce due to high processing temperatures and the associated engineering problems. Glass capillaries are much more temperature- and UV-resistant, but likewise mechanically not very stable. The recycling ability of glass, which is also possible with PMMA, is advantageous. Polycarbonates, on the other hand, are recyclable only with high energy expenditure and with quality losses.

For glazing systems with smaller thicknesses of 2–3 cm, aerogels are suitable. These are highly porous, open-pored solids consisting of more than 90% air and 10% silicate, with very low heat conductivity (λ = 0.02 W/mK). Aerogels are made of silica gel and can easily be poured into the cavity of double glazing. They are not inflammable, are easy to dispose of and to recycle. The significant disadvantage is that their light transmission is only around half that of capillary materials, and their sensitivity to water is also a problem. Water penetrating into the edge network of double glazing is absorbed by the aerogel material and the sensitive structure is broken by the capillary forces.

7.2.2.1 Construction principles of TWD systems

Transparent heat insulating systems are mainly used in two types of construction:

1. as a mullion-transom or element construction with framed TWD panel elements. To avoid dirtying of the TWD materials, the external covers usually consists of two highly transparent, iron-poor single glass panes. Element constructions are characterised by a higher degree of prefabrication with a corresponding cost-reduction potential; from the outside it is often impossible to distinguish between them and a mullion-transom construction installed on site.

Shading mechanisms such as blinds or shutters are preferably inserted between the outside window pane and the TWD material. Lamella type systems can also be used in front of the facade, and display high working reliability when few movements take place between open and closed status or when in an always-lowered status.
2. as a heat insulating compound system with frameless direct installation. The transparently plastered capillary structures are supplied with a fabric for attaching the plaster to the conventional insulation, and fastened to the external wall with a black adhesive that serves as an absorber.

7.3 Heat storage by interior building elements

Heat storage by the interior components is decisive for the degree of the useful energy in both passive solar use by windows and in transparent heat insulating systems. Only if solar gains do not lead to overheating of the interior can the heating demand be reduced. The heat storage capacity of components can be roughly estimated from the storage mass, the thermal capacity and the possible rise in temperature of the storage mass. Thus, for example, a solid concrete wall with a thickness d of 30 cm, a heat capacity c of 1 kJ/kgK and a gross density ρ of 2100 kg/m^3 can store, with a rise in temperature of 5°C, an amount of heat of 0.875 kWh per square metre of surface.

$$\frac{Q}{A} = \rho\, d\, c\, \Delta T = 2100\frac{\text{kg}}{\text{m}^3}\times 0.3\text{m}\times 1.0\frac{\text{kJ}}{\text{kgK}}\times 5\text{K} = 3150\frac{\text{kJ}}{\text{m}^2} = 0.875\frac{\text{kWh}}{\text{m}^2} \tag{7.18}$$

This view presupposes that the component is completely warmed or cooled to the temperature levels forming the basis of the calculation. This would presuppose very high heat transfer coefficients and high heat conductivities, which in practice is not the case. To what extent the storage capacity can be used depends, apart from the material values, primarily on the duration of a rise in temperature.

If, by dynamic calculation methods or by measurement, the amount of heat Q per surface A is determined which flows in a given period into the wall, then from this an effective thickness d_{eff} for the wall can be calculated, whose storage capability is fully used.

$$d_{eff} = \frac{Q}{A\, c\rho\Delta T} \tag{7.19}$$

During a three-hour rise in temperature, a concrete wall (with A = 1 m^2 surface) can, largely irrespective of its thickness, take up approximately 33 Wh for each Kelvin of temperature rise (with heat take-up on both sides). This corresponds to an effective thickness of approximately 5 cm. With a six-hour rise in temperature this value is approximately 9 cm.

If the heating up of a room is to be calculated, then for rough estimates the amount of heat Q flowing into the component can be calculated via the heat transfer coefficient h_i and the temperature difference between the component surface $T_{s,1}$ (at the beginning of a time step Δt) and the room air T_i.

$$Q = h_i\, A\, \Delta t\left(T_i - T_{s,1}\right) \tag{7.20}$$

After the time-step the new temperature of the component $T_{s,2}$ results from the stored amount of heat $Q_s = Q$:

$$T_{s,2} = T_{s,1} + \frac{Q_s}{A c \rho \, d_{eff}} \tag{7.21}$$

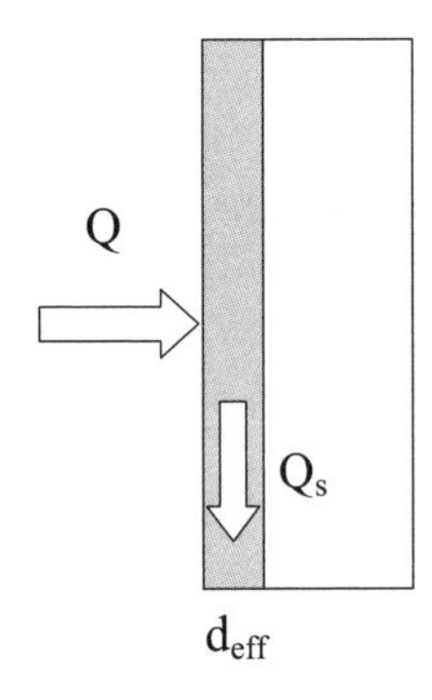

Figure 7.7: Amount of heat flowing into a component and the effective storage mass thickness d_{eff}.

This process is repeated with each time-step. For a more exact calculation of the temporally variable temperature distribution, an energy balance for a volume element of the storage mass must be created which leads to the classical thermal heat conduction equation. For simplicity, only one dimensional temperature distributions will be derived, i.e. from the air over the surface into the component depth.

For passive solar energy use the following boundary conditions play a role:

- the storage capacity of components during brief variations in temperature in the room caused by solar irradiance or air temperature modification,
- the potential for night cooling by utilisation of the periodic modification of the air temperature between day and night,
- the temperature amplitude and phase shift on the inner side of a transparently insulated wall.

In all cases heat is only absorbed or dissipated via the surface of the component. In the component interior there are no heat sources, so a very simple energy balance for each volume element results: from an entering heat flow $\dot{Q}_{in}$ by thermal conduction, part leads to the rise in temperature in the volume element (heat storage $\dot{Q}_{st}$), and the remainder is passed on by thermal conduction into the next element $\dot{Q}_{out}$.

$$\dot{Q}_{in} = \dot{Q}_{st} + \dot{Q}_{out} \tag{7.22}$$

The heat flow $\dot{Q}_{in}$ entering through surface A is, based on Fourier's law of thermal conduction, proportional to the temperature gradient at the point x_0.

$$\dot{Q}_{in} = -\lambda\, A \left.\frac{dT}{dx}\right|_{x_0} \tag{7.23}$$

The exiting heat flow $\dot{Q}_{out}$ at the point $x_0 + dx$ is, at constant heat conductivity λ, only different from $\dot{Q}_{in}$ if the temperature gradient has changed in the volume element, e.g. has become flatter during partial heat storage in the element.

$$\dot{Q}_{out} = -\lambda A \left.\frac{dT}{dx}\right|_{x_0+dx} \tag{7.24}$$

A Taylor series expansion of the temperature gradient at the point $x_0 + dx$ leads, if all higher-order members are ignored, to the following simplification:

$$\begin{aligned}\dot{Q}_{out} &= -\lambda A\left(\left.\frac{dT}{dx}\right|_{x_0} + \frac{d}{dx}\left(\left.\frac{dT}{dx}\right|_{x_0}\right)dx + \dots\right)\\ &\approx -\lambda A\left(\left.\frac{dT}{dx}\right|_{x_0} + \left.\frac{d^2T}{dx^2}\right|_{x_0} dx\right)\end{aligned} \tag{7.25}$$

The amount of heat $\dot{Q}_{st}$ stored in the volume element $dV = Adx$ is given by

$$\dot{Q}_{st} = \rho\, dV\, c\frac{dT}{dt} \tag{7.26}$$

Thus the energy balance Equation (7.22) leads to:

$$\begin{aligned}-\lambda\, A \left.\frac{dT}{dx}\right|_{x_o} &= \rho dVc\frac{dT}{dt} - \lambda\, A\left.\frac{dT}{dx}\right|_{x_o} - \lambda A \left.\frac{d^2T}{dx^2}\right|_{x_o} dx\\ \frac{\lambda}{\rho c}\frac{d^2T}{dx^2} &= \frac{dT}{dt}\end{aligned} \tag{7.27}$$

where $a = \dfrac{\lambda}{\rho c}\left[\dfrac{m^2}{s}\right]$ is termed the thermal diffusivity; it lies between 10^{-7} m^2/s for wood and 10^{-4} m^2/s for metals.

7.3.1 Component temperatures for sudden temperature increases

With periodic boundary conditions or with temperature-equalising processes, the differential equation can be solved by a product approach, with one function dependent only on time and the other only on place.

If a temperature jump on one side is given as a boundary condition, an approach with the Gauss error integral (or error function *erf (z)*) leads to a more general solution than the product approach, since the initial temperature distribution *F(x)* at the point in time $t = 0$ can assume any values. The temperature and time functions are, however, no longer separate.

$$T(x,t) = \frac{1}{\sqrt{\pi}} \frac{1}{\sqrt{4at}} \int_{-\infty}^{+\infty} F(\xi) \exp\left(-\frac{(\xi - x)^2}{4at} \right) d\xi \tag{7.28}$$

From this general approach, some solutions for simple boundary conditions can be represented analytically. Thus for a concrete floor slab, the change in temperature at a certain depth (as an indication of the utilisation of the heat storage) as well as the heat flow occurring through the surface $\dot{Q}_{in}$ and the stored heat $\dot{Q}_{st}$ as a function of the duration of the temperature jump at the surface can be examined. Far more relevant in practice is the case of a temperature jump in the room air, which is transferred via convection to the component surface. This situation is very complex in its mathematical derivation and is therefore discussed later, with solutions given.

The temperature distribution for a component with a constant initial temperature T_c and a temperature jump at the component surface to zero for $t > 0$ is directly calculable from the Gauss error integral. With the substitution

$$\eta = \frac{\xi - x}{\sqrt{4at}} \qquad d\xi = d\eta\sqrt{4at} \qquad F(\xi) = T_c$$

the result from Equation (7.28) is:

$$\frac{T(x,t)}{T_c} = \frac{2}{\sqrt{\pi}} \int_0^{z=\frac{x}{\sqrt{4at}}} \exp(-\eta^2) d\eta = erf\left(\frac{x}{\sqrt{4at}} \right) = erf(z) \tag{7.29}$$

Factor 2 results from splitting the integral from Equation (7.28) into two parts and the symmetry of the function $\exp(-\eta^2)$. Strictly, the solution only applies to a semi-infinitely expanded body, but can also be used for shorter time intervals (a few hours) for a finitely expanded wall.

The error function *erf (z)* can be approximated with an error $< 2.5 \times 10^{-5}$ with an exponentially dampened polynomial function of third order (Wong, 1997).

With the auxiliary variable $p = \dfrac{1}{1 + 0.47047 \times z}$ the approximation equation reads

$$erf(z) = 1 - \left(0.3480242 \times p - 0.0958798 \times p^2 + 0.7478556 \times p^3\right) \times \exp\left(-z^2\right) \quad (7.30)$$

From the values of the error function one can directly determine which temperature *T(x, t)* prevails at any point *x* and time *t*. If, on the other hand, the penetration of a given temperature ratio *T(x,t)* to the initial temperature T_c at time *t* or depth *x* is to be determined, Equation (7.30) must be solved iteratively for *z*, and *x* and *t* must be calculated from $z = x/\sqrt{4at}$. The graphical representation of the error function for direct reading of the z-value from the function value *erf(z)* = *T(x,t)/*T_c is represented below.

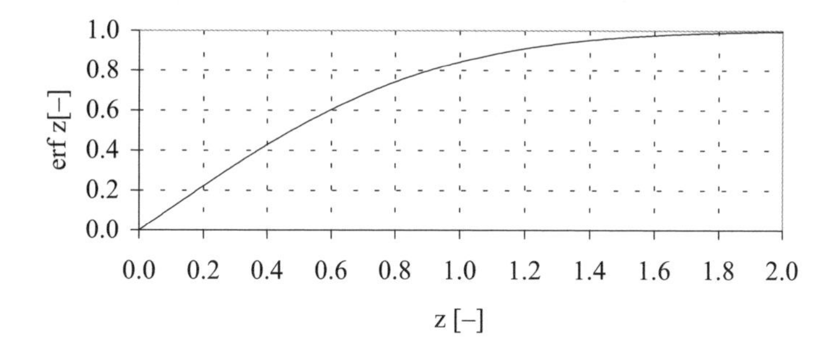

Figure 7.8: Error function *erf(z)* = *T(x,t)/*T_c as a function of $z = x/\sqrt{4at}$.

The temperature jump must always take place from the initial temperature T_{c0} to the jump temperature $T_j = 0$°C. If this is not the case ($T_j \neq 0$), a standardised initial temperature T_c is calculated from the temperature difference of the initial temperature T_{c0} and the jump temperature T_j: $T_c = T_{c0} - T_j$. In what follows, the calculations are always based on a standardised initial temperature T_c and a surface temperature of zero.

Example 7.1

On a 20°C concrete (or wooden) floor, a surface temperature jump to T_j = +30°C occurs. Over what period of time has the temperature at a depth of 20 cm risen by 5°C?

	Heat conductivity λ [W/mK]	*Density* ρ [kg/m³]	*Heat capacity* *c* [kJ/kgK]	*Thermal diffusivity* *a* [m²/s]
Concrete	1.28	2200	0.879	0.66×10^{-6}
wood	0.2	700	2.4	0.12×10^{-6}

In order to adapt the boundary condition to a temperature jump of the surface temperature to 0°C, the initial floor temperature T_{c0} = 20°C is replaced by the standardised initial temperature $T_c = T_{c0} - T_j$ = 20°C – 30°C = –10°C. A temperature rise *T(x,t)* of 5°C at a component depth of 20 cm thus corresponds to a temperature condition

$$\frac{T(x = 0.2m, t)}{T_c} = \frac{-5°C}{-10°C} = 0.5 = erf(z)$$

The iteratively calculated z-value (which can also be read off Figure 7.8) is $z = 0.477 = \frac{x}{\sqrt{4at}}$. From $t = \frac{x^2}{4a\,z^2}$ the result for the concrete ceiling is a temperature-rise time of 18.5 hours, and for the wooden ceiling 111 hours.

If the temperature ratio $T(x,t)/T_c = erf(z)$ is represented as a function of time, it can be seen how quickly a surface temperature jump propagates into the component. For $x > 0$ the temperature at $t = 0$ is at first the constant initial temperature T_c and the temperature ratio is unity. With increasing time, the component temperature approaches the surface temperature of zero, i.e. the temperature ratio tends towards zero.

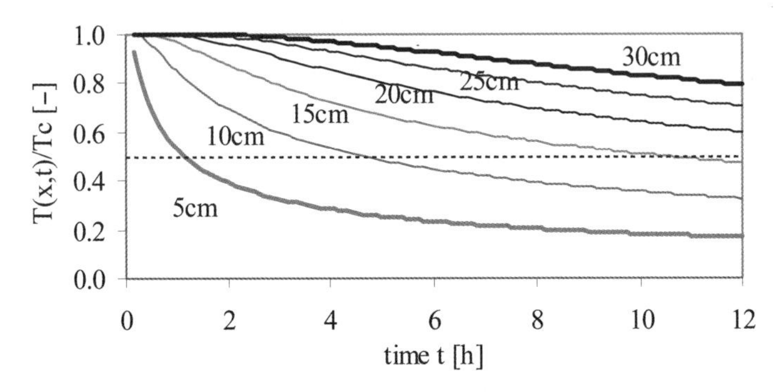

Figure 7.9: Error function $erf(z) = T(x,t)/T_c$ as a function of time for concrete floors with thermal diffusivity $a = 0.66 \times 10^{-6} m^2/s$.

After 12 hours, half of the surface temperature increase ($T(x,t)/T_c = 0.5$) is at a component depth of 16 cm for the concrete floor, which illustrates the limitation of the effectively usable storage capacity. Only with times >> 12 hours does the component temperature approach the surface value of zero ($T(x,t)/T_c \to 0$).

For wood with a lower thermal diffusivity a, a surface temperature jump clearly continues more slowly.

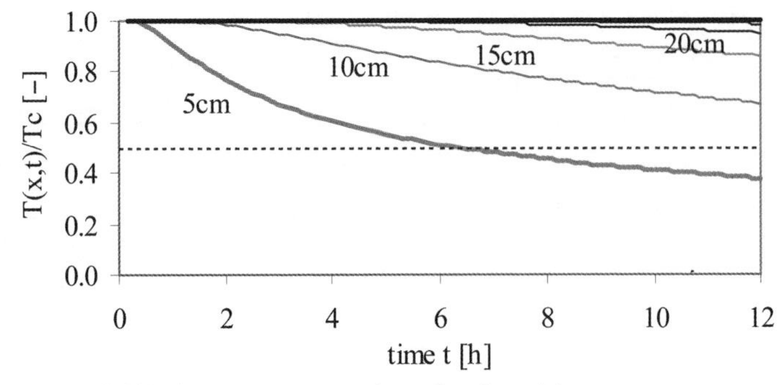

Figure 7.10: Temperature field in the component wood as a function of time.

After 12 hours, half of the surface temperature increase has only reached a depth of 7 cm. For component thicknesses over 20 cm, the temperature has not changed even after 12 hours.

The heat flow *dQ/dt* entering or leaving the surface *A* is proportional, based on Fourier's law, to the temperature gradient at the surface:

$$\frac{dQ}{dt} = -\lambda A \left.\frac{dT}{dx}\right|_{x=0} = -\lambda A T_c \frac{d}{dx} erf\left.\left(\frac{x}{\sqrt{4at}}\right)\right|_{x=0}$$
$$= -\lambda A T_c \frac{1}{\sqrt{4at}} \frac{2}{\sqrt{\pi}} \underbrace{\exp\left(-\frac{x^2}{4at}\right)}_{=1 \quad für\ x=0} = -A \underbrace{\sqrt{\lambda \rho c}}_{b} \sqrt{\frac{1}{\pi}} T_c \frac{1}{\sqrt{t}} \tag{7.31}$$

The heat flow is proportional to the so-called heat penetration coefficient $b = \sqrt{\lambda \rho c}$ and falls with $1/\sqrt{t}$.

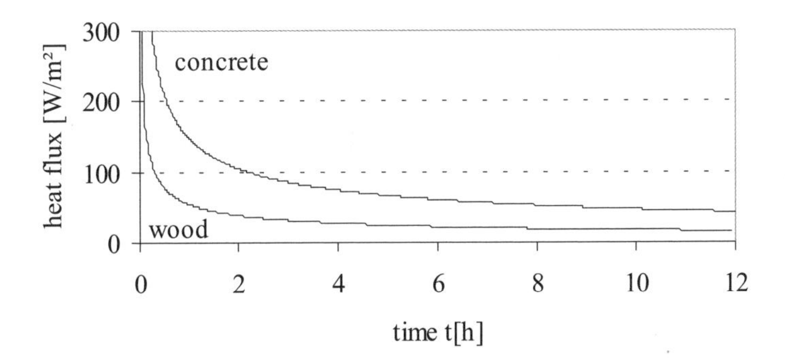

Figure 7.11: Heat flux as a function of time for a concrete and a wooden floor with a 10 K temperature jump.

The integration of the heat flow over time results in the total amount of heat penetrating into the component when there is a surface temperature jump.

$$\frac{Q}{A} = \int \frac{dQ}{A} = \int_0^{t_0} \left(-\sqrt{\frac{\lambda \rho c}{\pi}} T_c \frac{1}{\sqrt{t}} \right) dt = -\frac{2}{\sqrt{\pi}} \underbrace{\sqrt{\lambda \rho c}}_{b} \sqrt{t_0} T_c \tag{7.32}$$

Since the surface temperature for $t > 0$ has been set by definition to 0°C, for T_c the standardised initial temperature must be used again.

Example 7.2

Calculation of the amount of heat penetrating a component with a temperature jump at the surface of 10K for t_0 = 12h for the two floors from Example 7.1.

The heat penetration coefficient $b = \sqrt{\lambda \rho c}$ for concrete is $1.573\ \text{kJ/(m}^2\text{K}\sqrt{\text{s}})$, and for wood it is $0.579\ \text{kJ/(m}^2\text{K}\sqrt{\text{s}})$. Thus within 12 hours an amount of heat of 3689 kJ/m^2 = 1.02 kWh/m^2 is brought into the concrete ceiling and 0.38 kWh/m^2 into the wooden ceiling.

The amount of energy stored within a component is, like the heat flow, directly proportional to the heat penetration coefficient b and to the temperature jump ΔT at the surface, but it rises with the root of time.

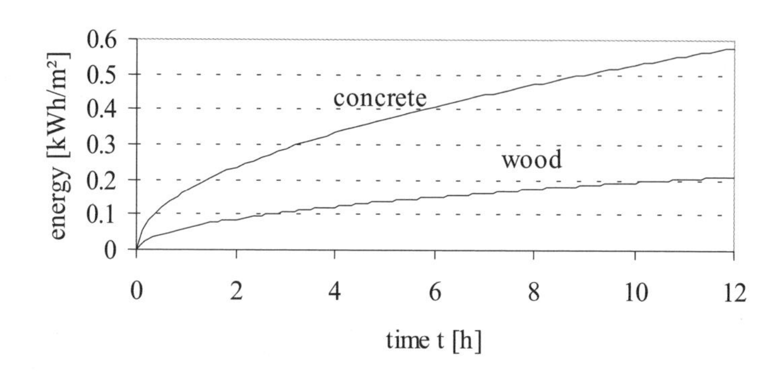

Figure 7.12: Amount of energy led into the component with a rise in temperature at the surface of 10K.

Usually, however, it is not the surface temperature but the air temperature which is known. Between the air and the surface temperature change there is a phase shift and a dampening of the amplitude. At a given air temperature T_o and a given heat transfer coefficient h between the air and the surface, the temperature field $T(x,t)$ can be calculated as follows (Gröber, 1988):

$$\frac{T(t,x)}{T_c} = erf\left(\frac{x}{\sqrt{4at}}\right) + \exp\left(at\left(\frac{h}{\lambda}\right)^2 + \frac{h}{\lambda}x\right)\left(1 - erf\left(\frac{x}{\sqrt{4at}} + \frac{h}{\lambda}\sqrt{at}\right)\right) \tag{7.33}$$

At the surface $x = 0$, therefore, the following temperature appears:

$$T(t, x=0) = T_c \exp\left(at\left(\frac{h}{\lambda}\right)^2\right)\left(1 - erf\left(\frac{h}{\lambda}\sqrt{at}\right)\right) \tag{7.34}$$

Example 7.3

Calculation of the temperature of a concrete ceiling at the surface and at a depth of 5 cm for t = 1 h and t = 10 h with parameters from Example 7.1, if the air temperature, rather than the surface temperature, jumps to 30°C. The heat transfer coefficient h is 8 W/m^{2}K.

First the surface temperature is calculated for $x = 0$ using Equation (7.33). For $t = 1\ h$ the temperature at the surface is:

$$T(t=1\,h, x=0)/T_c = 0.73$$

with
$$\exp\left(at\left(\frac{h}{\lambda}\right)^2\right) = \exp\left(\underbrace{0.66\times10^{-6}\frac{m^2}{s}\times1\,h\times3600\frac{s}{h}\times\left(\frac{8\frac{W}{m^2K}}{1.28\frac{W}{mK}}\right)^2}_{0.0928}\right) = 1.097$$

and
$$erf\left(\frac{h}{\lambda}\sqrt{at}\right) = erf(0.304) = 0.33$$

With an air temperature jump of 10 K, the ratio of the surface temperature to the initial body temperature T_c after one hour is still at 73%, i.e. with the selected boundary condition for heating, the surface temperature has increased by $(1-0.73)\times 10\ K = 2.7\ K$. After t = 10 h the temperature ratio is 43.8%. The surface temperature has then increased by 5.6 K.

At a depth of 5 cm, the temperature ratio after t = 1 h is 0.903, i.e. the temperature has only increased by 1 K. After 10 hours the temperature at a depth of 5 cm has increased by 4.33 K.

With air temperature modifications, the potential for heat storage with a limited duration of the temperature jump is clearly smaller than with direct impact of the temperature jump on the surface. After 12 hours the surface has only taken up 60% of the air temperature jump, i.e. the effective storage capacity sinks by 40%.

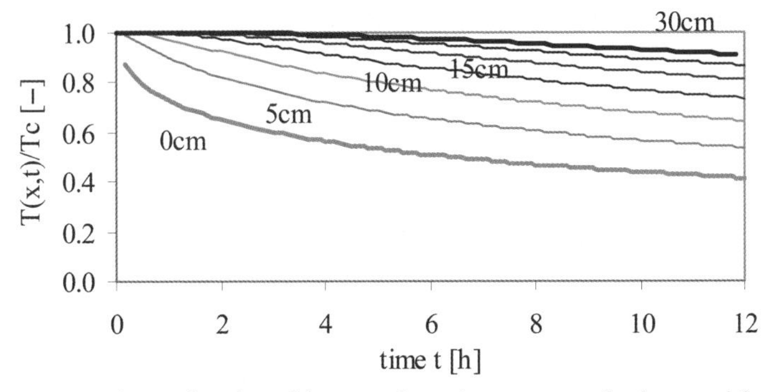

Figure 7.13: Temperature ratio as a function of time at a given air temperature for the material concrete.

The surface temperature changes during an air temperature jump essentially depend on the relation of the heat transfer coefficient h_i to the heat conductivity λ of the component. Component surfaces of materials with low heat conductivity clearly assume the air temperature faster (i.e. $T(x = 0,t)/T_c$ becomes zero) than components which conduct heat well. Deep in the component, on the other hand, a temperature jump of air continues only very slowly with poorly conducting materials.

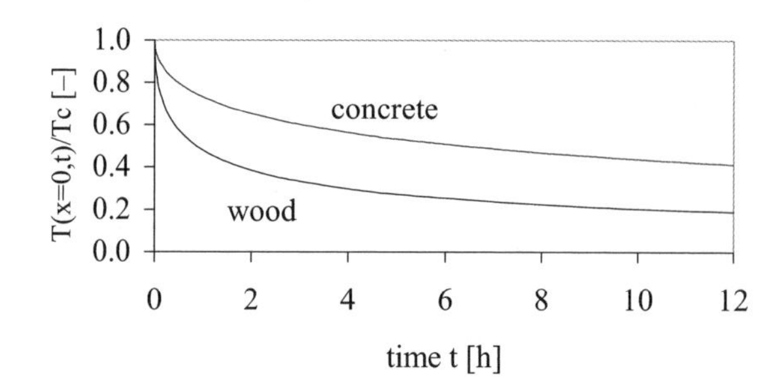

Figure 7.14: Relation of the surface temperature to the initial body temperature during an air temperature jump of 10 K.

The heat flow density into the component can be obtained either from:

$$\frac{\dot{Q}}{A} = h_i \left(T_o - T_{x=0}\right) \qquad \text{or} \qquad \frac{\dot{Q}}{A} = -\lambda \frac{\partial T}{\partial x}\bigg|_{x=0}$$

with the common solution for an air temperature jump T_o to zero:

$$\frac{\dot{Q}}{A} = -h_i T_c \exp\left(\left(\frac{h}{\lambda}\right)^2 at\right)\left(1 - erf\left(\left(\frac{h}{\lambda}\right)\sqrt{at}\right)\right) \tag{7.35}$$

Since the temperature difference of the surface $T(x = 0)$ and air T_o (here zero) is larger at the concrete surface than at the wood surface, the larger heat flows and stored amounts of energy occur there.

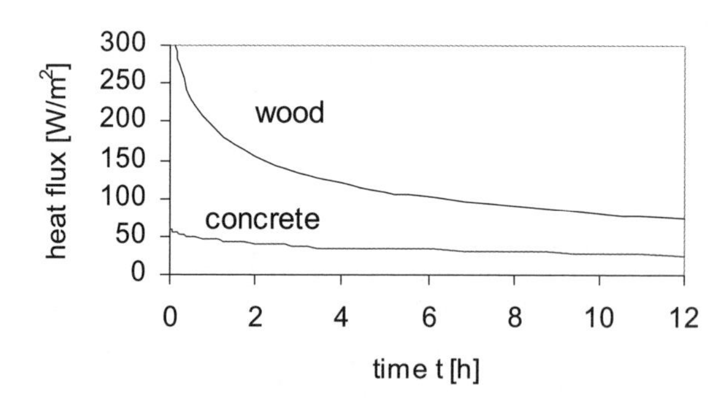

Figure 7.15: Heat flux into wood and concrete at an air temperature jump of 10 K.

To calculate the amount of heat Q for each unit area A which has flowed into the component till the point in time t_1, the heat flux density $\dot{Q}/A$ must be integrated over the time. A solution to this problem is probably very complex. It is simpler to calculate the heat flow density in smaller time intervals and total afterwards.

$$\frac{Q}{A}=\frac{1}{A}\int_0^{t_1}\dot{Q}dt\approx\frac{1}{A}\sum_1^n\dot{Q}_i\Delta t \quad \text{with}=\frac{t_1}{\Delta t} \tag{7.36}$$

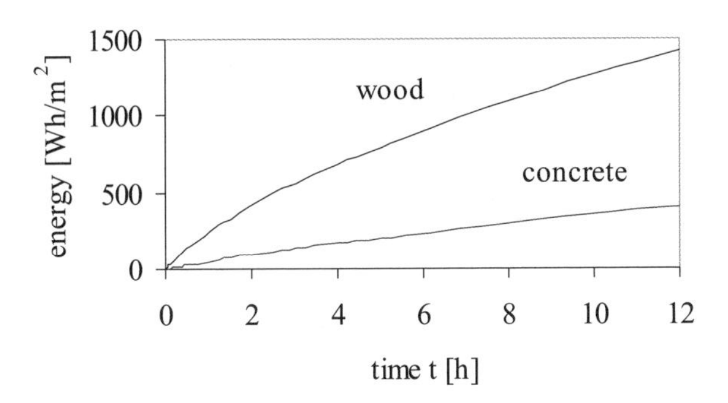

Figure 7.16: Amount of heat stored in the materials on an air temperature jump of 10 K.

7.3.2 *Periodically variable temperatures*

Apart from temperature jumps, for practical applications periodic changes in temperature caused by the external climate are of particular importance. The outside temperature T_o and the irradiance converted to a fictitious sol-air temperature can be approximated as periodic functions of time t with a period t_0 of 24 hours and an amplitude T_{om}.

The analytic solution of the thermal conduction equation is, as with the temperature jump, simpler for a periodic boundary condition at the surface. Firstly, therefore, the material temperatures of a semi-infinitely expanded component with periodically varying surface temperatures $T_s(t)$ will be calculated, and only afterwards will they be generalised to the boundary condition of the air temperature. With a period duration t_0 and an amplitude T_{sm}, the surface temperature $T_s(t)$ is set as follows:

$$T_s(t)=T_{sm}\cos\left(\frac{2\pi}{t_0}t\right) \tag{7.37}$$

Proceeding from a product approach for the temperature field $T(t,x)=\varphi(t)\psi(x)$, for the time function $\varphi(t)$ the complex exponential function $\exp(ipt)=\cos(pt)+i\sin(pt)$ is selected as the periodic function. With this approach for the time function and $d\varphi(t)/dt=ip\exp(ipt)$, the thermal conduction equation $a\frac{d^2T}{dx^2}=\frac{dT}{dt}$ becomes

$$a\frac{d^2\Psi(x)}{dx^2}\exp(ipt) - ip\psi(x)\exp(ipt) = 0$$
$$\Leftrightarrow \frac{d^2\Psi(x)}{dx^2} - i\frac{p}{a}\psi(x) = 0 \tag{7.38}$$

with the solution
$$\psi(x) = C\exp\left(x\sqrt{-i\frac{p}{a}}\right) \tag{7.39}$$

and
$$T(t,x) = C\exp(ipt)\exp\left(x\sqrt{-i\frac{p}{a}}\right) \tag{7.40}$$

which can be split into a real and an imaginary part. After some rearrangements and use of the surface boundary condition at $x = 0$, the constant of the imaginary solution becomes zero and the real temperature field is

$$T(x,t) = T_{sm}\exp\left(-x\sqrt{\frac{\pi}{at_0}}\right)\cos\left(\frac{2\pi}{t_0}t - x\sqrt{\frac{\pi}{at_0}}\right) \tag{7.41}$$

The wavelength x_l of the cosine function results in $x_l = 2\sqrt{\pi a t_0}$ from the relationship $x_l\sqrt{\pi/(at_0)} = 2\pi$, and the propagation rate of the wave is $v = x_0/t_0 = 2\sqrt{\pi a/t_0}$. The subsiding exponential function dampens the amplitude of the wave with rising x. The temporal phase shift t_x of the temperature wave at component depth x compared to the surface $x = 0$ is

$$\frac{2\pi}{t_0}t_x = x\sqrt{\frac{\pi}{at_0}} \Rightarrow t_x = \frac{x}{2}\sqrt{\frac{t_0}{a\pi}} \tag{7.42}$$

A periodic change in temperature at a component surface is seen in transparently insulated components in which the absorption of solar radiation on the wall surface leads to a periodic change in temperature.

With Equation (7.41) the periodic change in temperature within the component can be calculated with exponentially dampened amplitude and period duration t_0. The temperature field $T(x,t)$ is related for illustration purposes to the amplitude of the surface temperature fluctuation T_{sm}. Even at just 5 cm component depth, the surface amplitude of a concrete wall is reduced by 30%. The phase shift between the maximum surface temperature and the temperature at a depth of 5 cm is 1.5 h.

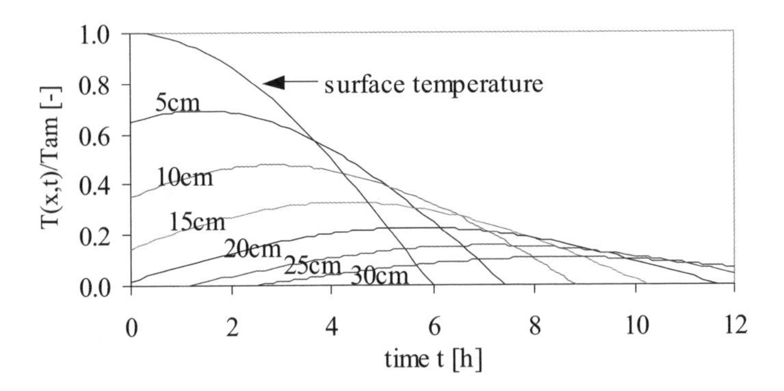

Figure 7.17: Variation in temperature $T(x, t)$ normalised to the amplitude T_{sm} around the average value as a function of time and component depth x in 5 cm steps for a concrete building component.

Example 7.4

On a solid wall with transparent thermal insulation, the absorber temperatures reaches 50°C during the day and falls to 10°C at night. To which value has the amplitude dropped after 24 cm (38 cm) wall thickness and after how many hours has the amplitude reached $x = 0.24$ m (0.38 m)? The thermal diffusivity a of the wall is 0.66×10^{-6} m²/s.

The phase shift between the maximum surface temperature and maximum temperature at 24 cm (38 cm) wall thickness is

$$t_x = \frac{0.24m}{2}\sqrt{\frac{24h \times 3600\frac{s}{h}}{0.66 \times 10^{-6}\frac{m^2}{s} \times \pi}} = 6.8h$$

and 10.8 h with 0.38 m wall thickness. The amplitude ratio is

$$\frac{T(x,t)}{T_{sm}} = \exp\left(-0.24m\sqrt{\frac{\pi}{0.66 \times 10^{-6}\frac{m^2}{s} \times 24h \times 3600\frac{s}{h}}}\right) = 0.17$$

for the 24-cm wall and 0.06 for the 38-cm wall. The temperature amplitude, which amounts to 20 K around the average value of 30°C, is dampened to 3.4°C and 1.2°C respectively.

The heat flow into a component with a periodic temperature boundary condition is again calculated, using Fourier's law, from the temperature gradient at the surface:

$$\frac{dQ}{dt} = -\lambda A \frac{dT}{dx}\bigg|_{x=0} = -\lambda A \frac{d}{dx}\left(T_{sm} \exp\left(-x\sqrt{\frac{\pi}{at_0}}\right)\cos\left(\frac{2\pi}{t_0}t - x\sqrt{\frac{\pi}{at_0}}\right)\right)\bigg|_{x=0}$$

$$= -\lambda A T_{sm}\left[\begin{array}{l} \underbrace{\exp\left(-x\sqrt{\frac{\pi}{at_0}}\right)}_{1}\times -\sqrt{\frac{\pi}{at_0}}\times\cos\left(\frac{2\pi}{t_0}t - \underbrace{x\sqrt{\frac{\pi}{at_0}}}_{0}\right) \\ +\underbrace{\exp\left(-x\sqrt{\frac{\pi}{at_0}}\right)}_{1}\times -\sin\left(\frac{2\pi}{t_0}t - \underbrace{x\sqrt{\frac{\pi}{at_0}}}_{0}\right)\times -\sqrt{\frac{\pi}{at_0}} \end{array}\right]\Bigg|_{x=0} \tag{7.43}$$

$$= \lambda A T_{sm}\sqrt{\frac{\pi}{at_0}}\left(\cos\left(\frac{2\pi}{t_0}t\right) - \sin\left(\frac{2\pi}{t_0}t\right)\right)$$

The heat flow $\dot{Q}/A$ is proportional to the surface temperature amplitude T_{sm}.

Example 7.5

A building with concrete ceilings is to be passively cooled by night ventilation. The surface temperatures can be approximated with a cosine function with an average value of 22°C, a maximum deviation of ±5°C and a period duration $t_0 = 24$ h.

The heat flow removed, based on Equation (7.43), is a maximum of –67 W/m² with a phase shift to the temperature of $t_0/8$, i.e. 3 hours.

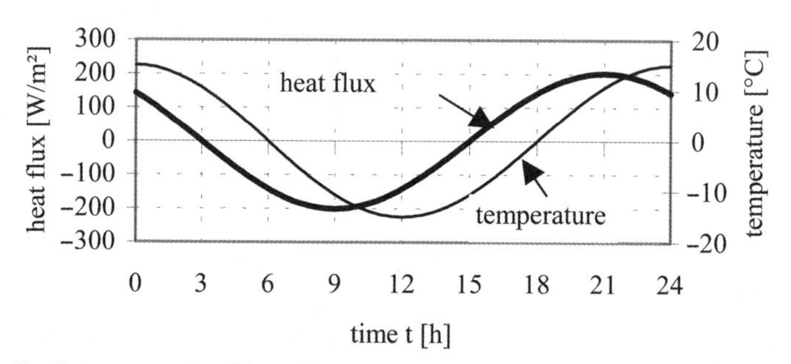

Figure 7.18: Heat flux into a concrete ceiling with a periodic surface temperature at amplitude $T_{sm} = 5$°C.

The integration of the heat flux results in the surface-related energy Q/A. If the stored or removed amount of heat is to be calculated, the integration limits t_1 and t_2 must be selected in such a way that only positive or negative heat flows are integrated. From Example 7.5 it can be seen that between the maximum heat flow and maximum surface temperature a phase shift of π/4 exists, i.e. one-eighth of a period.

$$\frac{Q}{A} = \lambda T_{sm}\sqrt{\frac{\pi}{at_0}}\int_{t_1}^{t_2}\left(\cos\left(\frac{2\pi}{t_0}t\right)-\sin\left(\frac{2\pi}{t_0}t\right)\right)dt$$
$$= \lambda T_{sm}\sqrt{\frac{\pi}{at_0}}\frac{t_0}{2\pi}\left(\sin\left(\frac{2\pi t_2}{t_0}\right)-\sin\left(\frac{2\pi t_1}{t_0}\right)+\cos\left(\frac{2\pi t_2}{t_0}\right)-\cos\left(\frac{2\pi t_1}{t_0}\right)\right) \tag{7.44}$$

The stored amount of heat of a half period results from the integral of the positive heat flows with a lower integration limit of $t_1 = 5/8\ t_0$ and an upper limit $t_2 = 5/8\ t_0 + t_0/2$, the released amount of heat from the integral being between $t_1 = t_0/8$ and $5/8\ t_0$.

Example 7.6

Calculation of the energy removed in a half period (12 h) by night cooling with a surface temperature amplitude of 5 K.

From Equation (7.44) the result is an amount of energy removed by night cooling of –0.51 kWh/m² at a lower integration limit of $t_1 = t_0/8$.

Normally it is not the surface temperature T_s which is known, but only the air temperature T_o with amplitude T_{om} and the heat transfer coefficient h between the air and the surface. The analytic solution corresponds to the solution with the surface temperature as a boundary condition from Equation (7.41), with the amplitude dampened by a factor η_0 and a phase shift ε_0 occurring due to thermal resistance between the air and the surface.

$$T(x,t) = T_{om}\eta_0 \exp\left(-x\sqrt{\frac{\pi}{at_0}}\right)\cos\left(\frac{2\pi}{t_0}t-\left(\varepsilon_0+x\sqrt{\frac{\pi}{at_0}}\right)\right) \tag{7.45}$$

with
$$\eta_0 = \sqrt{\left(1+2\sqrt{\frac{\pi}{(h/\lambda)^2 at_0}}+2\frac{\pi}{(h/\lambda)^2 at_0}\right)^{-1}}$$

and
$$\varepsilon_0 = \arctan\left(\left(1+\sqrt{\frac{(h/\lambda)^2 at_0}{\pi}}\right)^{-1}\right)$$

At $x = 0$ the surface temperature T_s is obtained with amplitude dampening η_0 and phase shift ε_0.

$$T(x=0,t) = T_s = T_{om}\eta_0 \cos\left(\frac{2\pi}{t_0}t-\varepsilon_0\right) \tag{7.46}$$

Example 7.7

Calculation of the surface temperature amplitude and the phase shift ε_0 for the concrete ceiling in Example 7.5, with a heat transfer coefficient h = 8 W/m^{2}K. The air temperature is to have an amplitude of ±5 K around an average value of 22°C.

With

$$\frac{\pi}{(h/\lambda)^2 at_0} = \frac{\pi}{\left(8\frac{W}{m^2K}/1.28\frac{W}{mK}\right)^2 0.66\times10^{-6}\frac{m^2}{s}\times 24h\times 3600\frac{s}{h}} = 1.41$$

the result is

$$\eta_0 = \sqrt{\frac{1}{1+2\sqrt{1.41}+2\times1.41}} = 0.4$$

The maximum temperature amplitude at the surface is now only $T_s = 5K\times0.4 = 2K$, and thus the removable energy falls by 60%! The phase shift is

$$\varepsilon_0 = \arctan\left(\left(1+\sqrt{\frac{1}{1.41}}\right)^{-1}\right) = 28.5°$$, i.e. in a 24 h period scarcely 2 h.

7.3.3 *Influence of solar irradiance*

If, in addition to air temperature fluctuations, the irradiance on a component surface is also to be considered, the use of a simple energy balance model is recommended, with which the short-wave solar irradiance is converted into a so-called sol-air temperature. With the help of the sol-air temperature, the analytic solutions of the thermal conduction equation already discussed can then be used.

The heat flow supplied to a component surface consists of the absorbed irradiance αG plus the heat flow transferred by the air (temperature T_o) to the surface (T_s) with the heat transfer coefficient h. This supplied heat flow is combined in a simple model, ignoring temperature-dependent modifications of the heat transfer coefficient h, into a purely temperature-dependent heat flow described by the sol-air temperature T_{So}.

$$\alpha G + h(T_o - T_s) = h(T_{So} - T_s)$$
$$T_{So} = T_o + \frac{\alpha G}{h} \tag{7.47}$$

With the sol-air temperature T_{So}, the heat storage in components can then be calculated with the solutions of the thermal conduction equation already considered.

Example 7.8

Calculation of the sol-air temperature and the energy stored in a concrete floor, with periodically varying irradiance at an amplitude of 500 W/m^2 and period t_0 = 24 h transmitted through the windows, an absorption coefficient of the floor of $\alpha = 0.6$, a room air and ceiling average temperature T_o = 20°C and a heat transfer coefficient h = 8 W/m^{2}K. The maximum sol-air temperature is:

$$T_{So,\max} = 20°C + \frac{0.6 \times 500 \frac{W}{m^2}}{8 \frac{W}{m^2 K}} = 57.5°C$$

Via the fictitious sol-air temperature, first the surface temperature at the concrete surface is calculated, and from this, using Equation (7.44), the stored energy per square metre of surface.

Using Example 7.7, $\eta_0 = 0.4$ and thus the maximum air temperature fluctuation of 57.5°C – 20°C = 37.5°C is reduced to a maximum surface temperature fluctuation of $T_{a,max} = \eta_0 (T_{So,max} - T_o)$ = 15°C. The energy stored during a half period is 1.54 kWh/m^2.

Altogether, however, an irradiance of $\int_0^{t_0/2} G dt = G_{\max} \int_0^{t_0/2} \sin\left(2\pi \frac{t}{t_0}\right) dt = \frac{G_{\max} t_0}{\pi} = 3.8 \frac{kWh}{m^2}$ is transmitted through the glazing during the half period, and of it 60%, i.e. 2.3 kWh/m^2, is absorbed by the concrete floor. The difference between the absorbed irradiance and stored heat has been transferred over the heat transfer coefficient h directly to the room air.

8 Lighting technology and daylight use

8.1 Introduction to lighting and daylighting technology

Lighting technology deals primarily with the supply of sufficient, glare-free lighting for workplaces and dwellings. Light, however, also assumes the important function of orientation in the interior and time, and enables reference to the exterior. These qualities are supplied predominantly by daylight and contribute crucially to visual comfort.

The human eye is optimally adapted to the visible spectral range of solar radiation, so that with short-wave solar irradiance there is a higher luminous efficiency per Watt of power than with most types of artificial light. Efficient daylight use thus contributes directly to reducing energy consumption, in particular in administrative buildings (Dudda, 2000).

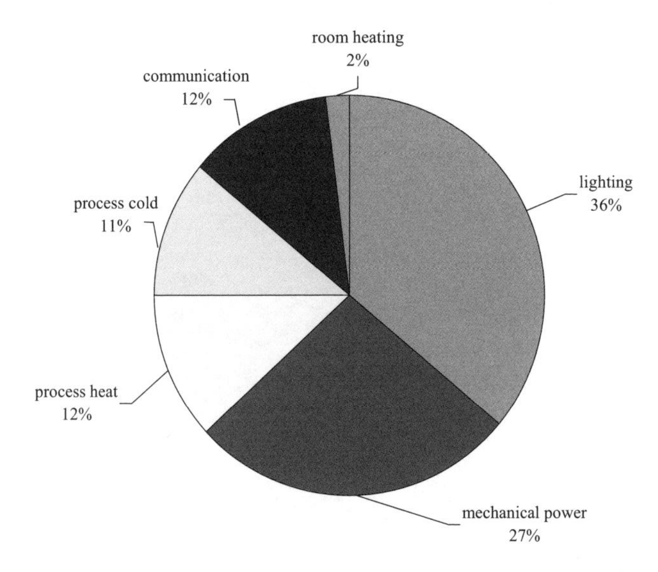

Figure 8.1: Average electricity consumption and lighting contribution for small consumers (trades, service sector and public sector) in Germany, 1998.

The average proportion of lighting in electricity consumption is 36% in an administrative building, compared to only about 5% in the industrial sector. Due to the high luminous efficiency of daylight, the internal thermal loads due to lighting are reduced, and thus also the problem of summer overheating in offices.

Daylight is predominantly used via conventional window openings. Special components are available for glare and sun protection, as well as for light distribution to the depth of the room. In particular in computer workplaces, glare shielding assumes a crucial role in daylight technology. Glare shielding systems should be based on the criteria of brightness reduction, light permeability, visual contact outward and light guidance far into the room. Textile anti-glare blinds reduce the brightness by their partially translucent layers, but cannot redirect the light. Also, with absorption and reflection glass, the decrease in glare luminance is proportional to the light permeability. With glare-reducing blinds, the user has more control. Transparency and visual contact outward can be improved by perforation.

Daylight-guiding elements within the upper window section (mirror-type lamellas, prism systems, etc.) also contribute to the reduction of glare in the window area and to improved illumination-in-depth of side-illuminated rooms. With an overcast sky, adequate illumination cannot be ensured even just 3 m deep into a 3 m high room with a window up to the ceiling. A light-guiding element of 20 cm shifts the area of sufficient illumination to a depth of 4.5 m.

The light guidance of direct radiation to the ceiling as far into the room as possible can be achieved either by fixed mirror-lamellas, prism systems, translucent insulation materials or the like. Cheapest and most widely used are mirror-lamellas divided into two parts, which guide light within the upper section of the window, while the lower area remains closed to reduce glare. Apart from open lamella systems, specially reflecting lamella profiles are also available for the space between panes, consisting of two mirror profiles opposite one another. The light reflected upwards by the lower mirror profile is, with a steep angle of incidence, reflected out from the pane by the mirror in front (sun protection); with a flat angle of incidence it is channelled inwards by the second mirror at the back. While the profiles mainly have a sun-protection function with a high sun position, a light-guiding effect can be obtained when the sun is low. Lamellas between non-ventilated glass panes, however, often cause high glass surface temperatures and thus add to the cooling load of the building.

Prism-profile panels use the total reflection of the light either to block it (sun-protection function) or to channel it. They are usually inserted between the panes. Prism systems are translucent and therefore only used in the skylight area or the overhead area. Sun-protection prisms reduce the sky brightness and thus the glare problem by a factor of 100 even with an overcast sky.

8.1.1 Daylighting of interior spaces

The illuminance level during daylight use in the interior is typically between 2% and 5% of the exterior illuminance, corresponding to a mean illuminance of 200–500 lux (lx) in the interior. The relation of the interior illuminance E_i to the horizontal exterior illuminance E_o is termed the daylight coefficient D.

$$D = \frac{E_i}{E_o} \tag{8.1}$$

A room lit by daylight, for which no perceptible difference in brightness between outside and inside is to exist, must have a daylight coefficient of at least 10%, i.e. illuminances of 1000–3000 lx or more. The visual sensitivity of the eye is almost constant in relation to further brightness increases. The illuminance E is defined by the ratio of the light flux Φ [lumen] and the illuminated surface area A [m²].

$$E = \frac{\Phi}{A} \quad \left[\frac{lm}{m^2} = lx \right] \tag{8.2}$$

In residential buildings, the minimum daylight requirement is essentially characterised by the avoidance of the impression of a dark room, corresponding to daylight coefficients of around 0.9% halfway into the room. For offices in which all workstations are near windows, an illuminance of 300 lx is sufficient; in offices with computer workstations 500 lx is necessary, and for open-plan offices or drawing workstations, 750 lx is required.

Example 8.1

Calculation of the mean illuminance for a 3 m high office of surface area 4 m × 5 m, with a window area of 9 m^2 with 40% total transmission (including the framework proportion). The outside available illuminance on the vertical window area on an overcast day is to be 4500 lux (lx).

Light flux onto the entire window area: $4500 \text{ lx} \times 9 \text{ m}^2 = 40500 \text{ lm}$

Transmitted light flux into the office: $40500 \text{ lm} \times 0.4 = 16200 \text{ lm}$

Light flux related to a square metre of office surface: $\frac{16200 \text{ lm}}{20 \text{ m}^2} = 810 \text{ lx}$

It is not the quantity of light which is the problem, but the unfavourable distribution of the daylight. Purely side-illuminated rooms are characterised by an almost exponential fall in the daylight coefficient, so near to windows glare problems occur, and with increasing room depth illuminances are too low.

For the use of daylight, glazings in the upper window section which enable good deep illumination of the room are particularly favourable. A centrally arranged 1 m high window in a 3 m high room leads to an exponential fall in the daylight coefficient with increasing room depth. An arrangement of the same window height just below the ceiling, on the other hand, increases the daylight coefficient in the room depth. The highest daylight coefficients are obtained with a window front over the entire room height.

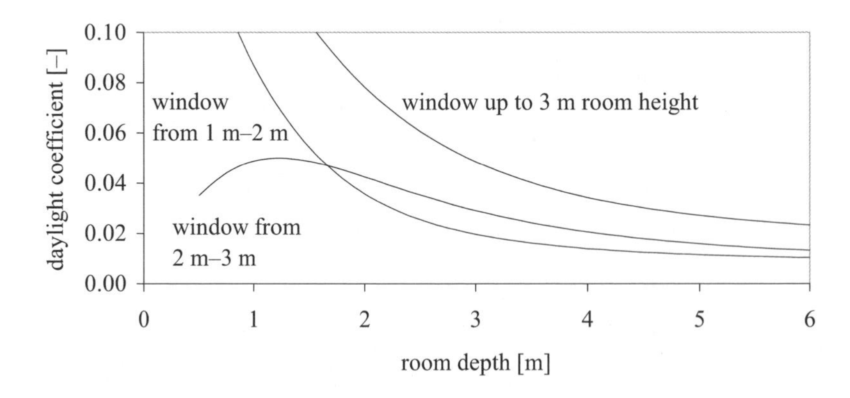

Figure 8.2: Daylight coefficient in the 3 m high side-illuminated room for a window from the floor up to the ceiling or windows in each case 1 m height centrally or within the area below the ceiling.

8.1.2 Luminance contrast and glare

Apart from the supply of sufficient illuminance, the main issues in administrative buildings are problems with workstations which must be illuminated, glare-free and with low contrast. Disturbance effects arise with very uneven distribution of the luminance, which in the worst cases impair the visual function (physiological or disability glare). The contrast between the object and the background luminance is reduced due to scattered luminance entering the eye sideways and the eye functions, such as distinction between nearby objects or contrast sensitivity, are reduced. The luminance is defined by the light flux, which is radiated from a surface into a spatial angle. At luminance levels above 10^4 cd/m^2 absolute glare occurs, where objects can no longer be distinguished within the visual field.

With psychological or discomfort glare a sensation of disturbance is caused without a real reduction in the eye's capability. The contrast relation between object luminance and surrounding luminance, which is experienced as pleasant, depends on the absolute height of the surrounding luminance and thus on eye adaptation: the higher the surrounding luminance, the lower the subjectively perceived brightness of an object luminance. This relative glare due to too large luminance contrasts in the visual field, which occurs most frequently in lighting technology, can therefore be reduced by raising the average luminance.

Average luminances of a monitor are approximately 100 cd/m^2, while the figure for a window is about 4500 cd/m^2, even with an overcast sky. The luminance contrast of 1:45 is so far over the acceptable contrast of 1:15 that glare reduction measures have to be taken. Ideal luminance contrasts between the visual task and darker immediate surroundings are only 3:1, with more distant surroundings 10:1 and should not exceed values over 20–40:1.

8.2 Solar irradiance and light flux

Optical radiation is part of the electromagnetic radiation with wavelengths above 1 nm (upper boundary of Röntgen radiation) up to 1 mm (lower boundary of radio waves). The radiated power P transported by the electromagnetic waves can be calculated using the Poynting vector $\vec{S}$ as a cross-product of electrical and magnetic field vectors $\vec{E}$ and $\vec{H}$, integrated over a closed surface A surrounding the radiating source.

$$\vec{S} = \vec{E} \times \vec{H}$$
$$P = \oint_A \vec{S}\, d\vec{A} \tag{8.3}$$

The solar irradiance covers a wide spectral range from about 0.3–4 μm due to the high temperature of the sun's surface. The measured radiated power per square metre is the irradiance G [W/m²] and is a so-called radiometric quantity. As the human eye is only sensitive to a small spectral range of 0.38–0.78 μm, the radiometric units have to be converted to eye sensitivity weighted photometric quantities. The light flux Φ [lumen] as a photometric quantity is the sensitivity weighted power. If the light flux is determined per square metre of surface, the illuminance E_v [lumen/m²] is obtained.

The lighting requirements at workstations are between 300 and 1000 lumen per square metre depending on the visual aspect of the work (EN 12464: lighting of workspaces in interior spaces). The illuminance describes the light flux Φ onto the work surface which results from the irradiated power, be it solar radiation or artificial light, weighted with the spectral sensitivity of the eye.

8.2.1 Physiological–optical basics

The human eye is an almost spherical object of 24 mm diameter and about 26 mm length. The pigmented iris functions as a lens with varying opening diameters between 2–8 mm, depending on the object's distance and average luminance. The maximum sensitivity of the retina cells during the day is 555×10^{-9} m (555 nm), i.e. in the green colour area, falling to zero in the short-wave area below 380 nm and in the long-wave area above 780 nm. This spectral area is called visible light. In some international norms the spectral boundaries for the visible part are rounded to 400 nm to 800 nm.

About 6.5 million cone cells, responsible for colour vision, are concentrated in the centre of the retina and contain three different dye stuffs with maximum spectral sensitivities at 419 nm, 531 nm and 558 nm. During illumination the dyes chemically change (isomerisation) and need around six minutes for regeneration. The rod cells, which are responsible for night vision, have a maximum spectral sensitivity at 496 nm, are extremely light-sensitive due to very large numbers (120 million) and coupling of different cells (up to 32 cells on one nerve end). The pigment regeneration time is around 30 minutes.

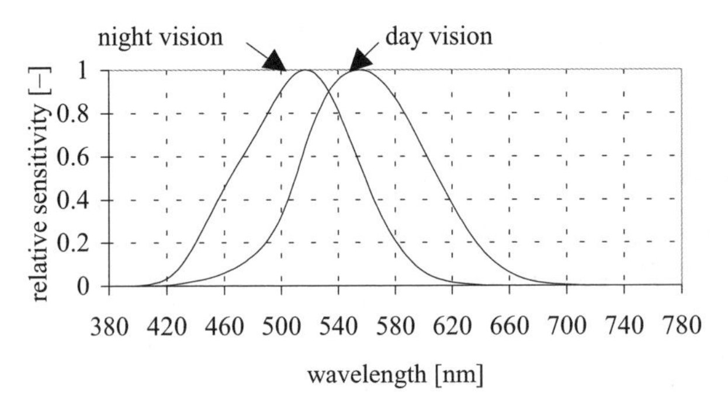

Figure 8.3: Relative spectral sensitivity $V(\lambda)$ of the eye for day vision with retina cells termed cones and for night vision with rods.

8.2.2 Photometric radiation equivalent

To convert an energetic radiant flux Φ_e (index *e* for energetic) in units of Watt into a light flux Φ_v (index *v* for visual) in lumen, the relative eye sensitivity $V(\lambda)$ at a given wavelength λ must be known, as well as an absolute conversion factor between the units, called the photometric radiation equivalent k_{max}. For each wavelength λ, the light flux $\Phi_{v,\lambda}$ is converted from the spectral radiating power $\Phi_{e,\lambda}$:

$$\Phi_{v\lambda} = k_{\max} V(\lambda) \Phi_{e\lambda}$$
$$k_{\max} = \frac{maximum\ light\ flux[lm]}{radiative\ power[W]} \tag{8.4}$$

The conversion factor k_{max} results from the definition of the photometric SI-unit candela [cd], which describes the light flux Φ per spatial angle Ω (in steradiant [sr]), i.e. the luminous intensity I:

$$I = \frac{\Phi}{\Omega} \qquad [cd] = \left[\frac{lm}{sr}\right] \tag{8.5}$$

For the definition of the luminous intensity I in candela, a black cavity emitter (platinum with a melting temperature of 2044.9 K) was historically used, whose spectral radiant emittance can be calculated using Planck's law of radiation. A candela was defined as the light flux per spatial angle, which 1/60 cm^2 of the surface of the black cavity emitter emits. The emitted power of this surface, calculated using Planck's law of radiation, is 1/673 W.

Today a candela is defined as the light flux of a monochromatic radiation source with the somewhat smaller power of 1/683 W, which radiates with a frequency of 540×10^{12} Hz, i.e. 555 nm, into a spatial angle of one steradiant. An emitted power of 1 W per steradiant thus results in a luminous intensity of 683 cd, corresponding to 683 lumen per steradiant.

Since the wavelength 555 nm corresponds to the maximum relative eye sensitivity, the maximum photometric radiation equivalent is today k_{max} = 683 lm/W.

Example 8.2

Calculation of the light flux $\Phi_{v,\lambda}$ for a light source of 10 W power with the wavelength λ = 633 nm (red helium-neon laser) and with λ = 588 nm (yellow resonance line of the sodium vapour low pressure lamp). The relative spectral sensitivity is V (633nm) = 0.25 or V (588nm) = 0.77. Based on Equation (8.4) the light flux is

$$\Phi_{v,633nm} = 683\frac{lm}{W} \times 0.25 \times 10W = 1707.5lm \quad \text{or} \quad \Phi_{v,588nm} = 683\frac{lm}{W} \times 0.77 \times 10W = 5258lm.$$

To determine the light flux and finally the illuminance on a surface of any radiation source, the entire spectrum must be weighted with the spectral sensitivity of the eye.

A single value of the photometric radiation equivalent for a given spectrum (of the sun or a lamp) is obtained by converting the energetic radiant flux for each wavelength into a light flux, integrating it over the visible area of the spectrum and standardising the value on the integrated total radiation flow of the source of light. A source of light is thus characterised by the following radiation equivalent:

$$k = \frac{\Phi_v}{\Phi_e} = \frac{k_{\max} \int_{380nm}^{780nm} \Phi_{e\lambda} V(\lambda) d\lambda}{\int_0^\infty \Phi_{e\lambda} d\lambda} \tag{8.6}$$

For an evenly overcast sky a radiation equivalent of k = 115 lm/W results, based on DIN 5034. The radiation equivalent varies, however, with cloud thickness, the vapour content, the height of the sun etc., and can be between 90 and 120 lm/W. The diffuse radiation of the clear sky can assume values over 140 lm/W; for the direct component values, between 50 and 120 lm/W have been measured. If one relates the light flux to a square metre of recipient surface, the illuminace E is obtained with the unit lumen per square metre (abbreviated to lux [lx]).

$$E = \frac{\Phi_v}{A} \quad \left[\frac{lm}{m^2} = lx\right] \tag{8.7}$$

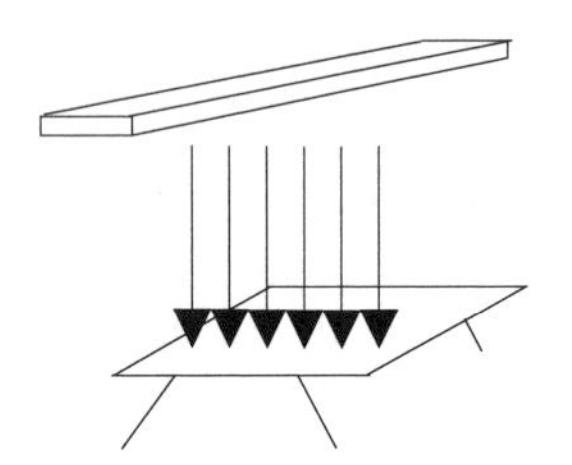

Figure 8.4: Light flux and illuminance on a horizontal surface.

Table 8.1: Photometric radiation equivalent

Light source	*Photometric radiation equivalent k* [lm/W]
evenly overcast sky	115
diffuse radiation of the clear sky	140
direct component of the clear sky	50–120

During full exposure to the sun (1000 W/m^2), with a photometric radiation equivalent of 120 lm/W, an exterior lighting strength of 120 000 lx is achieved on a horizontal surface.

$$\Phi_v = k\Phi_e = 120\frac{\text{lm}}{\text{W}} \times 1000\frac{W}{m^2} = 120\,000\frac{lm}{m^2} = 120\,000 lx$$

The annual average value of the exterior lighting strength is around 10 000 lx during the day.

8.2.3 *Artificial light sources*

50% of all artificial light sources are fluorescent lamps, consisting of a glas tube with two wolfram electrodes with an emitting surface at the ends. The gas filling is a mixture of rare gases such as argon or krypton (at a pressure of 70 Pa) and Hg (at 1 Pa pressure). The Hg atoms emit due to the electric discharge in the ultraviolet region (mainly at 185 and 254 nm) and excite fluorescent material on the tube walls (halogen phosphate or others) to give a continuous visible spectrum. Fluorescent tubes are not well suited for external applications as the light flux decreases to 25% at –10°C of its normal value at 20°C. Compact fluorescent lights are folded fluorescent lights with lifetimes over 10 000 h. Fluorescent lights have photometric radiation equivalents of 50–88 lm/W, clearly lower than daylight, and their power is obviously supplied by electricity. Electrical lightbulbs with hot wolfram wires provide a very low light flux of 6–16 lumen per Watt power. Halogen lights are slightly better, as evaporated wolfram from the hot wire no longer deposes on the light bulb walls. Furthermore, the pressure and operating temperatures are higher within the glass tube. At low power levels, light-emitting diodes start to provide good luminous efficiencies at a wide spectral range.

Table 8.2: Photometric radiation equivalent of artificial light sources

Light source	*Description*	*Power P* [W]	*Photometric radiation equivalent k* [lm/W]
(electric) light bulb	Wolfram glowing wire 2800K	15–200	6–16
halogen light	Wolfram wire at 3000K with higher gas pressure and halogen addition, quartz glass	15–200	8–20
fluorescent lamp	argon/krypton + Hg filled	18–58	50–88
compact fluorescent lamps	as above	5–55	50–88
sodium vapour lamp	widening of double emission line at 589/90 nm at high pressure	180	150
Light emitting diodes		40–80 mW	20–100
LED White	AlInGaN/Phosphor	20 mA, 4V	20
LED Green	506 , 530 , 571 nm, AlInGasN		30, 54, 14
LED Red	611 , 658 nm , AlInGaP	20 mA, 2V	102, 38

8.3 Luminance and illuminance

The light flux onto a surface and thus the illuminance E_v, defined as light flux per square metre, can in the simplest case be calculated, at a given solar irradiance, directly from the radiating power and the photometric radiation equivalent. For the solar irradiance one usually proceeds from an isotropic diffuse energy distribution. Since luminance distributions rising to the zenith are mostly used in lighting engineering, the light flux onto a randomly inclined surface must be calculated from the contributions of the luminance of different spatial angle areas of the sky's diffuse radiation. In the interior too, the light flux onto a recipient surface A_r consists of the total of the light fluxes of radiation-emitting or radiation-reflecting sender surfaces A_s at different spatial angles.

The task now is to integrate the light fluxes received by a surface over all spatial angle areas and thus to calculate the illuminance E for different luminance distributions. For this, a spatial angle $d\Omega$ is constructed by a two-dimensional surface element dA, which is at distance r from the sending or receiving surface.

$$d\Omega = \frac{dA}{r^2} \tag{8.8}$$

The unit of the three-dimensional spatial angle is steradiant [sr] and corresponds to the definition of a two-dimensional angle in arc measure [rad], which is constructed by a circular arc with radius r. A simple spatial angle is, for example, given by a surface A_{sp} on a sphere, which is calculated as a function of the sphere radius r and of the height h of the cut-out sphere segment.

$$\Omega = \frac{A_{sp}}{r^2} = \frac{2\pi r h}{r^2} = \frac{2\pi r^2 (1-\cos\theta)}{r^2} = 2\pi(1-\cos\theta) \tag{8.9}$$

Here the angle θ describes half the opening angle of the spherical cone. The full spatial angle of a hemisphere with half an opening angle $\theta = 90°$ is therefore 2π, and of a sphere 4π.

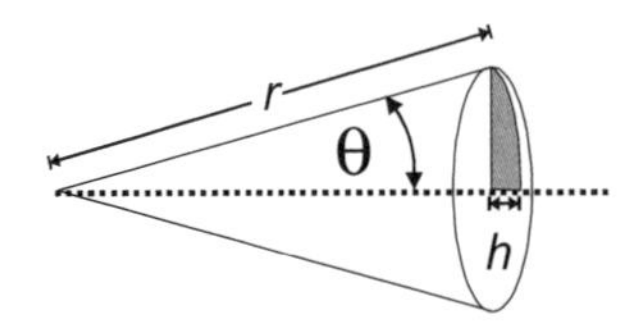

Figure 8.5: Spherical cone spatial angle.

If the light flux of a luminescent or radiation-reflecting surface is to be calculated from a certain spatial angle, first the orientation of the surface relative to the observed radiant emittance direction must be determined. If the radiant emittance is considered not as perpendicular to the sender surface element dA_s, but at an angle θ_s, effectively only the smaller surface $dA_s \cos\theta_s$ is visible.

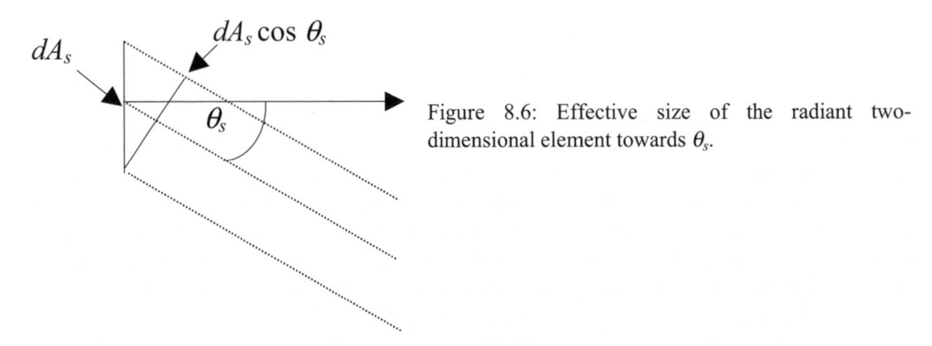

Figure 8.6: Effective size of the radiant two-dimensional element towards θ_s.

The light flux $d\Phi$, which a two-dimensional area dA_s sends at angle θ_s into a spatial angle $d\Omega_1$, is termed luminance L and describes the brightness of the surface.

$$L = \frac{d^2\Phi}{d\Omega_1 dA_s \cos\theta_s} = \frac{dI}{dA_s \cos\theta_s} \quad \left[\frac{lm}{sr\, m^2} = \frac{cd}{m^2}\right] \tag{8.10}$$

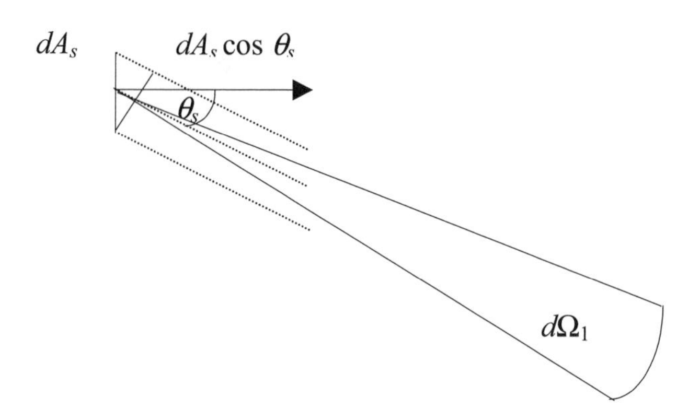

Figure 8.7: Luminance of a two-dimensional area dA_s radiating at an angle θ_s into a spatial angle $d\Omega_1$.

The light flux $d\Phi$ [lm] into a spatial angle $d\Omega$ [sr] can be replaced by the luminous intensity I. In many cases the luminance is independent of the angle θ_s, i.e. the surface appears equally bright independent of the viewing angle. Such surfaces are termed Lambert radiators. From Equation (8.10) it follows that the luminous intensity dI in Lambert radiators must decrease with the cosine of the angle θ_s: $I(\theta) = I(0)\cos\theta_s$. Rough, diffusely reflecting surfaces such as gypsum walls, paper etc. behave in good approximation like Lambert radiators.

In order to obtain the light flux of a sender surface A_s into a spatial angle Ω_1, integration must take place at a given brightness of this surface over the differential spatial angles $d\Omega_1$, with consideration given to the effective surface size through the cosine of the angle θ_s between the surface and the respective spatial angle element.

$$\Phi = \int_{\Omega_1} \int_{A_s} L dA_s \cos\theta_s d\Omega_1 \tag{8.11}$$

The spatial angle $d\Omega_1$ is constructed by the receiving two-dimensional surface dA_r, for example the work surface, or even the pupil opening surface of the eye. Of course the receiving surface dA_r also need not be perpendicular to the spatial angle regarded in each case, so effectively only $dA_r \cos\theta_r$ is available for radiation reception.

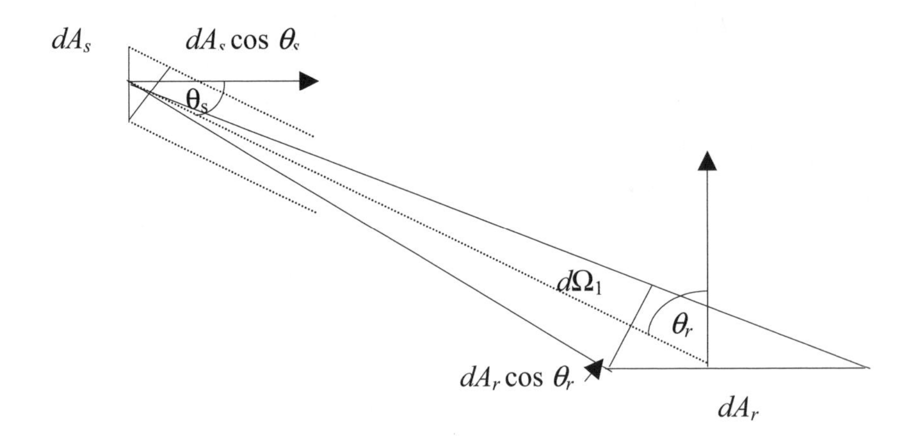

Figure 8.8: Light flux of a radiating two-dimensional surface dA_s onto a recipient surface dA_r.

The light flux falling from the radiating sender surface A_s onto the receiving surface A_r thus results in:

$$\Phi_{s\to r} = \int_{\Omega_1}\int_{A_s} L_s dA_s \cos\theta_s d\Omega_1 = \int_{A_r}\int_{A_s} L_s dA_s \cos\theta_s \frac{dA_r \cos\theta_r}{r^2} \tag{8.12}$$

If, conversely, the view is from the receiving surface A_r from the light flux over a spatial angle Ω_2 which covers the radiating surface A_s, for reasons of energy conservation the received light flux must correspond to the emitted light flux and the so-called Basic Law of Photometry is obtained:

$$\Phi_{r\leftarrow s} = \int_{\Omega_2}\int_{A_r} L_s dA_r \cos\theta_r d\Omega_2 = \int_{A_s}\int_{A_r} L_s dA_r \cos\theta_r \frac{dA_s \cos\theta_s}{r^2} = \Phi_{s\to r} \tag{8.13}$$

The cosine of the angle θ_r refers to the normal of the receiving surface dA_r and the spatial angle $d\Omega_2$, which is constructed by a differential two-dimensional surface area dA_s.

The illuminance E_r on the receiving surface dA_r for any sender surfaces dA_s with luminance L_s results from Equation (8.13).

$$E_r = \frac{d\Phi_r}{dA_r} = \int_{\Omega_2} L_s \cos\theta_r d\Omega_2 = \int_{A_s} L_s \cos\theta_r \frac{dA_s \cos\theta_s}{r^2} \tag{8.14}$$

Example 8.3

Calculation of the illuminance on a horizontal work surface of 1 m^2 illuminated by a 1 m^2 vertical window with a luminance of 4500 cd/m^2 at an average angle of 45° and 3 m distance (without consideration of the photometric boundary distance).

If the differentials are set as differences, Equation (8.14) is simplified to:

$$E_r \approx \frac{\Delta\Phi_r}{\Delta A_r} = L_s \cos\theta_r \frac{\Delta A_s \cos\theta_s}{r^2}$$

Both the angle of the radiating and of the receiving surface normal are 45° to the spatial angle.

$$E_r = \frac{\Delta\Phi_r}{\Delta A_r} = 4500\frac{\text{cd}}{\text{m}^2}\cos 45°\frac{1\text{m}^2\cos 45}{(3\text{m})^2} = 250lx$$

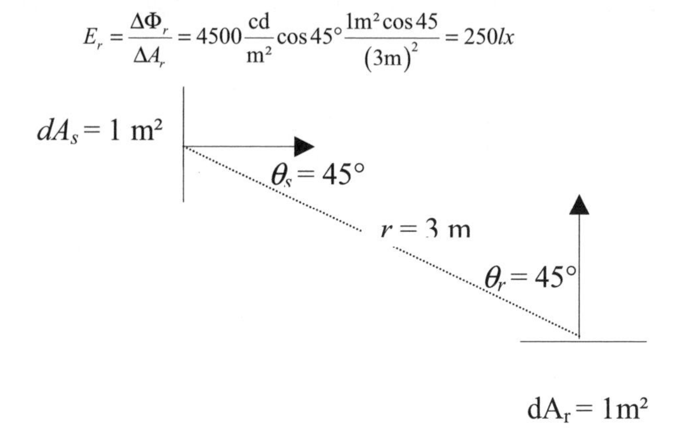

8.3.1 Luminance and adaptation of the eye

The luminance of sources of light or radiation-reflecting surfaces vary over a very wide range of values.

Table 8.3: Luminances of sources of light.

Light source	*Luminance* [cd/m²]
Sun, values depending on sun height	600 000 to 1 600 000 000
Clear sky	2000–12 000
Overcast sky	1000–6000
Lightbulb	20 000–50 000
Compact fluorescent lamp	9000–25 000
Candle flame	7000
Paper in a well lit office	250
Computer screen	20–200
Lower limit of light sensitivity	10^{-5}

The human eye adapts to the mean outside brightness L_o in the visual field. This is termed adaptation, and takes place via modification of the pupil surface A_p, which expands with falling brightness from 2 mm diametre to a maximum of about 8 mm during brightness adaptation (Henschel, 2002). The pupil surface, which varies as a logarithmic function of the luminance, enables adjustment of the penetrating light flux by a factor of 16.

The illuminance reaching the eye is let through with transmittance τ_p and strikes the retina. Brightness-sensitivity is determined by the number of photons which strike a two-dimensional element of the retina, i.e. by the luminance on the retina. This is obtained from the mean luminance of the visual field, which is seen by the spatial angle $d\Omega_1$ of the pupil surface A_p at a distance r between the lens and retina $d\Omega_1 = dA_p / r^2$. Between the spatial angle and the surface normal of the retina recipient surface dA_r, the angle θ_r is zero, so the cosine term for the recipient surface is omitted.

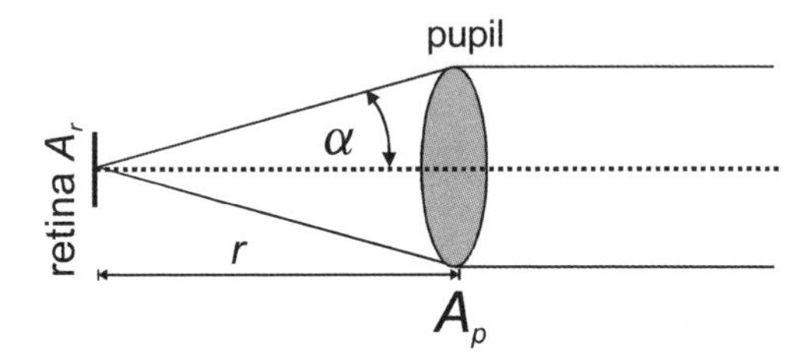

Figure 8.9: Cross-section of the eye with distance r between the lens and retina.

$$E_{retina} = \frac{d\Phi}{dA_r} = \int_{\Omega_1} L_o \tau_p \, d\Omega_1 = L_o \tau_p \frac{A_p}{r^2} = \text{const}\, L_o A_p \tag{8.15}$$

Besides the opening area of the pupil (A_p), which depends on the adaptation status of the eye, the brightness sensitivity is therefore determined solely by the luminance striking the eye. The luminance is thus the most important quantity in lighting engineering.

8.3.2 Distribution of the luminous intensity of artificial light sources

The luminance of lamps or lights is rarely constant in all directions in space. Luminous intensity distributions are indicated by the manufacturers in polar diagrams in different sections. Usually the absolute luminous intensity in candelas refers to a fixed lamp light-flux.

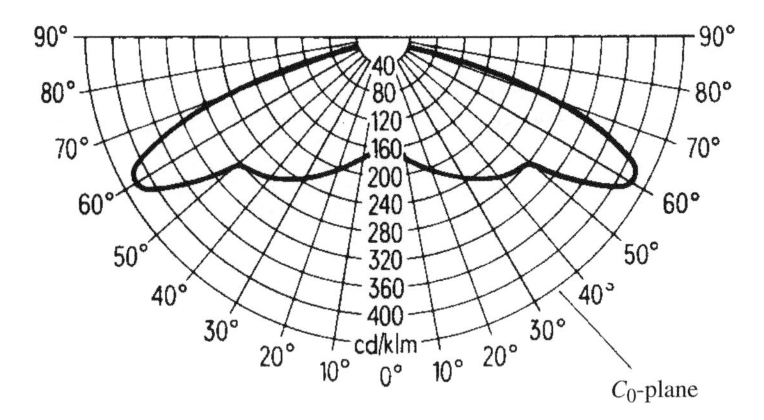

Figure 8.10: Light distribution curve in the C_0-plane, which is perpendicular to the radiating surface (in candelas per Kilolumen).

If in Equation (8.14) the luminance of the sender surface is replaced by the luminous intensity per unit area with

$$L_s = \frac{d^2\Phi_s}{d\Omega_1 dA_s \cos\theta_s} = \frac{dI(\theta)}{dA_s \cos\theta_s} \tag{8.16}$$

the result for the illuminance is

$$E_r = \frac{d\Phi}{dA_r} = \int_{\Omega_2} L_s \cos\theta_e d\Omega_2 = \int_{A_s} \frac{dI(\theta_s)}{dA_s \cos\theta_s} \cos\theta_r \frac{dA_s \cos\theta_s}{r^2} = \int_{A_s} \frac{dI(\theta_s)}{r^2} \cos\theta_r \tag{8.17}$$

with θ_r as the angle between the receiving surface and solid angle $d\Omega_2$.

With the usual constancy of the luminous intensity over the radiating surface A_s, the photometric distance law is obtained:

$$E_r = \frac{d\Phi}{dA_r} = \frac{I(\theta_s)}{r^2} \cos\theta_r \tag{8.18}$$

However, the functional characteristic of the illuminance, which decreases in inverse proportion to the square of the radius, only applies starting from the so-called photometric minimum distance, which is about 10 times as large as the largest linear dimension of the lighting surface.

Example 8.4

Calculation of the illuminance on an 80 cm high work surface lit by a Lambert emitting lamp at a height of 2.5 m and a lateral distance of 0.5 m with a luminous intensity *I(0)* = 500 cd.

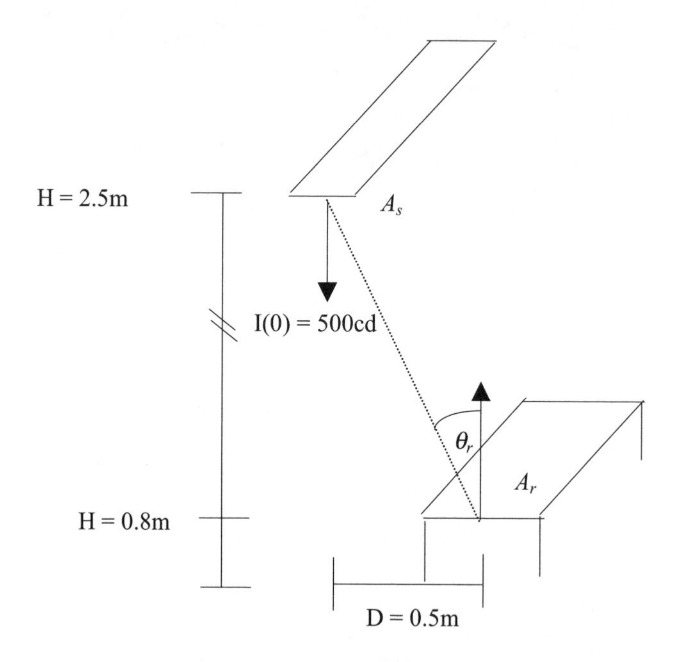

The angle θ_r between the recipient-surface normal and the light is obtained from the lamp height and lateral distance,

$$\theta_r = \arctan\left(\frac{0.5m}{1.7m}\right) = 16.4°$$

the distance is 1.77 m. The luminous intensity of the light toward the recipient surface is

$$I(0)\cos\theta_s = 500\ cd \times \cos 16.4 = 480\ cd$$

so the result is an illuminance of

$$E = \frac{I(\theta_s)\cos\theta_r}{r^2} = \frac{480\ cd \times \cos(16.4°)}{(1.77\ m)^2} = 147\,lx$$

8.3.3 *Units and definitions*

An overview of the definitions and units in lighting technology is given in Table 8.4.

Table 8.4: Units in lighting technology.

Photometric unit	*Symbol*	*Definition*	*Unit*
Luminous energy	Q_v	$Q_v = \int \Phi_v dt$	lm s
Luminous flux	Φ_v	–	lm
Luminous exitance	M_v	$M_v = \frac{d\Phi_v}{dA_1}$	lm m^{-2}
Luminous intensity	I_v	$I_v = \frac{d\Phi_v}{d\Omega}$	cd
Luminance	L_v	$L_v = \frac{d\Phi}{d\Omega dA_1 \cos\varepsilon_1}$	cd m^{-2}
Illuminance	E_v	$E_v = \int_{2\pi\, sr} L_v \cos\varepsilon_1 d\Omega$	lx
Luminous exposure	H_v	$H = \frac{dQ}{dA_2}$	lx s

Artificial and daylight sources can be characterised by the following efficiencies.

Table 8.5: Efficiency definitions in lighting technology.

Efficiency	*Symbol*	*definition*	*Unit*
Radiative efficiency	η_e	Φ_e/P	– (W/W)
Luminous efficiency	η_v	Φ_v/P	lm/W
Photometric radiation equivalent	K	$\frac{\Phi}{\Phi_e}$	lm/W
Optical efficiency	O	$\int \frac{d\Phi}{d\lambda} d\lambda \Big/ \Phi_e$	–
Visual efficiency	V	$\frac{\Phi}{k_m \Phi_e}$	–

8.4 Sky luminous intensity models

To calculate the daylight distribution in a room, models of the luminous intensity distribution of the sky are required. From the luminous intensity distribution, the effective surface and the solid angle, the light flux onto any recipient surfaces, for example windows, can then be calculated.

As the simplest model, the illuminance of the sky is set as constant, corresponding to an isotropic sky model. The density of light is calculated using Equation (8.14):

$$E_r = \frac{d\Phi_r}{dA_r} = \int_{\Omega_2} L_s \cos\theta_r d\Omega_2$$

The solid angle $d\Omega_2$ is selected using Equation (8.9) as a spherical cone with half opening angle θ, with $\theta = 0$ corresponding to the surface-normal of a horizontal recipient surface. With a horizontal recipient surface, the opening angle is equal to the angle of incidence θ_r and the following density of light results:

$$E_r = \int_{\Omega_2} L_s \cos\theta_r d\left(2\pi\left(1-\cos\theta_r\right)\right) = 2\pi L_s \int_0^{\pi/2} \cos\theta_r \sin\theta_r d\theta_r = 2\pi L_s \left.\frac{\sin^2\theta_r}{2}\right|_0^{\pi/2} = \pi L_s$$

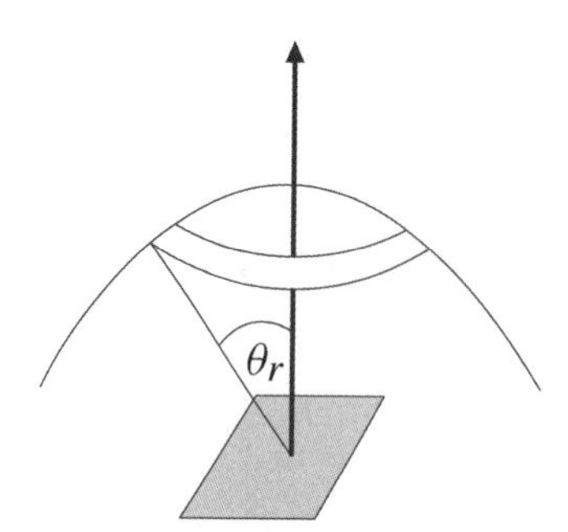

Figure 8.11: Projection of the sky dome with constant luminance onto a horizontal recipient surface.

The sky luminance can thus be calculated from the horizontal irradiance, if the radiating power is converted via the photometric radiation equivalent into a density of light.

Example 8.5

Calculation of the isotropic luminance of the sky for a horizontal irradiance of the overcast sky of 100 W/m^2 and a photometric radiation equivalent of 115 lm/W.

The illuminance on the horizontal surface is 11 500 lm/m^2. The illuminance then produces a sky luminance of

$$L_s = \frac{E_r}{\pi} = 3660 \ \frac{\text{cd}}{\text{m}^2}$$

Apart from the isotropic luminance distribution, standardised distributions of the international lighting engineering commission (CIE) for a clear and an overcast sky are used. The overcast, so-called Moon and Spencer sky, is characterised by a rise in the luminance L_α with the elevation angle α, with the zenith luminance L_z assuming three times the value of the horizon luminance,

$$L_\alpha = \frac{1}{3} L_z (1 + 2\sin\alpha) \tag{8.19}$$

and the zenith luminance with an overcast sky is a function of the sun height angle α_s:

$$L_z = \frac{9}{7}\pi (300 + 21000\sin\alpha_s) \tag{8.20}$$

For a horizontal surface, the conical solid angle can be used again and the illuminance can be represented as a function of the angle of incidence θ_r on the horizontal (which corresponds to the zenith angle) instead of the elevation angle α.

$$\begin{aligned} E_{r,h} &= \int_0^{\pi/2} \left(\frac{1}{3} L_z (1 + 2\cos\theta_r) \right) \cos\theta_r d\left(2\pi (1 - \cos\theta_r)\right) \\ &= \frac{2\pi}{3} L_z \int_0^{\pi/2} (1 + 2\cos\theta_r) \cos\theta_r \sin\theta_r d\theta_r \\ &= \frac{2\pi}{3} L_z \left(\int_0^{\pi/2} \cos\theta_r \sin\theta_r d\theta_r + \int_0^{\pi/2} 2\cos^2\theta_r \sin\theta_r d\theta_r \right) \\ &= \frac{2\pi}{3} L_z \left(\frac{1}{2}\sin^2\theta_r \Big|_0^{\pi/2} + 2\left(-\frac{1}{3}\cos^3\theta_r \Big|_0^{\pi/2} \right) \right) = \frac{L_z \pi}{3}\left(1 + \frac{4}{3}\right) = L_z \frac{7}{9}\pi \end{aligned} \tag{8.21}$$

For vertical surfaces, the solid angle must be selected in such a way that the angles of incidence on the vertical, which change with the azimuth, can be taken into account. For example, two-dimensional elements on a ring zone of the sky hemisphere can be used, indicated as a function of the zenith and azimuth angles.

In spherical polar coordinates, a two-dimensional surface in the sky hemisphere with radius r is given by the product of the sides $r\, d\theta$ on the meridian and $r \sin\theta\, d\gamma$ on the parallel circle, with θ corresponding to the zenith angle θ_z. The solid angle $d\Omega$ is thus:

$$d\Omega = \frac{r \sin\theta_z d\theta_z\ r d\gamma}{r^2} = \sin\theta_z d\theta_z\ d\gamma \tag{8.22}$$

and the solid angle of a ring zone

$$\Omega_z = \int_{\gamma=0}^{2\pi} \int_{\theta_1}^{\theta_2} \sin\theta d\theta d\gamma = 2\pi\left(\cos\theta_2 - \cos\theta_1\right) \tag{8.23}$$

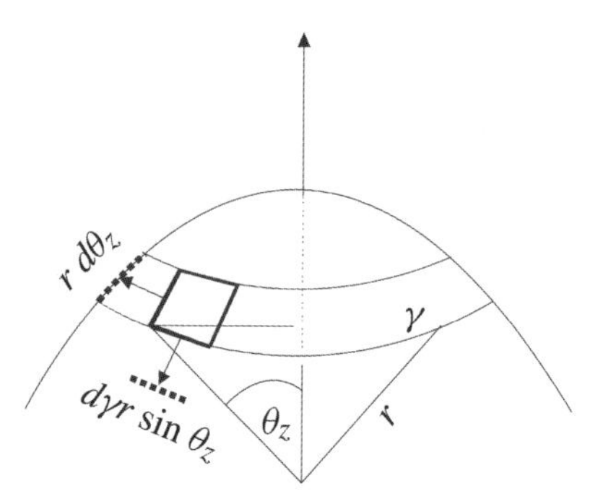

Figure 8.12: Solid angle element of a ring zone on a sphere.

The angle of incidence θ_r between the solid angle element of the sky with the coordinates zenith angle θ_z and azimuth γ and the vertical surface normal (angle of inclination β = 90°) is calculated with the known sun-position equations. Since the luminance does not depend on the azimuth, the vertical surface can be arbitrarily oriented in azimuth direction. For simplification, the surface azimuth $\gamma_s = 0$ is selected.

$$\cos\theta_r = \cos\theta_z \cos\underbrace{\beta}_{90^\circ} + \sin\theta_z \underbrace{\sin\beta}_{1} \cos\left(\gamma - \underbrace{\gamma_s}_{0}\right) = \sin\theta_z \cos\gamma \tag{8.24}$$

The integration over the solid angle is carried out with the integration limits of the zenith angle from 0 to $\pi/2$ and of the azimuth of $-\pi/2$ to $+\pi/2$, since only half the sky is seen from the vertical surface.

$$\begin{aligned}
E_v &= \int_{d\Omega} L(\theta_z)\cos\theta_r d\Omega = \int_{-\pi/2}^{\pi/2} \int_0^{\pi/2} \left(\frac{1}{3} L_z \left(1+2\cos\theta_z\right)\right) \underbrace{\sin\theta_z \cos\gamma_{sky}}_{\cos\theta_r} \underbrace{\sin\theta_z d\theta_z d\gamma}_{d\Omega} \\
&= \frac{1}{3} L_z \left(\int_{-\pi/2}^{\pi/2} \int_0^{\pi/2} \sin^2\theta_z d\theta_z \cos\gamma d\gamma + \int_{-\pi/2}^{\pi/2} \int_0^{\pi/2} 2\cos\theta_z \sin^2\theta_z d\theta_z \cos\gamma d\gamma \right) \\
&= \frac{1}{3} L_z \left(\left(\underbrace{\frac{1}{2}\theta_z}_{\pi/4} - \underbrace{\frac{1}{4}\sin(2\theta_z)}_{0} \right)\Bigg|_0^{\pi/2} \underbrace{\sin\gamma\Big|_{-\pi/2}^{\pi/2}}_{2} + 2 \underbrace{\left(\frac{1}{3}\sin^3(\theta_z)\right)\Bigg|_0^{\pi/2}}_{1/3} \underbrace{\sin\gamma\Big|_{-\pi/2}^{\pi/2}}_{2} \right) \\
&= \frac{1}{3} L_z \left(\frac{\pi 2}{4} + 2\frac{1}{3}2 \right) = L_z \pi \underbrace{\left(\frac{1}{6} + \frac{4}{9\pi} \right)}_{0.308}
\end{aligned} \tag{8.25}$$

In contrast to the isotropic sky model, in which from a vertical surface exactly half of the horizontal illuminance is seen, in the Moon and Spencer sky model only 40% of the horizontal value is obtained.

$$\frac{E_{r,v}}{E_{r,h}}=\frac{L_z\pi\left(\frac{1}{6}+\frac{4}{9\pi}\right)}{L_z\pi 7/9}=0.4 \tag{8.26}$$

The solid angle element of the spherical ring zone $d\Omega=\sin\theta_z d\theta_z\, d\gamma$ can (as an alternative to the spherical cone) also be used to calculate the horizontal illuminance, with of course the same results being obtained as with the conical solid angle. This sphere ring zone solid angle is used to calculate daylight coefficients, since in this way integration can take place over limited azimuth ranges of windows.

$$\begin{aligned}E_h &= \int_{-\pi/2}^{\pi/2}\int_0^{\pi/2}\underbrace{\left(\frac{1}{3}L_z\left(1+2\cos\theta_z\right)\right)}_{L}\underbrace{\cos\theta_z}_{\cos\theta_r}\underbrace{\sin\theta_z d\theta_z d\gamma}_{d\Omega}\\ &= \frac{L_z}{3}\left(\int_{-\pi/2}^{\pi/2}d\gamma\left(\frac{1}{2}\sin^2(\theta_z)\right)\Big|_0^{\pi/2}+\int_{-\pi/2}^{\pi/2}d\gamma_{sky}\left(-\frac{2}{3}\cos^3(\theta_z)\right)\Big|_0^{\pi/2}\right)\\ &= L_z\frac{7}{9}\pi\end{aligned} \tag{8.27}$$

8.5 Light measurements

The human eye is capable of comparing illuminances of adjacent surfaces with an accuracy of about 2%. This was quantitatively first used in the so-called grease photometer of Bunsen, where luminous intensities were determined by shining two light sources from opposite sides on a paper screen partly covered with grease. If one light source has a known light intensity I_1, the other light intensity I_2 can be determined by varying the distance d from the paper surface, until no difference in illuminance E is detected by the eye.

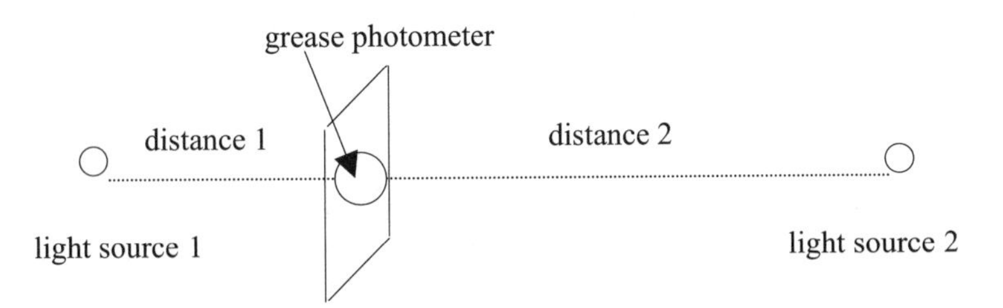

Figure 8.13: Determination of light intensities by illuminance comparison on a greasy screen.

$$E = \frac{I_1}{d_1^2} = \frac{I_2}{d_2^2}$$

Today's photometric receivers measure light or radiation either thermally (thermocouples or thermopiles) or by photo-electric techniques. The most important sensors are silicium photodiodes, which at short-circuit current operation linearily measure irradiance over about 6 orders of magnitude. Very low light fluxes can be measured with a photomultiplier tube.

Charged coupled devices within CCD cameras have up to 2048 × 2048 individual sensors organised in a matrix, where electrons are generated within the sensor pixels proportionally to the illuminance level and time: $n_{electrons} \propto \int E dt$. At a frequency of 20 MHz the whole matrix can be analysed within fractions of a second. The dynamic range of CCD receivers is very large. CCD cameras can be used to measure luminance or spectral distributions of large areas.

To provide the correct visible spectrum to the detector, either integral filtering, combining a range of different filters completely covering the sensor surface (f_1-error is 1.5%), or partial filtering, combining different filters only partially covering the sensor (1% f_1-error) can be used. The f_1-error is an energy-weighted error using the spectral sensitivity of the eye and a standardised light source (at 2856 K) divided by the energy-weighted measured curve of the filters. If the incident light flux is not perpendicular to the detector surface, additional reflection losses occur, which can be compensated by round domes or additional transmitting surfaces on the detector sides. Further measurement errors are due to linearity, temperature coefficients etc.. The best photometric receivers have a total error of 3% (highest accuracy class) up to 20% (lowest accuracy class).

8.6 Daylight distribution in interior spaces

With the asymmetrical luminance distribution of the Moon and Spencer sky, the daylight distribution in the interior can now be calculated. First, as an overview, an illustration will be made of which zenith angle areas the main part of the light-flux falls from, onto a vertical or horizontal glazing surface, in order to make simple estimations of the depth illumination of rooms.

Using Equation (8.25) the illuminance on a vertical glazing for a zenith angle range of $\theta_{z,1}$ to $\theta_{z,2}$ can be calculated as follows (angles in arc measure):

$$E_v\Big|_{\theta_{z,1}}^{\theta_{z,2}} = \frac{1}{3} L_z \left(\left(\theta_{z,2} - \theta_{z,1}\right) - \frac{1}{2}\left(\sin\left(2\theta_{z,2}\right) - \sin\left(2\theta_{z,2}\right)\right) + \frac{4}{3}\left(\sin^3\left(\theta_{z,2}\right) - \sin^3\left(\theta_{z,1}\right)\right) \right) \quad (8.28)$$

The luminance falls with rising zenith angle, but the average angles of incidence onto the vertical surface become smaller and the solid angles of the spherical ring zones with constant zenith angle steps (for example 15°) likewise become larger with rising zenith angle. All three effects taken together lead to the fact that despite the highest luminance in the zenith, the light-flux from the high zenith angle areas is most relevant.

Zenith angle [°]	*Luminance ratio* L/L_z [–]	*Normalised vertical illuminance* E_v/L_z [sr]
0–15	0.99	0.01
15–30	0.95	0.07
30–45	0.86	0.17
45–60	0.74	0.24
60–75	0.59	0.26
75–90	0.42	0.21

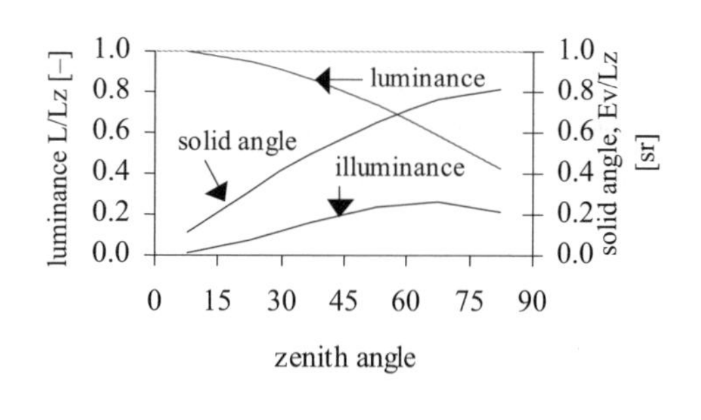

Table 8.6 and Figure 8.14: Reduction in luminance, and increase in the solid angle as well as in the vertical illuminance based on Equation (8.28), with the zenith angle.

The maximum of the vertical illuminance, here standardised on the zenith luminance L_z, comes from the zenith angle interval of 60–75°. The total of the vertical illuminance over all zenith angle areas results in

$$\sum \frac{E_v\big|_{\theta_{z,1}}^{\theta_{z,2}}}{L_z} = 0.986 = \pi\left(\frac{1}{6}+\frac{4}{9\pi}\right)$$

so again the result from Equation (8.26) is obtained. The angle of incidence is, based on Equation (8.25), given by $\cos\theta_r = \sin\theta_z \cos\gamma$, it is thus not constant for a given zenith angle interval, as the azimuth γ varies from $-\pi/2$ to $+\pi/2$ respectively. The average angle of incidence onto the vertical surface falls from 83.7° in the zenith angle interval of 0–15° to 51.3° for the interval of 75–90°.

When light passes through the window, the curve shifts to even higher zenith angles, since the reflection losses for steeply incident light are large. As a rule of thumb for the design of window openings it follows that for sufficient illumination in depth, at least the lower 30° of the sky should be seen. After all, from these zenith angle intervals (60–90°) originate (0.21 + 0.26)/0.986 = 0.48, i.e. 48% of the entire illuminance of the sky hemisphere.

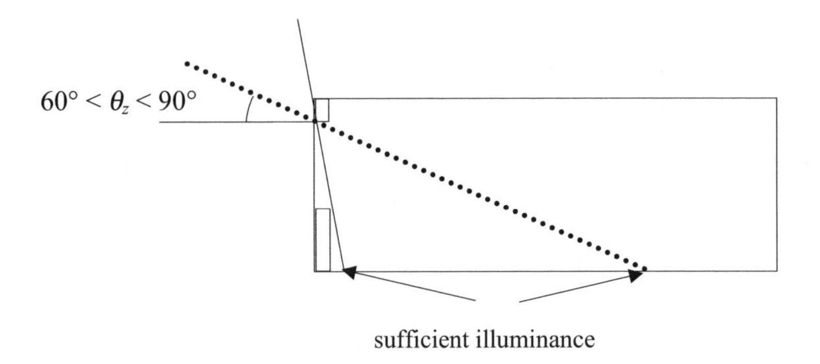

Figure 8.15: Sufficient illuminance of a room.

High-positioned windows without obstructions therefore result in good depth illumination. Narrow, high windows up to the ceiling are, photometrically, clearly better than broad strip windows. With horizontal skylights the illuminance is calculated from different zenith angle areas using Equation (8.21):

$$E_{r,h}\Big|_{\theta_{z,1}}^{\theta_{z,2}} = \frac{2\pi}{3} L_Z \left(\frac{1}{2}\left(\sin^2\theta_{z,2} - \sin^2\theta_{z,1}\right) - \frac{2}{3}\left(\cos^3\theta_{z,2} - \cos^3\theta_{z,1}\right) \right)$$

The maximum of the horizontal illuminance standardised on the zenith luminance is now in the zenith angle area 30–45°. From the zenith angle area 0–30° comes a total of 31% of the entire illuminance.

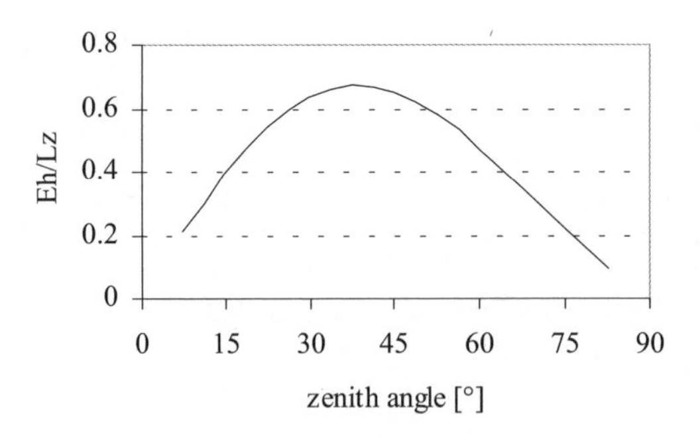

Figure 8.16: The quantity of light striking the horizontal glazing with an overcast sky, depending on the zenith angle.

Illuminating a room with skylights is sufficient if no more than a zenith angle area of 0–30° is cut off.

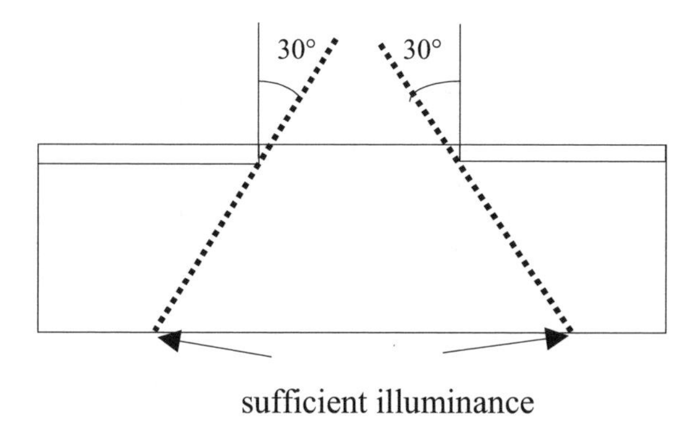

Figure 8.17: Sufficient lighting of a room using skylights.

8.6.1 Calculation of daylight coefficients

The daylight coefficient D as the relation of interior illuminance to exterior illuminance based on Equation (8.1) is defined as standard for two measuring points at a height of 0.85 m, and 1 m away from the side walls. The daylight coefficient consists of a skylight proportion D_{sky}, a proportion of diffuse reflection of shading obstructions D_{sh} and an interior reflection proportion D_r.

$$D = D_{sky} + D_{sh} + D_r \tag{8.29}$$

Since about 1920, graphic methods have been used to determine the proportion of the sky D_{sky} seen from the work surface and of the shaded proportion D_{sh}, which only reflects light. The so-called Waldram diagram contains the projection of the sky dome onto a horizontal surface and takes into account the luminance increase to the zenith. Shading buildings obstructing the horizon are included with their solid angle and reflection coefficient. The more complex inter-reflections in the interior (interior reflection proportion D_r) were only later included in the calculation of the daylight coefficient and are today usually calculated in a simplified way by the so-called "split flux method", in which only sky light reflections from the interior floor and the lower wall sections plus reflection of ground reflected light from the ceiling and the upper wall sections are regarded separately.

The daylight coefficient is determined first in dependence on the room geometry from the raw dimensions of the window openings (index r) and afterwards multiplied by light-reducing factors (window transmittance τ, a framework proportion factor k_1, a dirt factor k_2 and a correction factor for non-vertical incident k_3.).

$$D = \left(D_{sky,r} + D_{sh,r} + D_{r,r}\right)\tau k_1 k_2 k_3 \tag{8.30}$$

For the calculation of the sky light proportion seen from the point of reference P, first the effective elevation angle α_w of the window's upper edge and the lateral delimitations of the window by a left and right azimuth angle γ_{wl} and γ_{wr} have to be determined.

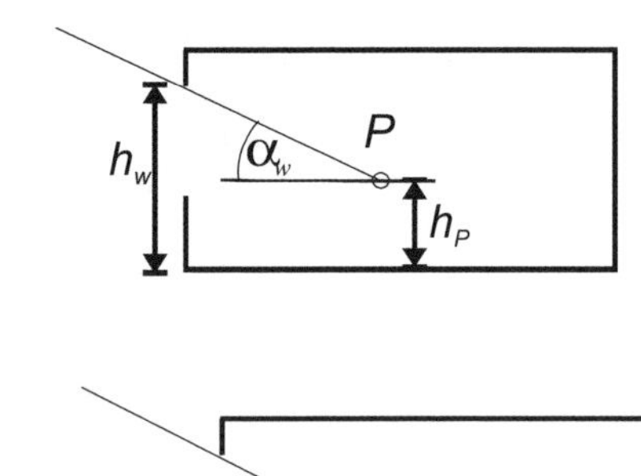

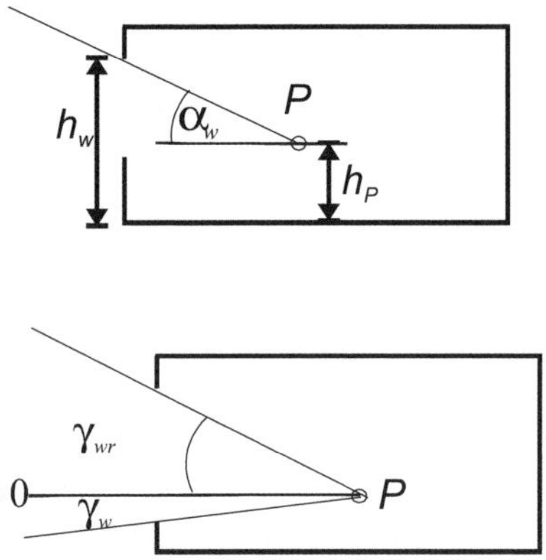

Figure 8.18: Geometrical characteristics of the window in cross-section and plane view.

To determine the effective elevation angle α_w, first determine the maximum elevation angle $\alpha_{w,\max}$ from window height h_w (upper edge) and the shortest distance between the window and point of observation, and then calculate the lateral reduction of the elevation angle with the azimuth γ.

$$\alpha_w = \arctan\left(\tan\alpha_{w,\max}\cos\gamma\right) \tag{8.31}$$

To take obstructions into account, the obstruction elevation angles $\alpha_{sh}(\gamma)$ must be known as a function of the azimuth. The illuminance on a horizontal surface in the unshaded exterior is $E_{h,o} = L_Z\pi 7/9$, based on Equation (8.27). The illuminance on a horizontal surface in the interior can likewise be calculated, taking into account the reduced height and azimuth angles of the window:

$$E_{h,i} = \frac{L_z}{3}\left(\int_{-\pi/2}^{\pi/2} d\gamma \left(\frac{1}{2}\sin^2(\theta_z)\right)\Bigg|_{\theta_{z,1}}^{\theta_{z,2}} + \int_{-\pi/2}^{\pi/2} d\gamma \left(-\frac{2}{3}\cos^3(\theta_z)\right)\Bigg|_{\theta_{z,1}}^{\theta_{z,2}}\right) \tag{8.32}$$

The zenith angle is replaced by the window height angle $\alpha_w = \pi/2 - \theta_z$, with

$$\cos(\theta_z) = \sin(\alpha_w),\ \ \sin(\theta_z) = \cos(\alpha_w) \quad \text{and} \quad \sin^2(\theta_z) = \cos^2(\alpha_w) = 1 - \sin^2(\alpha_w).$$

The delimitation angles for the luminance integral are the left and right azimuth angle of the window opening γ_{wl} and γ_{wr}, the lower elevation angle is the obstruction elevation angle α_{sh} (corresponding to the larger zenith angle $\theta_{z,2}$) and the upper elevation angle is the window height angle α_w (corresponding to $\theta_{z,1}$).

$$E_{h,i} = \frac{L_z}{3}\left(\int_{\gamma_{wl}}^{\gamma_{wr}} \left(\left(\frac{1}{2}\left(\sin^2(\theta_{z,2}) - \sin^2(\theta_{z,1})\right)\right) - \frac{2}{3}\left(\cos^3(\theta_{z,2}) - \cos^3(\theta_{z,1})\right)\right) d\gamma\right)$$
$$= \frac{L_z}{3}\left(\int_{\gamma_{wl}}^{\gamma_{wr}} \left(\left(\frac{1}{2}\left(-\sin^2(\alpha_{sh}) + \sin^2(\alpha_w)\right)\right) - \frac{2}{3}\left(\sin^3(\alpha_{sh}) - \sin^3(\alpha_w)\right)\right) d\gamma\right) \quad (8.33)$$

Thus the daylight coefficient results in:

$$D_{sky,r} = \frac{E_{h,i}}{E_{h,o}}$$
$$= \frac{\frac{L_z}{3}\left(\int_{\gamma_{wl}}^{\gamma_{wr}} \left(\left(\frac{1}{2}\left(-\sin^2(\alpha_{sh}) + \sin^2(\alpha_w)\right)\right) - \frac{2}{3}\left(\sin^3(\alpha_{sh}) - \sin^3(\alpha_w)\right)\right) d\gamma\right)}{L_z \frac{7}{9}\pi} \quad (8.34)$$
$$= \frac{3}{7\pi}\int_{\gamma_{wl}}^{\gamma_{wr}} \left(\frac{2}{3}\left(\sin^3\alpha_w(\gamma) - \sin^3\alpha_{sh}(\gamma)\right) + \frac{1}{2}\left(\sin^2\alpha_w(\gamma) - \sin^2\alpha_{sh}(\gamma)\right)\right) d\gamma$$

The externally reflected proportion $D_{sh,r}$ results as a function of the obstruction angles and the reflection coefficient ρ_{sh} of the obstruction (typically 20%), by integration from the elevation angle $\alpha = 0$ up to the obstruction elevation angle α_{sh} as well as over the azimuth angles of the obstruction $\gamma_{sh,l}$ and $\gamma_{sh,r}$. In the German standard DIN 5034 the external reflection proportion is reduced at a flat rate by a factor of 0.75.

$$D_{sh,r} = 0.75\rho_{sh}\frac{3}{7\pi}\int_{\gamma_{sh,l}}^{\gamma_{sh,r}} \left(\frac{2}{3}\sin^3\alpha_{sh} + \frac{1}{2}\sin^2\alpha_{sh}\right) d\gamma \quad (8.35)$$

The interior reflection proportion calculated using the split flux method is calculated depending on the surface-weighted reflection coefficient of the floor and of the wall lower part ρ_{fw} (without window walls, wall lower part to height of window centre) as well as on the corresponding surface-weighted reflection coefficient of the ceiling and wall upper sections ρ_{cw} (likewise without window walls). In contrast, the average reflection coefficient of the room $\bar{\rho}$ includes all walls.

$$D_{r,r} = \frac{\sum b_w h_w}{A_{room}} \frac{\bar{\rho}}{1-\bar{\rho}^2}\left(f_{up}\rho_{fw} + f_{low}\rho_{cw}\right) \quad (8.36)$$

A_{room}: total room confinement surface [m^2]
b_w, h_w: Window width and height [m]

The upper window factor f_{up} describes the integrated luminance of the Moon and Spencer sky model on the vertical surface, depending on an average obstruction angle α

(obstruction elevation angle in arc measure, measured from the window centre). The lower window factor f_{low} takes into account the diffuse radiation reflected by the floor.

$$\begin{aligned} f_{up}(\alpha) &= 0.3188 - 0.1822\sin\alpha + 0.0773\cos(2\alpha) \\ f_{low}(\alpha) &= 0.03286\cos\alpha' - 0.03638\alpha' + 0.01819\sin(2\alpha') + 0.06714 \end{aligned} \tag{8.37}$$

with $\alpha' = \arctan(2\tan\alpha)$.

Example 8.6

Calculation of the daylight coefficient of a side-illuminated room without obstructions, with window transmittance $\tau = 0.65$, a glazing proportion of 80%, a dirt factor $k_2 = 0.9$ (low contamination) and a factor $k_3 = 0.85$ to take account of the non-vertical incidence angle of the irradiance.

Room geometry:

Width B:	4 m
Depth T:	6 m
Height H:	3 m
Height of window upper edge h_w:	2.5 m
Height of window bottom edge h_{wb}:	0.85 m
Width of window b_w:	4 m

Reflection coefficients of the surfaces:

ρ_{floor}:	0.3
$\rho_{ceiling}$:	0.7
ρ_{wall}:	0.5
$\rho_{windows}$:	0.15

From this results a surface-weighted reflection degree of

$$\bar{\rho} = \frac{\rho_{floor}A_{floor} + \rho_{ceiling}A_{ceiling} + \rho_{wall}A_{wall} + \rho_{window}A_{window} + \rho_{wall}A_{wall,window}}{\sum A}$$

$$= \frac{0.3\times 24m^2 + 0.7\times 24m^2 + 0.5\times 48m^2 + 0.15\times 6.6m^2 + 0.5\times 5.4m^2}{108m^2} = 0.48$$

Without obstructions the window factors are $f_{up} = 0.3961$ and $f_{low} = 0.1$. The interior reflection proportion thus becomes $D_{r,r} = 0.008$, less than 1%!

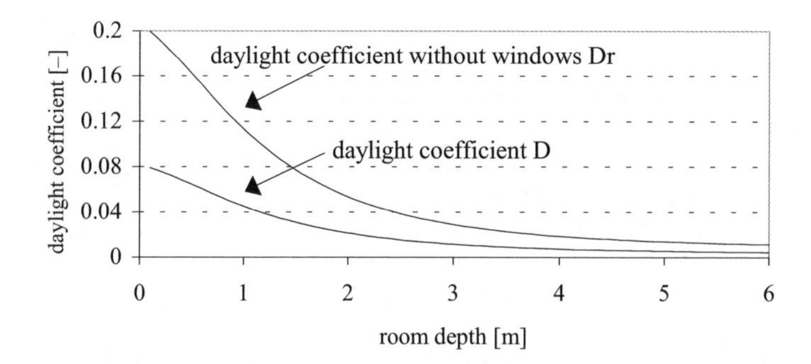

Figure 8.19: Daylight coefficient of the side-illuminated room.

If the reflection coefficient of the walls is increased to 0.7, the interior reflection proportion rises to 1.28%. Raising the window's upper edge to room height (3 m) increases the interior reflection proportion further to 1.7%.

If the reduction in the daylight coefficient by transmittance, framework proportion etc. is taken into account (factor 0.4!), at a depth of 4 m a daylight coefficient of about 1% is obtained.

References

Adnot, J., Final report "Energy efficiency of room air-conditioners", Study for the Directorate General for Energy DG XVII, contract 4.1031/D/97.026, 1999.

Aguiar, R. and Collares-Pereira, M. "Statistical properties of hourly global irradiance", *Solar Energy* **48**, 3, pp. 157–167, 1992.

Al-Amouri, A., "Aufbau einer Wärmeübertragerdatei zur Charakterisierung und Auswahl von Wärmeübertragern", Diss. 1994, TU Dresden.

Albring Industrievertretung GmbH, Alsbach-Hähnlein, "Produktunterlagen Mycom-Adsorber: Technik und Daten", 2001.

Altfeld, K., "Exergetische Optimierung flacher solarer Lufterhitzer", VDI Fortschrittsberichte Reihe 6, Nr. 75, VDI Verlag, 1985.

Berger, R., "Auslegung und Konstruktion einer Versuchs-DEC-Anlage in der Firma Klingenburg GmbH", Diplomarbeit 391, Technische Universität Dresden 1996.

Bloem, J.J., Zaaiman, W. and van Dijk, D., "Electric and thermal performance assessment of hybrid photovoltaic systems using the PasLink Test Facility". In *Proceedings of the 14th European Photovoltaic Solar Energy Conference*, Barcelona, 1997.

Bosnjakovic, F., Vilicic, M and Slipcevic, B., "Einheitliche Berechnung von Rekuperatoren", VDI-Forschungsheft 432, Band 17, 1951.

Bourseau, B., "Réfrigération par cycle à absorption-diffusion" *Int. J. Refrig.* **9**, 1986.

Braun, P.O., Goetzberger, A., Schmid, J. and Stahl, W., "Transparent insulation of building facades", *Solar Energy* **49**, No. 5, 1992.

Breembroek, G. and Lazáro, F., "International heat pump status and policy review 1993–1996", IEA Heat Pump Centre, 1999.

Brunger, P. and Hooper, F.C., "Anisotropic sky radiance model based on narrow field of view measurements of shortwave radiance", *Solar Energy* **51**, 1993.

Deschamps, G., "European Energy consumption", Meeting of the Enerbuild network in Malmö, Sweden, www.enerbuild.net, 2001.

Diekmann, B. and Heinloth, K. "Energie", Teubner Verlag, 1997.

Dudda, Ch., "Energie- und Kosteneinsparung durch innovative Beleuchtungssysteme", Sechstes Symposium Innovative Lichttechnik in Gebäuden, Staffelstein, OTTI Kolleg, 2000.

Duffie, J. and Beckmann, W., "Solar Engineering of thermal processes", John Wiley & Sons, 1980.

Eicker, U., "Mit transparenter Wärmedämmung von der Altbausanierung zum Niedrigenergiehaus", dB 9/96.

Eicker, U. and Huber, M., "Wohnungslüftungsanlagen mit solarer Nachheizung in Niedrigenergiehäusern", Tagungsband achtes Symposium Thermische Solarenergie, Otti-Technologiekolleg, 1998.

Eicker, U., Huber, M., Schürger, U., Schumacher, J. and Trinkle, A., "Komponenten- und Anlagenverhalten solar betriebener sorptionsgestützter Klimaanlagen", KI-Luft- und Kältetechnik, 10/2002.

Erbs, D.G., Klein, S.A. and Duffie, J.A., "Estimation of the diffuse irradiance fraction for hourly, daily and monthly average global irradiance", *Solar Energy* **28**, 4, pp. 293–304, 1982.

European Commission Communication, "The competitiveness of the construction industry", COM(97) 539, Chapter 2, 1997.

Franzke, U., "Chancen der solar unterstützten Klimatisierung in Deutschland", Tagungsband der FhG-ISE, "Solar unterstützte Klimatisierung von Gebäuden mit Niedertemperaturverfahren", 1995.

Fuentes, M.K., "A simplified thermal model for flat-plate photovoltaic arrays", Sandia Report SAND85-0330-UC-63, Albuquerque, N.M., 1987.

Gassel Programmauszug TRNSYS Type 107, 1998.

Gassel, A., "Betriebserfahrungen mit einer solar beheizten Adsorptionskältemaschine", Dresdner Kolloquium Solare Klimatisierung, ILK Dresden, 2000.

GBU mbH, Bensheim, "Hinweise zur Planung und Einsatzvorbereitung der Adsorptionskältemaschine", 1998.

Glück, B., "Zustands- und Stoffwerte (Wasser, Dampf, Luft), Verbrennungsrechnung", Verlag für Bauwesen, Berlin, 1991.

Goetzberger, A., Voß, B. and Knobloch, J., "Sonnenenergie: Photovoltaik", Teubner Verlag, 1997.

Gordon, J.M, Reddy, T.A., "Time series analysis of daily horizontal solar irradiance", Solar Energy **41**, 2, pp. 215–226, 1988.

Grammer, K.G. and Amberg, "Planungsunterlagen Luftkollektoren", company product information, 2000.

Granados, C., "Solar Cooling in Spain – Present and Future", Workshop Forschungsverbund Sonnenenergie, 1997.

Gregorig, R., "Wärmeaustauscher", Band 4, Verlag H.R. Sauerländer & Co. Frankfurt am Main, 1959.

Gröber, E.G., "Die Grundgesetze der Wärmeübertragung", Springer Verlag, 1988.

Grossman, G., "Solar-powered systems for cooling, dehumidification and air-conditioning", *Solar Energy Journal*, **72**, pp. 53–62, 2002.

Hahne, E. *et al.* "Solare Nahwärme – ein Leitfaden für die Praxis", BINE Informationspaket, 1998.

Hartwig, H., "Thermotrope Schichten als Regelungssysteme zur Tageslichtnutzung", 6. Symposium Innovative Lichttechnik in Gebäuden, Staffelstein, 2000.

Heinrich, J. and Franzke, U., "Sorptionsgestützte Klimatisierung", C.F. Müller Verlag, 1997.

Heinrich, J. "Energieeinsparung durch sorptionsgestützte lufttechnische Anlagen", C.F. Müller Verlag, 1999.

Henning, H.-M., Fraunhofer Institut ISE, Freiburg, personal communication, 2001.

Henning; H.-M., "Regenerierung von Adsorbentien mit solar erzeugter Prozeßwärme", Fortschrittsberichte VDI Reihe 3, Nr. 350, VDI Verlag, 1994.

Henschel, H.-J., "Licht und Beleuchtung-Theorie und Praxis der Lichttechnik", Hüthig Verlag, Heidelberg, 2002.

Hering, E., Martin, R. and Stohrer, M., "Physik für Ingenieure", Springer Verlag, Berlin, 1997.

Iqbal, M., "An introduction to solar irradiance", Academic Press, 1983.

Kast, W., "Adsorption aus der Gasphase", VCH Verlagsgesellschaft Weinheim, 1988.

Kerskes, H., Heidemann, W. and Müller-Steinhagen, H., "Über das thermische Verhalten von Kombianlagen", Zwölftes Symposium Thermische Solarenergie, Staffelstein, 2002.

Knaupp, W., ZSW Stuttgart, "Untersuchungen zur photovoltaischen Anlagentechnik im Rahmen des Photovoltaik-Testgeländes Widderstall", Abschlußbericht BMBF, FKZ 032 9048A, 1993.

Kotsaki, E., "European Solar Standards CEN/TC 312 Thermal solar systems and components, EN ISO 9488 – Solar Energy – Vocabulary, EN 12975/12976/12977 – Thermal Solar systems and components", Refocus-ISES, Elsevier Advanced Technology, June 2001.

Kübler, R. and Fisch, N., "Wärmespeicher", BINE Informationsdienst, 1998.

Ladener, H., "Solaranlagen", Ökobuch Verlag, 1994.

Lazzarin, R.M., "Experimental report on the reliability of ammonia-water absorption chillers", *International Journal of Refrigeration*, **19**, No. 4, S.247, 1996.

Lehrbuch der Klimatechnik, Band 3: Bauelemente, C.F.Müller Verlag, Karlsruhe, 1983.

Liu, B. and Jordan, R., "The interrelationship and characteristic distribution of direct, diffuse and total solar irradiance", *Solar Energy*, **4**, 1960.

Lokurlu, 2. Symposium Solares Kühlen in der Praxis, University of Applied Sciences, Stuttgart, 2002.

Merker, G.P. and Eiglmeier, C., ``Fluid- und Wärmetransport, Wärmeübertragung", Teubner Verlag, Stuttgart, 1999.

Meyer, O., "Optimierung von Niedrigenergiehäusern", Diplomarbeit am FB Bauphysik, Fachhochschule für Technik Stuttgart, Schellingstr. 24, 70174 Stuttgart, 1995.

Milow, B. and Hennecke, K., "Solarthermische Kraftwerke und Prozesswärme – Aktivitäten im Bereich des Forschungsverbunds Sonnenenergie", Zwölftes Symposium Thermische Solarenergie, Staffelstein, 2002.

Ministry for Transport and Buildings Germany, "Entwurf der Energieeinsparverordnung", Tagungsbeitrag zum 10. Symposium thermische Solarenergie, Staffelstein, May 2000.

Munters Euroform GmbH, Aachen, Abteilung HumiCool, Product documentation for adsorption cooling.

Otten, W., "Simulationsverfahren für die nichtisotherme Ad- und Desorption im Festbett auf der Basis der Stoffdaten des Einzelkorns am Beispiel der Lösungsmitteladsorption", Fortschrittsberichte VDI Reihe 3, Nr. 186, Düsseldorf, VDI Verlag, 1989.

Palz, W. and Zibetta, H., "Energy payback time of photovoltaic modules", *International Journal of Solar Energy*, **10**, pp. 211–216, 1991.

Pauschinger, T., "Solaranlagen zur kombinierten Brauchwassererwärmung und Raumheizung", 7. Symposium thermische Solarenergie, Regensburg, 1997.

Perez, R., Ineichen, P., Seals, R., Michalsky, J. and Stewart, R., "A new simplified version of the Perez diffuse irradiance model for tilted surfaces", *Solar Energy*, **39**, 1987.

Peuser, F., Croy, R. and Wirth, H., "Erfahrungen mit Regelungen für thermische Solaranlagen im Programm Solarthermie 2000", Teilprogramm 2, Tagungsband 10. Symposium thermische Solarenergie, Staffelstein, 2000.

Pukrop, D., "Zur Modellierung großflächiger Photovoltaik-Generatoren", Shaker Verlag, 1997.

Quaschning, V., "Regenerative Energiesysteme", Hanser Verlag, 1998.

Quaschning, V., "Simulation der Abschattungsverluste bei solarelektrischen Systemen", Verlag Dr. Köster, Berlin 1996.

Recknagel, Sprenger, and Schramek, "Taschenbuch für Heizung + Klimatechnik", Oldenbourg Verlag, 2001.

Reichelt, J., "Wo steht die Kältetechnik in Deutschland und weltweit?", *Kälte- & Klimatechnik*, 10/2000.

Sauer, D.U., "Untersuchungen zum Einsatz und Entwicklung von Simulationsmodellen für die Auslegung von Photovoltaik-Systemen", Diplomarbeit TH Darmstadt, Institut für angewandte Physik, 1994.

Schirp, W., "Die DAWP macht weiter von sich reden", Wärmetechnik, Nr. 4, 1993.

Schmidt, H. and Sauer, D.U., "Wechselrichterwirkungsgrade – praxisgerechte Modellierung und Abschätzung", *Sonnenenergie*, **4**, 1996.

Schumacher, J., "Digitale Simulation regenerativer elektrischer Energieversorgungssysteme", Dissertation Universität Oldenburg, 1991.

Schwab, A., "Zur Wärmeübertragung bei Mischkonvektion in luftdurchströmten Bauteilen", *Bauphysik*, **24**, 6, 2002.

Seeberger, P., "Passiv-Bürohaus Lamparter Weilheim Teck: Konzept und Betriebserfahrungen", Proceedings of conference "Passiv-Haus 2002'', Böblingen, www.solarbau.de, 2002.

Shah, R.K. and London, A.C., "Laminar flow forced convection in ducts", *Advances in heat transfer*, Academic Press, New York, 1978.

Sodha, M.S., Mathur, S.S., Macik, M.A.S., and Kaushik, S.C., "Reviews of Renewable Energy Resources", Wiley Eastern Limited – New Delhi, First Edition, 1983.

Spencer, J.W., "Fourier series representation on the position of the sun", *Search*, **2** (5), 1971.

Staiß, F., "Photovoltaik, Technik, Potentiale und Perspektiven der solaren Stromerzeugung", Vieweg Umweltwissenschaften Verlag, 1996.

Steinemann, U. *et al.*, "SIA Empfehlung V382/2: Kühlleistungsbedarf von Gebäuden", Zürich, SIA 1992.

Stryi-Hipp, G., "European Solar Thermal Market", Refocus-ISES, Elsevier Advanced Technology, June 2001.

Tan, H.M. and Charters, W.W.S., "An experimental investigation of forced convective heat transfer for fully developed turbulent flow in a rectangular duct with asymetric heating", *Solar Energy*, **13**, 1970.

Truschel, S. "Passivhäuser in Europa", Diploma thesis department of Building Physics, University of Applied Sciences, Stuttgart, 2002.

University of Bochum, Prof. Unger, Software Sunorb 1.02, www.lee.ruhr-uni-bochum.de/nes/for/sunorb.html, 1999.

VDI Wärmeatlas, VDI publications, Düsseldorf, 1994.

Versluis R., Bloem J.J. and Dunlop E.D., "An energy model of hybrid photovoltaic building facades". In *Proceedings of the 14th European Photovoltaic Solar Energy Conference*, Barcelona, 1997.

Vollmer, K. "Thermische Charakteristik und Energieertrag von hinterlüfteten PV-Fassaden", Diplomarbeit FH Stuttgart, Fachbereich Bauphysik, 1999.

Wagner, A., "Transparente Wärmedämmung an Gebäuden", BINE Informationsdienst, 1998.

Weiß, W., Suter, J-M. and Letz, T., "Solare Kombianlagen im europäischen Vergleich, Ergebnisse der IEA SHC task 26", Zwölftes Symposium Thermische Solarenergie, Staffelstein, 2002.

Wittkopf, H., Becker, H. and Jödicke, D., "Pilkington E-Control- das variable Sonnenschutzglas der neuen Generation", Tagungsband Glaskon, München, 1999.

Wong, S.S.M. "Computational methods in physics and engineering", World Scientific Publishing Singapore, 1997.

Zimmermann, M., "Handbuch der passiven Kühlung", EMPA ZEN, Dübendorf, Switzerland, 1999.

Zimmermann, M. and Andersson, J., "Low energy cooling – case studies buildings", EMPA ZEN, Dübendorf, Switzerland, 1998.

Index